USING STATISTICS

Kenneth J. Travers

William F. Stout

James H. Swift

Joan Sextro

Addison-Wesley Publishing Company
Menlo Park, California • Reading, Massachusetts
Wokingham, Berkshire • Amsterdam • Don Mills, Ontario • Sydney

This book is published by the Addison-Wesley Innovative Division.

Design, composition, and illustrations: The Bookmaker, San Jose, California

Cover design: John Edeen

Printed in the United States of America. Published simultaneously in Canada.

ISBN-0-201-20070-8
ISBN-0-201-20676-5
ABCDEFG-VH- 8987654

Introduction

Using Statistics is intended for a one- or two- semester course for students who have successfully completed at least one year of high school algebra. The book has been developed over several years as a cooperative project between the Department of Mathematics and the College of Education at the University of Illinois. *Using Statistics* can serve as an introductory course in statistical concepts, a course in mathematics for general education, or as a supplementary reference for innovative and challenging topics in computer mathematics. For a one-semester course, Chapter One through Six would be appropriate, with Chapters Seven and Eight as optional material. Chapters One through Twelve are appropriate for a two-semester course, with Chapters Thirteen and Fourteen as optional material.

A grasp of the elements of statistics is essential for every educated citizen. This book presents these concepts in a simplified manner, minimizing the symbolism and formulas typically utilized in such a course but preserving the integrity of the fundamental concepts. A typical presentation of a concept involves the use of an interesting example, the conduct of an experiment or activity, the collection of data, and discussion of the data. The progression is usually from concrete experience to abstract concept formulation. A five-step procedure for solving probabilistic and statistical problems was devised at the University of Illinois and utilized successfully with students at a wide variety of ability levels.

Overview of Content

The book begins with elementary topics in descriptive statistics and in Chapters Three and Four presents key concepts of probability in a highly intuitive, experience-based manner, relying heavily upon laboratory activities. Chapter Five, Making Statistical Decisions, provides an elementary introduction to hypothesis-testing.

The chi-square statistic is discussed in Chapter Six and is done by first discussing a "D" statistic. The D statistic has proved in our experimental classroom work to be easier to utilize to present the notion of correspondence between expected and obtained outcomes than is the chi-square. The other novel and pedagogically important feature of this chapter is that the distributions are developed initially from chi-square values generated as class activities. The curve-smoothing idea is utilized in order to arrive at theoretical chi-square distributions and tables.

Bivariate statistics is introduced in Chapter Seven in an informal way, using the idea of a simple rule which relates two measures. There is sufficient material in this chapter to cover the basic notion of linear regression. However, more able students could cover Chapter Eight, as well, which leads to the standard formulations of the regression and correlation coefficients, but from the point of view of searching for the line that minimizes the error of prediction (estimation) of one variable from the other.

Chapter Nine first considers the transformation of scores, then deals with the uniform and normal distribution. Concepts of measurement (Chapter Ten) are introduced from the point of view of sampling, and form a basis for the more detailed consideration of the Central Limit Theorem in Chapter Eleven.

Chapter Twelve deals with the elements of hypothesis testing, which are further expanded upon in Chapter Thirteen. At this point, students should be well equipped with basic statistical concepts for general education (such as is needed for intelligent reading of newspapers or viewing television) as well as for the further study of statistics required in the fields of business, economics, physics, chemistry—indeed, most disciplines in today's world of study and work.

Chapter Fourteen, on Monte Carlo Methods, is intended for enrichment, and provides unusually rewarding opportunities for computer applications.

The Role of the Hand Calculator

The authors regard the hand calculator as an extremely useful tool in the statistics classroom and laboratory, and hope (expect) that liberal use will be made of this tool as the book is taught. However, the book has been written so that, as far as possible, the book is not calculator-dependent.

The Role of the Computer

The computer is an important component of the contemporary statistics classroom. However, access to the computer still appears to be a problem for many. Therefore, emphasis on the computer is minimized within the text. However, supplementary materials will provide especially useful programs and propose computer investigations (such as generating certain sampling distributions).

Experimental Trials

As already indicated, this book makes heavy use of experimental trials in obtaining solutions which are estimates of theoretical values. One clear advantage of this approach, we believe, is that students become thoroughly acquainted with the distinction between statistic and parameter. They are continually obtaining statistics that vary from student to student and soon learn that these sample estimates can be improved by larger samples or increasing the number of trials.

We recommend that teachers require individual students to generate only a small number (say, 5) of trials, in order to demonstrate a grasp of the correct procedure. Then the trials are pooled as a class activity to obtain, say, 150 trials from a class of 30 students. This procedure has been successfully used by the authors in trying out the materials as they have been written.

Key Problems and Special Interest Features

Two features of this book, designed to stimulate interest in statistics and demonstrate the wide variety of situations in which statistics is used, are Key Problems and Special Interest Features. Each chapter is introduced by a Key Problem—a real world example—which is understandable and illustrates a central concept treated in the chapter. Special Interest Features are used throughout the text to help maintain a lively flow of ideas and further elaborate, at times in a humorous or curiosity-prodding way, upon the major ideas in the chapter.

Contents

Chapters

Appendices

CHAPTER 1

Descriptive Statistics

TELLING IT LIKE IT IS

KEY PROBLEM

Will the Real Author Please Stand Up?

Finding out who actually wrote a play, a novel, or a will can be a very important problem to solve. For example, when the billionaire Howard Hughes died, several different wills came to light. It was left to the courts to decide which was written by Howard Hughes and which were forgeries. As another example, some scholars doubt that William Shakespeare actually wrote his plays. Sir Walter Raleigh and Sir Francis Bacon are among those suggested as the real author.

For this first key problem in the book, we will show how statistics can be used to help solve a famous problem in U.S. history having to do with the *Federalist Papers*. These papers were written anonymously during 1787–1788 by Alexander Hamilton, John Jay, and James Madison to persuade persons in the state of New York to ratify the U.S. Constitution. They contain many details about the framing of the Constitution because, for example, James Madison took extensive notes at the Constitutional Convention.

The authorship of 70 of the papers is agreed upon. There are 12 about which there is disagreement as to whether they were written by Hamilton or Madison. But by looking at use of the words *while* and *whilst*, researchers found the following information, as shown in Figure 1.1.[1]

FIGURE 1.1

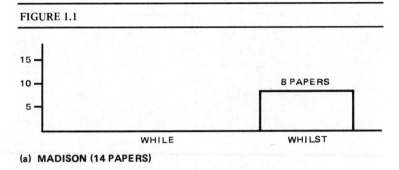

(a) MADISON (14 PAPERS)

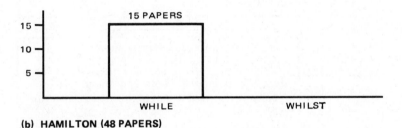

(b) HAMILTON (48 PAPERS)

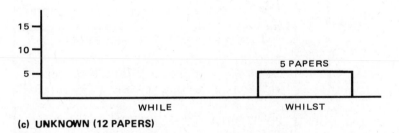

(c) UNKNOWN (12 PAPERS)

Figure 1.1 (a) shows that in the 14 papers known to be written by Madison, *whilst* is used in 8 of them, and *while* is never used. Figure 1.1 (b) shows, on the other hand, that in the 48 papers written by Hamilton, *while* is used 15 of the papers, and *whilst* is never used.

The investigators then went to the 12 papers of disputed authorship. They found that *whilst* was used in 5 of the papers and *while* was not used at all. Therefore, it seems reasonable to conclude that these 5 papers were written by Madison.*

What cautions should be kept in mind as we make this conclusion?

*Notice that we cannot be sure that Madison *never* used *while* in his other writing, apart from the *Federalist Papers*. It has been found that he did, in fact, use the word *while* on two occasions in other papers.

1

1.1   WORKING WITH DATA

Statistics are used daily in newspapers, on television, and in ordinary conversation. Before we give a definition of what statistics are, let's look at some examples of how they are used. Each of the following items was in a newspaper, but could just as well have been a news story on the radio or television.

HELP!

People may complain that police take a long time to respond to calls for help. But *Time* (April 3, 1978) reported that victims often delay in calling the police. A study in Kansas City, Missouri, showed that for 1000 victims of major crimes, it took a median (a kind of average—see Chapter 2) time of 6 minutes 17 seconds to call the police. Once police were called, it was reported that a car was sent in an average of 2 minutes 50 seconds and arrived at the scene in an average of 5 minutes 34 seconds.

WHAT'S IN A NAME?

Robert Weiszman of Idaho decided to change his name. His reason, according to the Lewiston *Morning Tribune* (July 15, 1975) had to do with an experiment by his brother, who sent out job applications. Ten of the applications used the name of Weiszman and 10 (to the same people) used the name of Smith. The applications sent out under his own name received no responses. Those with the name Smith received 8 responses.

SUNNY-SIDE UP?

The *Chicago Tribune* on January 20, 1976, carried a full-page advertisement by the National Commission on Egg Nutrition reporting on research in relationships between eggs in the diet and risk of heart disease. In one experiment, two groups of men were involved. One group was on a high-egg diet and the other was on a low-egg diet. The advertisement says, "The results (after eight weeks) indicate that there were no significant differences in average serum cholesterol levels between any two time periods in either groups."

THE MEDIUM AND THE MESSAGE

Temple University's Institute for Survey Research examined the kind and amount of fear possessed by American children, as relative to the amount of television watched. It reported that the highest level of fear is found in children who watch 4 hours or more of television daily. (*Chicago Daily News*, March 3, 1977.)

WHERE DOES ALL THE MONEY GO?

Statistics Canada reported in the *Market Research Handbook* (April 1979) that the largest chunk of a Canadian's paycheck (17%) goes to paying personal taxes. The next biggest bites are for food (15%) and shelter (14%). About 9% of the family income goes to savings.

Each of these news clips involves information in the form of numbers—length of time for police to arrive, numbers of responses to job applications, amounts of cholesterol in the body, levels of fear in children, and so on. These pieces of information are called *data*. The word *data* is plural (we say *data are . . .*), but hardly anyone uses the singular form *datum*, which refers to only one number or piece of information! In each of the news items, conclusions can be drawn from the data presented.

Statistics is the science of gathering, organizing, and drawing conclusions from data. In this book, we will study some of the basic ideas in statistics and learn how to use these ideas to help solve problems in the world around us.

THE LONG AND SHORT OF AIRPLANE TRAVEL

Several thousand people were asked which in-flight service they considered most important for *short* flights (less than 2 hours). See Table 1.1.

TABLE 1.1

SERVICE	PERCENT
Carry-on baggage compartments	48
Flight attendant attitude/behavior	31
Beverage service	8
Food service	7
Other	6
Total	100

For *long* flights (3 hours or more), the answers in Table 1.2 were obtained.

TABLE 1.2

SERVICE	PERCENT
Food service	39
Flight attendant attitude/behavior	34
Carry-on baggage compartments	10
In-flight entertainment	7
Other	10
Total	100

We notice that for long flights, food service becomes an important factor to passengers, and compartments for carry-on baggage become less important. But we also notice that the attitude and behavior of the flight attendants remains a key factor for both short and long flights.[2]

Many factors have to be taken into account when decisions are made. One of the important advantages of statistics is that it helps to clarify information and present it in a way that is easy to grasp.

EXERCISES

1   Infant mortality refers to the number of infants that die for each 1000 live births. The following data are for the years 1920–1970 in the United States (rounded to the nearest whole number).[3]

	1920	1930	1940	1950	1960	1970
Infant Mortality	86	65	47	29	26	20

What could help explain the decrease in infant mortality during this time period?

2   Life expectancy is the number of years, on the average, that a person born at a certain time could expect to live. This is based on records of death during that period.[4]

	1900	'10	'20	'30	'40	'50	'60	'70
Life Expectancy (nearest year)	47	50	54	60	63	68	70	71

Here we see an increase in life expectancy over the time period. What factors might explain this? If you were to speculate upon life expectancy figures from now until the year 2100, what do you think they would be?

3   The *New York Times* reported on June 28, 1979, on the density of millionaires (number per 1000 persons) for each of the 50 States in the United States. The figures

reported here are rounded to the nearest whole number.

State	Millionaires per 1000 Persons
Idaho	27
Maine	8
North Dakota	7
Nebraska	7
Minnesota	6
Indiana	5
Wisconsin	4
Iowa	4
New Jersey	4
Connecticut	3

Can you explain what would account for this high proportion of millionaires in these states? Why, do you think, states like California, New York, or Illinois are not on this "top 10" list? (See Section 1.6 for a further discussion of this.)

4   From a newspaper, find three examples of how data are used in different areas (sports, medicine, politics, and so on).

5   By reading magazines and newspapers or by viewing television news, find three ways in which statistics are used.

6   Interview a professional person (such as a medical doctor, lawyer, or economist) and determine if and how that person uses statistics.

7   The distance from Earth to the sun is called the astronomical unit. Here are values for the astronomical unit (A.U.) as obtained by astronomers in the time period 1895 to 1961.[5]

Source and Date	A.U. (Millions of Miles)
Newcomb, 1895	93.28
Hinks, 1901	92.83
Noteboom, 1921	92.91
Spencer Jones, 1928	92.87
Spencer Jones, 1931	93.00
Witt, 1933	92.91
Adams, 1941	92.84
Brower, 1950	93.00 (rounded from 92.977)
Rabe, 1950	92.91 (rounded from 92.9148)
Millstone Hill, 1958	92.87 (rounded from 92.874)
Jodrell Bank, 1959	92.88 (rounded from 92.876)
S.T.L., 1960	92.93 (rounded from 92.9251)
Jodrell Bank, 1961	92.96 (rounded from 92.960)
Cal.-Tech., 1961	92.96 (rounded from 92.956)
Soviets, 1961	92.81 (rounded from 92.813)

What would you consider to be a good estimate of how far it is from Earth to the sun, based on this information? Why do you think there are more decimal places reported in the values in the more recent measurements?

1.2   STEMS AND LEAVES

Making sense out of data is an important part of statistics. Suppose, for example, that the weights of a sample of 50 girls at Central High, rounded to the nearest pound, are:

132	135	114	153	135
138	120	106	131	132
105	115	127	103	100
117	114	125	137	110
137	129	98	128	112
122	88	100	119	121
103	118	152	148	125
114	139	89	100	92
92	132	132	107	146
99	134	108	101	89

With data jumbled up like this, it's hard to find the story being told. What is the lowest weight? The highest weight? What is the range (highest to lowest) of the weights? What weight do girls most commonly have?

In order to answer questions like these, we need to organize the data. One good way is to use the *stem-and-leaf* approach. Here is how it works.

We first notice from the data that the weights could easily be grouped by tens. That is, we can group the 80s together, the 90s together, and so on, using a table like this:

8
9
10
11
12
13
14
15

The numbers shown in the table stand for 80, 90, 100 and so on. These numbers are called *stems*. Now we can start listing the weights in the table. The first weight is 132 pounds. The stem is 13 so we go to row 13 and write 2 to the right of 13. The 2 is a *leaf*.

STEM	LEAF
8	
9	
10	
11	
12	
13	2
14	
15	

The next weight in the list (reading down the first column) is 138 pounds; then comes 105

pounds. The stem for 138 is 13 and that for 105 is 10. So we write the leaf of 8 next to the 13 and the 5 opposite the 10. Thus, the weights of 132, 138, and 105 are recorded as below:

STEM	LEAF
8	
9	
10	5
11	
12	
13	2, 8
14	
15	

Note that leaves are separated by commas.

The completed stem-and-leaf table for the weights of the 50 girls at Central High is shown in Table 1.3.

TABLE 1.3

STEM	LEAF
8	8, 9, 9
9	2, 9, 8, 2
10	5, 3, 6, 0, 8, 3, 0, 7, 1, 0
11	7, 4, 5, 4, 8, 4, 9, 0, 2
12	2, 0, 9, 7, 5, 8, 1, 5
13	2, 8, 7, 5, 9, 2, 4, 2, 1, 7, 5, 2
14	8, 6
15	3, 2

Let's notice an important feature of the stem-and-leaf table. It is easy to tell, for example, that many girls (12—count them!) have weights in the 130-pound range (from 130 to 139 pounds). There are 10 girls in the 110-pound range (from 110 to 119 pounds). We can also see that the lowest weight is 88 pounds and the greatest weight is 153 pounds. So, the *range* of all the weights is from 88 to 152 pounds; it is usually expressed as the difference 153 − 88 = 65 pounds.

Another important feature of the stem-and-leaf table is that one could use it to list all the original weights (that is, no original information is lost by using this table). For example, reading from the top row (the 8-stem), we have the weights 88 pounds, 89 pounds, and 89 pounds.

■ **Example—New Minted Coins:** A sample of 25 pennies from the U.S. Mint were weighed on a balance. Here are the results, in grams:

3.01	3.04	3.05	3.02	3.12
3.14	3.11	3.16	3.20	3.14
3.08	3.09	3.10	3.10	3.13
3.16	3.17	3.12	3.13	3.11
3.10	3.07	3.09	3.12	3.17

A stem-and-leaf table can be used to record these weights. We will get rid of the decimal points by multiplying each weight by 100. Then, when we read the table, we will have to remember to divide each weight by 100. See Table 1.4.

TABLE 1.4

STEM	LEAF
30	1, 8, 4, 9, 7, 5, 9, 2
31	4, 6, 0, 1, 7, 6, 0, 2, 0, 3, 2, 2, 4, 3, 1, 7
32	0

(*Remember:* Each weight has been multiplied by 100.)

There it is. We see that the range of measurements is rather small, from a low weight of 3.01 grams (remember to divide by 100) to a high weight of 3.20 grams.

There is a problem with this table, though. The data are scrunched together so much that we find it hard to get a good idea about the various weights in the 3.10 interval, for example.

We can overcome this difficulty by stretching out the table. We can divide each interval into two parts. For example, in the first row, we can write:

STEM	LEAF
30	1, 4, 2
3*	8, 9, 7, 5, 9

The * means that the interval is continued. So, the top row includes values from 3.00 to 3.04, and the second row includes the remaining values, in the interval 3.05 to 3.09.

Table 1.5 shows this expanded look.

TABLE 1.5

STEM	LEAF
30	1, 4, 2
3*	8, 9, 7, 5, 9
31	4, 0, 1, 0, 2, 0, 3, 2, 2, 4, 3, 1
3*	6, 7, 6, 7
32	0

(Values are multiplied by 100.)

We now get a much better look at the data and find that the interval 3.10 to 3.14 contains about one-half of all the weights (12 out of 25). ■

EXERCISES

1 The following scores were obtained by 50 students on a final exam in statistics. Construct a stem-and-leaf table of these data.

51	46	31	35	37	51	56	51	43	48	52
33	42	37	27	57	65	36	37	55	42	43
33	49	31	46	50	57	52	35	38	47	42
58	38	47	54	39	51	68	36	48	36	47
32	51	50	44	32	36					

2   The weights of a group of rats used in an agricultural research project range from 182 to 220 grams and are reported to the nearest gram. What stems would you use for recording these data?

3   The weights of the 40 rats in Exercise 2 are as follows:

```
206  202  200  215  191  193  196  202
204  191  190  215  188  196  190  182
205  192  194  201  207  205  203  211
198  206  203  192  215  195  210  216
206  208  190  197  210  220  220  211
```

Construct a stem-and-leaf table of these weights.

4   The stem-and-leaf table gives stopping distances (in meters) of ten test cars going 50 kilometers per hour when the brakes were applied. List the original data from which the table was made.

STEM	LEAF
6	4, 8, 0, 1, 4
7	5, 1, 3
8	2, 0

5   This stem-and-leaf table gives final scores on a statistics test for 20 students. From the table, write the score of each student.

STEM	LEAF
7	3, 5, 2, 4, 9
8	7, 0, 4, 4, 5, 3, 2, 8
9	6, 3, 6, 5, 2, 0, 1

6   From the stem-and-leaf table in Exercise 5, answer these questions about the final scores of the students on the statistics exam:

(a)   What was the lowest score on the exam?
(b)   What was the highest score on the exam?
(c)   How far apart were the lowest and highest scores?

7   The ages of the first 36 U.S. presidents at inauguration and death are given in Table 1.6.[6] Make a stem-and-leaf plot of ages at inauguration and another for ages at death. For which data is there more variation —age at inauguration or age at death?

TABLE 1.6

WHICH PRESIDENT	AGES	
	INAUG.	DEATH
1 Washington	57	67
2 Adams	61	90
3 Jefferson	57	83
4 Madison	57	85
5 Monroe	58	73
6 Adams	57	80
7 Adams	57	78
8 Van Buren	54	79
9 Harrison	68	68
10 Tyler	51	71
11 Polk	49	53
12 Taylor	64	65
13 Filmore	50	74
14 Pierce	48	64
15 Buchanan	65	77
16 Lincoln	52	56
17 Johnson	56	66
18 Grant	46	63
19 Hayes	54	70
20 Garfield	49	49
21 Arthur	50	57
22 Cleveland	47	71
23 Harrison	55	67
24 Cleveland	55	71
25 Mckinley	54	58
26 Roosevelt	42	60
27 Taft	51	72
28 Wilson	56	67
29 Harding	55	57
30 Coolidge	51	60
31 Hoover	54	90
32 Roosevelt	51	63
33 Truman	60	88
34 Eisenhower	62	78
35 Kennedy	43	46
36 Johnson	55	64

8   Using the information from Exercise 7, construct a new set of data, the difference in age between a president's age of death and his inauguration (that is, how long each president lived after his inauguration). For example, for Washington the number is 10 (years).

Construct a stem-and-leaf plot of these data. What is the most frequent number of years a president lived after inauguration? What is the shortest time? The longest time?

9   Table 1.7 gives data on British monarchs: Age of accession to the throne (AC) and age of death (DE).[7] Construct a stem-and-leaf plot of age of accession and another of age of death. Which set of data has more variation— age of accession to the throne or age at death?

10   Using the information from Exercise 9, construct the set of data on difference between age of death and age of accession to the throne. (How many years did a monarch live after becoming king or queen?) What is the range of these data? The least number of years lived? The greatest number?

11   Table 1.8 gives the number of home runs scored by leaders in the National League (NL) and the American League (AL) between 1921 and 1980.[8] Draw stem-and-leaf plots of these data. Which league has the greatest range in the number of home runs scored? In which league does there seem to be, on the average, more home runs scored?

12   Construct a stem-and-leaf table from the following data, which are the times in hours that each of 60 batteries lasted before needing to be recharged. (*Hint:* For 63.2, use a stem of 6 and leaf of 32. Record leaves as two digits. Then note that values are to be divided by 10.)

63.2	71.0	81.6	68.5	76.2	82.3
70.3	84.6	65.9	74.7	74.2	79.2
63.2	89.7	78.4	86.8	59.0	77.9
50.3	55.5	74.3	60.6	62.1	74.9
63.6	63.2	79.4	57.6	61.4	81.8
71.3	64.1	64.8	64.9	71.8	64.0
74.5	51.8	64.6	75.2	64.4	70.9
74.2	83.3	64.1	77.5	54.1	71.6
63.4	55.2	64.2	66.2	73.4	61.1
60.1	65.3	83.8	68.1	74.5	66.3

13   The length of a copper bar was measured by each of 30 persons in a physics class. Make a stem-and-leaf table of these data. (*Hint:* Use stems of 347, 348, and so on.)

35.12	35.08	35.15	35.10	35.02	35.02
34.95	34.93	34.99	34.93	35.01	35.13
34.95	34.79	34.93	35.07	34.87	35.11
35.20	35.00	35.10	35.06	34.86	34.95
35.14	35.01	35.00	34.95	35.20	35.04

TABLE 1.7

MONARCH	DE	AC	MONARCH	DE	AC
William 1	60	39	Mary 1	43	37
William 2	43	31	Elizabeth 1	69	25
Henry 1	67	32	James 1	59	37
Stephen	54	35	Charles I	48	25
Henry 2	56	21	Cromwell 1	59	54
Richard 1	42	32	Cromwell 2	86	54
John	50	32	Charles 2	55	30
Henry 3	65	9	James 2	55	4
Edward 1	68	33	William 3	51	39
Edward 2	43	23	Mary 2	32	27
Edward 3	65	15	Anne	49	37
Richard 2	34	12	George 1	67	54
Henry 4	47	32	George 2	77	44
Henry 5	37	26	George 3	81	22
Henry 6	49	1	George 4	67	58
Edward 4	41	19	William 4	71	65
Edward 5	13	13	Victoria	81	18
Richard 3	35	31	Edward 7	68	60
Henry 7	53	28	George 5	70	44
Henry 8	56	18	Edward 8	77	42
Edward 6	16	10	George 6	57	41

TABLE 1.8

YEAR	AL	NL	YEAR	AL	NL
1921	59	23	1951	33	42
1922	39	42	1952	52	37
1923	41	41	1953	43	47
1924	46	27	1954	32	49
1925	33	39	1955	37	51
1926	47	21	1956	52	43
1927	60	30	1957	42	44
1928	54	31	1958	42	47
1929	46	43	1959	42	46
1930	49	56	1960	40	41
1931	46	31	1961	61	46
1932	58	38	1962	48	49
1933	48	28	1963	45	44
1934	49	39	1964	49	47
1935	36	34	1965	32	52
1936	49	33	1966	49	44
1937	46	31	1967	44	39
1938	58	36	1968	44	36
1939	35	28	1969	49	45
1940	41	43	1970	44	45
1941	37	34	1971	33	48
1942	36	30	1972	37	40
1943	34	29	1973	32	44
1944	22	33	1974	32	36
1945	24	28	1975	36	38
1946	44	23	1976	32	38
1947	32	51	1977	39	52
1948	39	40	1978	46	40
1949	43	54	1979	45	48
1950	37	47	1980	41	38

1.3   GRAPHS: PICTURES OF DATA

Often in statistics, as in other parts of life, a picture is worth a thousand words. So statisticians take the opportunity to draw pictures, or graphs, of their data, whenever feasible. One such picture, the bar graph, is commonly used in newspapers and magazines. In statistics, this graph is called a *histogram* One axis of the graph tells about the measurements involved and the other axis shows how often each of the measurements occurred (that is, gives the frequency of the observations).

Histograms can readily be made from stem-and-leaf tables. Let's look at the table of weights given in Section 1.2, reproduced here as Table 1.9.

TABLE 1.9

STEM	LEAF
8	8, 9, 9
9	2, 9, 8, 2
10	5, 3, 6, 0, 8, 3, 0, 7, 1, 0
11	7, 4, 5, 4, 8, 4, 9, 0, 2
12	2, 0, 9, 7, 5, 8, 1, 5
13	2, 8, 7, 5, 9, 2, 4, 2, 1, 7, 5, 2
14	8, 6
15	3, 2

By simply enclosing each row (set of leaves) in a bar, we have a respectable histogram.

We can also count how many leaves there are for each stem and redraw our histogram showing the frequencies (how many leaves there are) along one axis of the graph. See Figure 1.2.

PROJECT CORNER

Locate a collection of 50 pennies, or other similar objects. Use a very accurate scale or balance to find the weight of each. Construct a stem-and-leaf table of your data.

FIGURE 1.2

STEM	LEAF
8	8, 9, 9
9	2, 9, 8, 2
10	5, 3, 6, 0, 8, 3, 0, 7, 1, 0
11	7, 4, 5, 4, 8, 4, 9, 0, 2
12	2, 0, 9, 7, 5, 8, 1, 5
13	2, 8, 7, 5, 9, 2, 4, 2, 1, 7, 5, 2
14	8, 6
15	2, 3

It is more common to use vertical bars instead of horizontal bars. In that case, our histogram of the weights of the 50 girls looks like the one in Figure 1.3.

■ **Example—Fill 'Er Up, Please:**   The horizontal axis often expresses numerical data, as in this example. Here, the vertical axis shows gasoline consumption in the United States in millions of barrels per day and the horizontal axis is time in years. See Figure 1.4.

One clear message is that a dip in consumption took place in 1974. This was caused by the embargo on oil imposed by the OPEC nations and the resulting shortages of supply.

Another useful graph is the *frequency polygon*. It can be drawn from a histogram by connecting midpoints of the tops of the bars as in Figure 1.5. Or, a frequency polygon can be drawn directly from a frequency or a stem-and-leaf table. You may recall that a

FIGURE 1.3

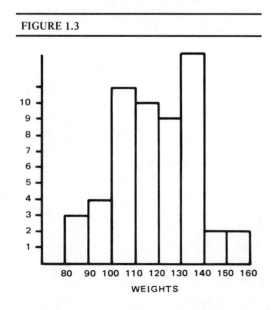

WEIGHTS

FIGURE 1.4

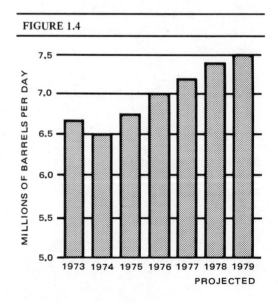

PROJECTED

FIGURE 1.5

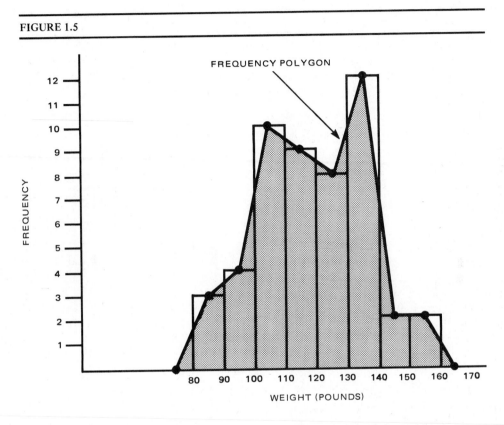

polygon is a closed figure consisting of straight lines. Perhaps you can see why this graph is given the name *frequency polygon.* ∎

EXERCISES

Unless otherwise noted, histograms should show frequencies rather than proportions.

1   Draw a histogram of the stem-and-leaf data given below. The data (from Section 1.2, Exercise 5) are the scores of 20 students on a statistics test.

STEM	LEAF
7	3, 5, 2, 4, 9
8	7, 0, 4, 4, 5, 3, 2, 8
9	6, 3, 6, 5, 2, 0, 1

2   Refer to the airline passenger data for long and short trips (Table 1.2, page 3). Draw a histogram of each set of data.

Use the graph in Figure 1.6 to answer Exercises 3, 4, and 5. The histogram shows the number of cases of polio in the United States from 1930 through 1956.[9] (Salk vaccine was first used on a large-scale basis in 1954.)

3   In what year was the highest number of cases of polio reported? About how many cases was this?

4   In what year was the lowest number of cases of polio reported? About how many cases was this?

FIGURE 1.6

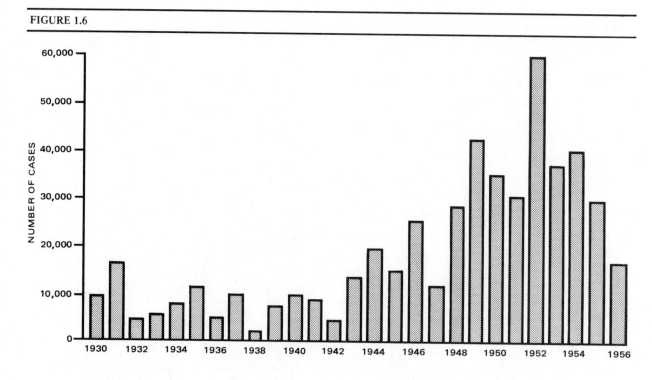

5   Construct a frequency polygon from the data given in the histogram of the number of cases of polio. Do you think the chances of getting polio were the same from year to year?

For each of the following questions, you should collect your own data.

6   Toss one coin 30 times. Prepare a table of outcomes and a histogram of the data in the table. Construct a histogram to show your results. (*Hint:* Your histogram will consist of only two bars.)

7   Do Exercise 6, above, tossing the coin 50 times.

8   From Exercise 6, above, calculate the proportion of heads and of tails obtained.

9   Roll a die (singular of dice) 30 times, recording in a table the frequency of each of the outcomes. For example, if the die falls with one dot showing, call the outcome 1. There are six different outcomes possible.

10   Draw a histogram of the data of Exercise 9.

11   Draw a frequency polygon of the data of Exercise 9.

12   Toss an ordinary thumbtack 40 times. Record whether it falls point up or point down. Find the proportion of times it falls up and the proportion of times it falls down.

13   Draw a histogram of the data given in Table 1.10. Then, by outside reading, try to determine the size of the catch of Blue Whales in Antartica today.

TABLE 1.10
CATCH OF BLUE WHALES IN THE ANTARCTICA[10]

YEAR	ESTIMATED NUMBER OF WHALES (IN THOUSANDS)
1931	28
1932	7
1933	10
1934	18
1935	16
1936	17
1937	15

14 Draw a frequency polygon of the data in Exercise 13.

15 Fingerprints can be classified according to the number of ridges between "loops" in the patterns. The number of ridges is called the *ridgecount* for a particular person. Suppose we have the following ridgecounts for the fingerprints of 20 persons.

189	181	205	210	198	207	201	185
188	192	186	189	192	194	215	205
213	207	213	220				

Draw a stem-and-leaf table from these data. Then draw a histogram and a frequency polygon.

1.4 PROPORTIONS AND PROBABILITIES

Table 1.11 gives some data obtained from a quality control board, which tested 20 fire alarms and found their *trip temperature* (the temperature at which the alarm is sounded).

We can now ask what *proportion* of the 20 alarms tested had a trip temperature of, for example, 93 degrees? In order to answer such

TABLE 1.11

TEMPERATURE	FREQUENCY	PROPORTION
85	1	.05
86	0	0.0
87	2	.10
88	3	.15
89	5	.25
90	4	.20
91	0	0.0
92	3	.15
93	2	.10
Total	20	1.00

(*Note:* Due to rounding, the total of the proportions may not always be exactly 1.00.)

a question, we have added a column to the right which tells what fraction, or proportion, of the 20 temperatures are a given value, such as 93 degrees. We see that there are two such temperatures, so the proportion is 2/20 or .10.

Proportions can be thought of as probabilities. Instead of asking what *proportion* of the alarms had a trip temperature of 93 degrees, we could ask a question in terms of probabilities. The question would be something like this: "If we pick a fire alarm at random, what is the *probability* that it has a trip temperature of 93 degrees?" We will consider what is meant by *at random* several times in this book. The idea of randomness is a very important idea in statistics. For now, we'll just say that at random means that every object (in this case, every fire alarm) has the same chance of being picked. Using symbols, we say

Probability (trip temperature of 93 degrees) = .10

Similarly,

Probability (trip temperature of 89 degrees) = .25

Probability will be discussed in more detail in Chapters Three and Four. For now, we will say only that probability is a way of saying *how likely* something is to happen.

From the table, we see that getting an alarm with a trip temperature of 85 degrees has a low probability. Getting an alarm with a trip temperature of 89 degrees has a higher probability.

Data from a stem-and-leaf table can be interpreted similarly.

Table 1.12 again gives the weights of the 50 girls. But this time we show the interval corresponding to each stem and the number of leaves for each.

TABLE 1.12

INTERVAL	FREQUENCY OF LEAVES	PROPORTION
80–89	3	.06
90–99	4	.08
100–109	10	.20
110–119	9	.18
120–129	8	.16
130–139	12	.24
140–149	2	.04
150–159	2	.04
Total	50	1.00

We can ask the question, "What is the probability that a girl chosen by chance will weigh between 100 and 109 pounds, for example?" We see that there are 10 girls in this interval. So, we say the probability is 10 out of 50. (We also notice that $10/50 = .20$.)

As a third example, suppose we toss a coin 100 times and get the results in Table 1.13.

We see that a head was obtained 48 out of 100 tosses. So the proportion of the 100

TABLE 1.13

OUTCOME	FREQUENCY	PROPORTION
Heads	48	$\frac{48}{100} = .48$
Tails	52	$\frac{52}{100} = .52$
Total	100	1.00

times that a head was obtained is .48 (48% of the time). This is also a probability.

Each of the examples we have given is obtained from looking at collected data and finding some proportion based on these data. These probabilities can be thought of as *experimental* probabilities, since they come from data produced by an experiment.

You may not think of recording the trip temperatures of fire alarms or weights of girls or tosses of coins as experiments. But in a sense they are. We will not deal with details here. But suppose a consumer group wants to know how safe certain fire alarms are. The group may obtain 30 such alarms and test them. Or suppose that we want to see whether a certain coin is fair. We can toss it lots of times and see how often we get heads.

These probabilities that come from data are represented by P. So, we write, based on our experiments,

$$P(\text{trip temperature of 93 degrees}) = .10$$
$$P(\text{weight between 100 and 109 pounds}) = .20$$
$$P(\text{heads}) = .48$$

There is another kind of probability, called *theoretical* probability. In the example of the coin, suppose that we continue to toss it, say, for 10,000 times. The toss of a coin can have two outcomes: Heads or tails (ignoring the possibility of its falling on its edge!).

If the coin is not biased, we expect neither outcome to occur more often, in the long run, than the other. Therefore, in 10,000 tosses, we would expect about 5000 heads and 5000 tails.

So, we have two equally likely outcomes, and thus the *theoretical* probability of heads is one-half, or .5. We write such probabilities with lowercase *p*.

$$p(\text{heads}) = .5$$
$$p(\text{tails}) = .5$$

This kind of probability is called theoretical because it has to do with ideal or assumed conditions. We *assume* that each outcome of the coin is equally likely. Or, we can *assume* that each outcome of a die is equally likely.

We can test our assumptions (our theory) by carrying out an experiment. We can toss a coin and see if we get heads about half of the time. If the coin is indeed fair, we would expect the experimental probability to get closer and closer to the theoretical probability as the number of tosses is increased.

Using symbols, we can write $P(\text{heads})$ gets closer to $p(\text{heads})$, for a fair coin, as the number of tosses gets larger and larger.

EXERCISES

1   A coin was tossed 200 times and the following results obtained:

	Frequency	
Heads	Tails	Total
103	97	200

Use these data to find:

(a) $P(\text{heads}) =$ _____ ,
(b) $P(\text{tails}) =$ _____ .

2   A six-sided die was rolled 120 times. Each of the sides was obtained as follows:

Side	Frequency
1	12
2	21
3	15
4	18
5	24
6	30
	120

The *experimental probability* of rolling a 1 is written as $P(1)$. Find each of these experimental probabilities using the given data.

(a) $P(1) =$ _____
(b) $P(3) =$ _____
(c) $P(6) =$ _____
(d) $P(\text{even number}) =$ _____
(e) $P(\text{odd number}) =$ _____

Do you think this die is fair? Why or why not?

3   A true die (a die that is not loaded) is rolled 300 times. About how many ones would you expect to get? How many sixes? Find these theoretical probabilities for a true die:

(a) $p(1)$
(b) $p(6)$
(c) $p(\text{even number})$

4   Toss a coin 50 times and record the number of heads and tails you get. Find $P(\text{heads})$ and $P(\text{tails})$ from your data. Compare your answer with that obtained in Exercise 1. Can you explain the differences, if any, obtained?

5   A new rocket is being tested. The same model of rocket is fired on 40 occasions and lift-off occurs on 35 occasions. What is $P(\text{lift-off})$ for this rocket?

6   The United Nations Demographic Year-book for 1977 reported that in 1975 the world's population was made up of 1,987,049,000 women and 1,979,956,000 men. Using these statistics, find:

(a) P(women)
(b) P(men)

Write your answer first as a fraction, then change it to a decimal, rounding to the nearest ten-thousandth.

7   A food corporation has developed a new product, product M. Tasters are recruited to sample the product and express their liking for it by marking from 1 (terrible) to 5 (fantastic). The table gives the results of 50 tasters.

Rating	Frequency
Terrible	2
Tolerable	18
OK	20
Good	7
Fantastic	3
	50

(a)   What proportion of persons rated product M as terrible?
(b)   What is the estimated probability that the product was rated as fantastic?
(c)   What is P(OK)?

8   J. E. Kerrich, while interned during World War II, tossed a coin 6000 times and obtained 3009 heads.[11] What is P(heads) for his experiment?

9   Roll a die 60 times and record the number of outcomes in a table like that in Example 2. Find these experimental probabilities from your data:

(a) $P(1)$
(b) $P(3)$
(c) P(odd number)
(d) P(even number)

10   The histogram in Figure 1.7 reports the results of tossing a coin 100 times. From this histogram, estimate:

(a) P(heads)
(b) P(tails)

FIGURE 1.7

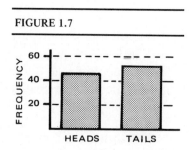

11   The frequency polygon in Figure 1.8 reports the results of rolling a die 60 times. From this graph, estimate:

(a) $P(1)$
(b) $P(4)$
(c) $P(6)$

FIGURE 1.8

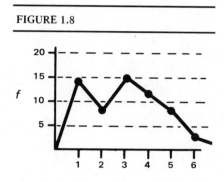

1.5   IT ALL ADDS UP
(Cumulative Frequencies and Proportions)

An important question often asked about data is: How many scores or values in the set are less than a given value? This kind of information is easily obtained from a frequency table by including an additional column. One of the columns, labeled *cumulative frequency*, tells how many scores in the set are less than or equal to that score.

As an example we will use the data in Table 1.14, which are test scores for 30 students in a statistics course.

TABLE 1.14

SCORE	FREQUENCY (f)	CUMULATIVE FREQUENCY (cf)
75	1	1
76	0	1
77	2	3
78	3	6
79	4	10
80	11	21
81	5	26
82	3	29
83	1	30
Total	30	

For example, a score of 78 has a cumulative frequency of 6. That is, 6 persons obtained a score of 78 *or less.* Similarly, 21 persons had a score of 80 or less, and so on. Thus cumulative frequencies provide a comparison of a score with others in the set.

Cumulative frequencies can be plotted on a graph, called a *cumulative frequency polygon.* A graph of the above data is given in Figure 1.9.

FIGURE 1.9

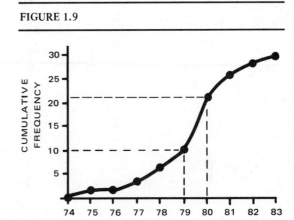

We now add another column, called *cumulative proportions.* This column gives us the proportion of persons with scores less than or equal to a given score. See Table 1.15.

TABLE 1.15

SCORE	f	P	cf	CUMULATIVE PROPORTION (cP)
75	1	.03	1	.03
76	0	0	1	.03
77	2	.07	3	.10
78	3	.10	6	.20
79	4	.13	10	.33
80	11	.37	21	.70
81	5	.17	26	.87
82	3	.10	29	.97
83	1	.03	30	1.00
	30	1.00		

A graph of cumulative proportion data is shown in Figure 1.10. This graph is called a *cumulative proportion polygon* (an old-fashioned name for this graph is *ogive*).

FIGURE 1.10

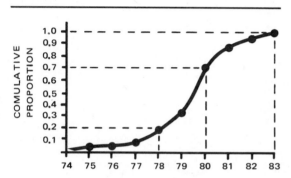

The cumulative proportion polygon is useful in showing the proportion of scores less than of equal to a given score. The above graph, for example, shows that .20, or 20%, of the persons obtained a score of 78 *or less*.

We can also think of cumulative proportions as probabilities. Take the example of the test scores. If one of the scores is picked at random, as in a lottery or raffle, what is the likelihood, or probability, that the score is 78 *or less*?

From Figure 1.10 we can read this probability as .20, as we have just noted.

We can also draw a graph of the proportions as given by the column labeled P in Table 1.15. This graph is given in Figure 1.11. It turns out that the estimated probability of getting a score less than or equal to 78 is approximated by the shaded region. (We assume that each of the 30 scores has equal chance of being picked.)

FIGURE 1.11

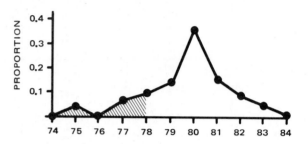

Thus we have

$$P(\text{score of 78 or less}) = .20$$

This can also be written

$$P(\text{score} \leqslant 78) = .20$$

Similarly,

$$P(\text{score of 80 or less}) = .70$$

Of course, this probability depends upon pulling out a score by chance as in a raffle. If you wanted a particular score and, for example, wrote it on a very large piece of paper, the probability of finding that score would differ from what we have calculated here. (Would it be greater or less?)

Let's go back one more time to the data on the weights of the 50 girls at Central High (see Table 1.16).

TABLE 1.16

STEM	LEAF
8	8, 9, 9
9	2, 9, 8, 2
10	5, 3, 6, 0, 8, 3, 0, 7, 1, 0
11	7, 4, 5, 4, 8, 4, 9, 0, 2
12	2, 0, 9, 7, 5, 8, 1, 5
13	2, 8, 7, 5, 9, 2, 4, 2, 1, 7, 5, 2
14	8, 6
15	2, 3

TABLE 1.17

STEM	LEAF	FREQUENCY	CUMULATIVE FREQUENCY
8	8, 9, 9	3	3
9	2, 9, 8, 2	4	7
10	5, 3, 6, 0, 8, 3, 0, 7, 1, 0	10	17
11	7, 4, 5, 4, 8, 4, 9, 0, 2	9	26
12	2, 0, 9, 7, 5, 8, 1, 5	8	34
13	2, 8, 7, 5, 9, 2, 4, 2, 1, 7, 5, 2	12	46
14	8, 6	2	48
15	2, 3	2	50
		Total 50	

We can add columns showing frequencies and cumulative frequencies, as in Table 1.17.

Let's interpret the frequency and cumulative frequency columns. Take the 8-stem. There are 3 leaves on this stem. The scores in the interval 80–89 are 88, 89, and 89. The cumulative frequency column also has a 3. That is, there are 3 leaves up until the end of the interval 80–89.

Now consider the 9-stem. There are 4 leaves on this stem. The cumulative frequency of 7 tells us that up until the end of interval 90–99 *there is a total of 7 leaves* (3 for the 8-stem and 4 for the 9-stem).

In order to show the correct meaning of the cumulative frequency data, we show the *end of each interval* for that particular cumulative frequency on the horizontal axis of the graph. So, on Figure 1.12 we see the values 89, 99, 109, and so on. Cumulative proportion polygons are drawn the same way.

FIGURE 1.12

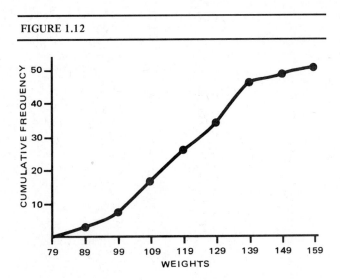

EXERCISES

1   This table shows the number of children in each of 50 households interviewed in a survey in Anabru.

N Children	f	Proportion
0	3	
1	8	
2	26	
3	10	
4	2	
5	0	
6	1	
	50	

(a)   Complete the column to show the proportion of families with each number of children.

(b)   Add a column and calculate the cumulative proportion for each number children.

(c)   Draw a cumulative proportion graph of these data.

(d)   Use your graph to find the proportion of families consisting of 2 or less children. Then find the proportion, from the graph, of families with 4 or less children.

2   The following gives the number of telephone calls, timed to the nearest minute, to a doctor's office.

Length of call (minutes)	1	2	3	4	5	6	7	8	9
Number of calls		18	29	23	15	8	2	1	3	1

(a)   Construct a histogram of these data, where the vertical axis gives proportions rather than frequencies.

(b)   Construct a cumulative proportion graph of the data.

(c)   From your graph, find:
   (i) the proportion of telephone calls that are 3 minutes or less;
   (ii) the proportion of telephone calls that are 4 minutes, or less;
   (iii) the length of the most frequent, or common, telephone call;
   (iv) the average length of a telephone call.

3   A six-sided die is rolled 120 times and the following outcomes are obtained.

Outcome	f	Proportion
1	15	
2	21	
3	23	
4	19	
5	17	
6	25	
	120	

(a)   Complete the table to show what proportion of the 120 rolls resulted in each of the 6 outcomes.

(b)   Add a column and calculate the cumulative proportions for the 6 outcomes.

(c)   Draw a cumulative proportion graph for these data.

(d)   From the graph, find these cumulative proportions:
   (i) The proportion of the rolls resulting in 3 or less.
   (ii) The proportion of the rolls resulting in 5 or less.
   (iii) The proportion of rolls resulting in more than 3.
   (iv) The proportion of rolls resulting in an odd number.

4   Find these estimated probabilities, using the outcomes of the 120 die rolls given in Exercise 3, above.

(a) $P(3 \text{ or less})$
(b) $P(5 \text{ or less})$
(c) $P(\text{more than 3})$
(d) $P(\text{even number})$
(e) $P(1 \text{ or } 2)$

1.6   TELLING IT LIKE IT *ISN'T*

About 20 years ago, there was a popular book entitled *How to Lie with Statistics.* Many people had a lot of fun with this book, because it certainly is possible to use statistics to give misleading or wrong impressions. Throughout our book, we give examples of how statistics might be misused. An important reason for studying statistics is to learn ways in which statistics is misused, so that we can read and make decisions more intelligently. Then, when you encounter statistics in newspapers, magazines, or on radio or television, you should be better able to interpret them correctly.

■ **Example 1—The Top 10 Millionaire States:** *The New York Times* reported on June 28, 1979 on the number of millionaires in the United States. The report first gave the numbers in terms of millionaires per 1000 of population for each state. When given this way, the top 10 states were as shown in Table 1.18.

TABLE 1.18

STATE	FREQUENCY PER 1,000 POPULATION
Idaho	26.7
Maine	7.7
North Dakota	6.9
Nebraska	6.6
Minnesota	5.7
Indiana	4.6
Wisconsin	4.1
Iowa	4.0
New Jersey	3.6
Connecticut	3.4

After looking at this report, one natural reaction is to remark that Idaho is a wealthy state—look at all its millionaires! However, it must be noted that many of these states also have a relatively small population and have no major cities.

Therefore, the data can be given another way and can present a rather different picture. The number of millionaires, reported state by state, can be given as shown in Table 1.19 for the new top 10 list.

TABLE 1.19

STATE	NUMBER OF MILLIONAIRES
New York	51,031
California	33,509
Illinois	31,131
Ohio	27,607
Florida	26,647
New Jersey	26,565
Indiana	24,345
Idaho	23,797
Minnesota	22,873
Texas	21,051

Notice that the number of states appearing in both lists is rather small: only New Jersey, Indiana, Idaho, and Minnesota appear twice, and they appear toward the end of the second list.

When data are presented, it is important to try and see how they are being used. As the "millionaire example" shows, data can be presented in different ways to tell different stories. ■

■ **Example 2—The Closing Gas Stations:** We have seen in this chapter that a picture, or graph, can be very helpful. But a graph also can be used to convey different messages with the same data. *Time,* in its August 22, 1977 issue stated in its story on closing gas stations that "nationwide, the number of stations has dropped from 226,000 in 1973 to 180,000 at present (1977)." Now, depending upon the impressions we wanted to give, we could use different versions of a graph. Figure 1.13(a) presents a picture of relative stability. After all, only a moderate decline is seen in a period of 4 years. Figure 1.13(b), on the other hand, shows a dizzying decline in the number of service stations in the United States. The data are the same, but the change of scale in the vertical axis of (b) presents a rather different picture. ■

■ **Example 3—Seeing Is Believing:**   Graphs can be used (or misused) in another way to create the impression intended by the author or artist. *Time* magazine (April 9, 1979) carried an article on the rising prices of crude oil from OPEC countries and the corresponding price increases of gasoline in the United States. The story was accompanied by two graphs, represented in Figure 1.14.[12]

Figure 1.14(a) shows a series of successively larger oil drums. The smallest, with dimension approximately .5 cm by .5 cm, is labeled with the price $2.41 per barrel, which was the price of light crude leaving Saudi Arabia on January 1, 1973. The largest, with dimension approximately 2 cm by 3.5 cm, is labeled with the price $13.34 per barrel, the price on January 1, 1979.

Figure 1.14(b) shows a series of service station gasoline pumps. The smallest, with dimension approximately .5 cm by 1.5 cm, is labeled with the price 37.3¢ per gallon, which was the average price of gasoline in the U.S. in 1973. The largest, with dimension approximately 2 cm by 5.0 cm, is labeled with the monthly averages of price per gallon for three months in 1979.

FIGURE 1.13

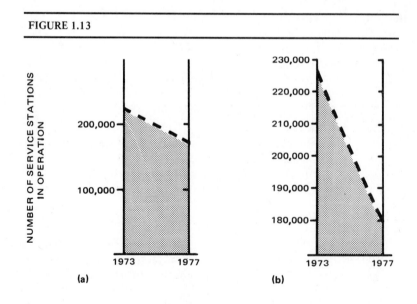

FIGURE 1.14

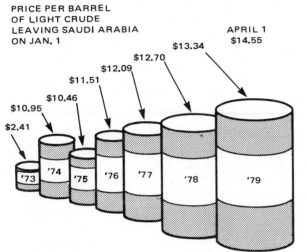

(a) IN THE BARREL

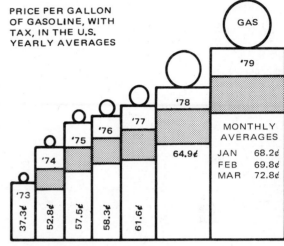

(b) AT THE PUMP

Although at the time that the article was published, there was great concern about inflation, there remains the question of how accurately those graphs represent the price situation. The drawing of the oil barrel for 1979, for example, is (in terms of area) over 25 times as large as the oil barrel for 1973, while the actual price increase was about sevenfold. An increase of 700% is indeed whopping, but it still is not a third of the increase suggested by the relative sizes of areas of the graphs.

What evidently happened is that the concern of the artist was focused upon the heights of the figures representing the price increase. The ratio of .5 centimeter (representing 1973) to 3.5 centimeters (representing 1979) is 1 to 7, an accurate depiction of the amount of price increase. ∎

EXERCISES

1   Conduct an analysis of the graphs representing the price increase for gasoline similar to the one done in this section for the price of oil.

2   Find examples from newspapers, magazines, and similar materials, of presentations of statistics that are subject to the kinds of misinterpretation presented in this section.

3   Try to identify other possible misuses of statistics. Keep in mind that this book will present others later.

Centers and Spreads

KEY PROBLEM

HELP WANTED

Company XYZ on the move is looking for a lab technician in its research division. Training on the job. Many employee benefits. Average salary at our company is $12,000. Call 555-3456 and ask for Mr. Phelps.

Jane was very interested in this advertisement and decided to check it out. When she did, she found the information about salaries shown in Table 2.1.

It is true that the average of the salaries is $12,000. (The total salary earned by the nine persons in the company is $60,000 + $8000 + $7000 + $7000 + $6000 + $5000 + $5000 + $5000 + $5000 = $108,000. Therefore, the average salary is $\frac{108,000}{9} = 12,000$.) But she also found that the salaries are bunched up at the low end of the scale. More than one-half of the people earn under $7,000. The salary of $12,000 quoted in the ad is not at all typical of the pay at XYZ.

In this chapter we will learn ways of describing characteristics of sets of data like the salaries at XYZ in meaningful ways.

TABLE 2.1

SALARY	NUMBER OF PERSONS
60,000	1
8,000	1
7,000	2
6,000	1
5,000	4

2.1 CENTERS
(Measures of Central Tendency)

One of the first things we usually want to know about a set of data such as salaries, prices of stereos, or number of home runs is, "About how much?" or "About how many?"

About how much do people earn at company XYZ?

About how much does a decent, but reasonaly-priced stereo cost?

About how many home runs per game would Reggie be expected to make next year?

When we ask, "About how much?" we probably want to know the value of a score at the middle, or center, or a set of data, such as we can see in a stem-and-leaf table. Table 2.2 is the stem-and-leaf table of the weights of the 50 girls at Central High (from Chapter One).

TABLE 2.2

STEM	LEAF
8	8, 9, 9
9	2, 9, 8, 2
10	5, 3, 6, 0, 8, 3, 0, 7, 1, 0
11	7, 4, 5, 4, 8, 4, 9, 0, 2
12	2, 0, 9, 7, 4, 8, 1, 5
13	2, 8, 5, 7, 9, 2, 4, 1, 7, 5, 2
14	8, 6
15	3, 2

From this table, we can see that the girls weigh about 120 pounds, on the average. This value of 120 pounds is somewhere in the center of the set of scores.

Here are the prices of 10 stereos (rounded to the nearest $5.00):

$100 $120 $125 $115 $105
$ 90 $100 $110 $120 $125

The stem-and-leaf table of these data is given in Table 2.3.

TABLE 2.3

STEM	LEAF
9	0
10	0, 5, 0
11	0, 5
12	0, 0, 0, 5

We can see from the table that the stereos cost an average of *about* $110. Again, this is a *center*, or middle value, of the data.

MEAN

The center of a set of data can be defined in different ways. The most common definition is the *average*, or *mean*.

The average or mean of a set of data is usually easy to find. Just add all the values and divide by the number of values.

■ **Example 1:**   Five tape decks are listed at these prices:

$100 $120 $200 $200 $450

The average price is:

$$\frac{\$100 + \$120 + \$200 + \$200 + \$450}{5} = \frac{\$1070}{5}$$

$$= \$214.00 \ ■$$

Finding the mean of a set of data: To find the average, or mean, of a set of data, add all the data and divide by the number of individual values in the set. A little more precisely, we can say that if a set consists of data values X and there are N such numbers, then the mean is

$$\overline{X} = \frac{\text{sum of } X}{N}$$

MEDIAN

Another way of looking at the idea of center is given in this next example.

■ **Example 2:**   Pete is interested in how he is doing in his statistics course. His instructor gives him the following results of the midterm exam.

You got a score of 62.
About half of the class got more than 62 on the test.

What does Pete know about how he is doing in the course, relative to the rest of the class? Let's suppose all the test scores were in a stem-and-leaf table, as in Table 2.4.

TABLE 2.4	
STEM	LEAF
3	8
4	5, 0, 1, 1, 3, 2
5	0, 0, 2, 1, 5
6	②, 5, 6, 6, 4, 7
7	1, 2, 8
8	2, 7, 7
9	0

We see that there are 25 students in Pete's class. Twelve students got a score lower than 62 and 12 got a score higher than 62. ■

The notion of a middle score is often used to express the concept of center, so this score is given a special name, *median*. (To help you remember the term, think of the median strip on a freeway—it divides the total road way in half.)

■ **Example 3:**   The following quiz scores were obtained by a class of 15 students. Find the median score on the quiz.

Scores obtained:

8, 5, 5, 9, 10, 2, 3, 3, 5, 5, 8, 7, 4, 5, 4

Solution:

Scores Arranged from Highest to Lowest

10
9
8
8
7
5
5
5 Middle score = median
5
5
4
4
3
3
2 ■

So far, our examples have involved an odd number of cases. If the number of scores is even, a slight adjustment is needed in the definition.

■ **Example 4:**   Sue is looking for a used car and finds six cars within her budget that interest her. The prices are:

$500 $650   <u>$700 $900</u> $925 $950
Middle Values

Solution:

Here, the halfway point in the price range she is considering is between $700 and $900. So, we find the average of $700 and $900; the median price is thus $800. ■

One other minor complication can arise. Let's think about this example of quiz scores. There is an even number of students (8), but the middle scores are the same. Here, the median is 7, since

$$7 = \frac{7 + 7}{2}$$

Quiz Scores

10
9
9
7
— Median score = 7
7
4
1
1

Finding the median of a set of data: To find the median of a set of scores, first arrange them from high to low.

Then: If there is an odd number of scores, find the score that has as many scores above it as below it.

If there is an even number of scores, take the two middle scores and find their mean (halfway point).

MODE

The mode (or modal score) of a set of data is another kind of center. For example, if we are looking for a suit and, in glancing around the shop, we notice a lot of price tags with $200 on them, we probably would not attempt either to add all the prices to find the mean, nor to list them from highest to lowest to calculate the median. We would be more likely to say, "We went shopping for a new suit yesterday, but do you know that they cost about $200?" What we are giving is our impression of the typical price of a suit based upon seeing lots of price tags for $200. This brings us to the concept of *mode.*

The most frequent or most common score in a set of scores is called the *mode.*

■ **Example 5:**   Table 2.5 shows the prices of 10 bicycles. What is the modal, or most common, price?

Solution:

TABLE 2.5	
STEM	LEAF
9	0
10	0, 5, 0
11	0, 5
12	0, 0, 0, 5
13	

We see that more bicycles are $120 than any other price (three bicycles cost $120). So, the modal price is $120. ■

A stem-and-leaf table is very helpful in locating the modal value of a set of data. Suppose the prices of the bicycles were as in Table 2.6.

TABLE 2.6

STEM	LEAF
9	0, 0, 0, 5
10	5
11	
12	
13	0, 5, 0, 0

In this case, two prices are more popular than the others. This set of data has two modes: $90 and $130. If a set of data has two modes, it is called *bimodal*.

EXERCISES

1   The prices of a liter of milk at six stores were 60¢, 63¢, 62¢, 68¢, 60¢, 62¢. Find the mean price of milk per liter.

2   The number of accidents, A, was observed at a busy intersection for eight weekdays:

$$3, \quad 2, \quad 0, \quad 2, \quad 5, \quad 0, \quad 1, \quad 2$$

Find $\bar{A}$, the mean number of accidents per day.

3   Samples of 25 flash bulbs were taken once a week for 5 weeks and tested. The number of defectives, D, found in each of the samples is shown below. What is $\bar{D}$, the mean number of defectives per sample?

Number of Defectives (D)

1
3
0
0
7

4   Carlos earned the following numbers of points playing basketball in each of 8 games. Find the mean points earned by Carlos.

$$21, \quad 10, \quad 14, \quad 5, \quad 8, \quad 19, \quad 19, \quad 12$$

5   The "normal temperature" for a particular day of the year is computed by finding the mean temperature for that day for the past 20 years. Here are July 1 temperatures for Anabru from 1960 until 1979, in degrees Celsius.

$$15 \quad 17 \quad 14 \quad 14 \quad 16 \quad 19 \quad 18 \quad 15 \quad 15 \quad 15$$
$$14 \quad 12 \quad 13 \quad 16 \quad 16 \quad 17 \quad 15 \quad 15 \quad 16 \quad 14$$

Estimate the mean temperature. Then calculate the mean, which is the normal temperature for Anabru for July 1.

6   The city council of McBride is investigating the efficiency of the fire department. The time taken by the fire department to respond to a fire alarm was surveyed. It was found that the response times in minutes for 15 alarms were:

1, 3, 2, 2, 1, 9, 4, 6, 1, 10, 1, 4, 5, 10, 1

What was the mean response time? What was the median response time?

7   The scores of seven persons on a statistics quiz were:

$$9, 9, 8, 6, 5, 4, 4$$

What was the median score on the quiz? What was the mean score?

8   Polly scored the following numbers of points in 8 basketball games:

$$11, 5, 8, 11, 10, 7, 12, 10$$

What median number of points did she score?

9   Mr. and Mrs. Rivera invited their grandchildren for Sunday dinner. The ages of their grandchildren are:

3, 5, 4, 5, 10, 5, 1

What is the modal age of the grandchildren? (That is, what is the mode of the grandchildren's ages?)

10   The numbers of runs batted in by 10 batters in the National League for 1982 were:

94, 93, 71, 58, 76, 37, 89, 36, 39, 29

What was the median number of runs batted in by these players?

11   The winning team in the Roller Bowler Bowling League got these scores:

190, 205, 201, 199, 220

What was the median score for the team?

12   Three coins were tossed 10 times. The number of heads on each toss was:

2, 0, 1, 1, 3, 1, 3, 2, 2, 1

What was the mean number of heads per toss? What was the median number of heads per toss?

13   Toss three coins 10 times and record the number of heads on each toss. Draw a histogram of your results. Then find the mean and median number of heads per toss.

14   Find the mode for the number of baskets scored by five players on a team:

17, 21, 21, 18, 24

15   What is the mode for this set of test scores?

31, 33, 29, 31, 40, 45, 33, 31

16   What is the modal response time for the fire department in Exercise 6 of this section? If the city council wishes to choose between reporting the mean, median, or mode of the response time, what would be the advantages of choosing each? (First, find all three centers.)

17   The word *mode* is related to a French word meaning *fashion*. In what way can the mode of a set of numbers be thought of as "fashionable"? For example, here is a frequency distribution of hair styles in the 1950s for a sample of 29 twelfth grade students:

Haircut	f
Sideburns	2
Duck cuts	15
Crew cuts	12

Use these data to help answer the question.

18   Four coins are tossed 20 times, with the following numbers of heads obtained on each toss:

2, 3, 1, 2, 2, 0, 1, 2, 2, 4,
3, 2, 2, 1, 1, 2, 2, 1, 3, 2

Make a frequency table of the number of heads obtained. Draw a histogram of the data, where the vertical axis gives the proportion of times each number of heads was obtained. What is the modal number of heads? What is the mean number of heads? What is the median number of heads?

19   Toss four coins 20 times and record the number of heads obtained on each toss. Draw a histogram of your data. What is the mode of the number of heads obtained per toss? How many heads are most likely to be obtained, based on your data, when four coins are tossed?

PROJECT CORNER

1.   In baseball, a batting average tells mean number of hits per time at bat. It is written as a three-place decimal. From the sports page of a newspaper, find the batting averages of the three top batters of a team and interpret each.

2.   From the sports page of the newspaper, find ways in which the center of a set of data (mean, median, or mode) are used. (Batting average is discussed in Problem 1.)

3.   Go through yesterday's newspaper and find five ways other than in sports in which the mean, median, or mode are used. Clip the articles and write a brief note explaining how the idea of the center of a set of data is used in each article.

2.2 PROPERTIES OF CENTERS

Here are the hourly wages of five workers:

$2.50, $3.00, $3.50, $4.00, $5.00

This set of wages has a median of $3.50 and a mean of $3.60.

Now let's see what happens if one of the workers gets a huge pay increase. We change $5.00 to $10.00, getting:

$2.50, $3.00, $3.50, $4.00, $10.00

This set of wages has a median of $3.50 and a mean of $5.75.

Notice first that the median remains at $3.50. One-half of the workers still earn more and one-half still earn less than $3.50. But the mean wage in the second example has increased so much that it is larger than the wages of four of the workers. We see here that the mean is influenced by large scores in the set. It also is influenced by small scores in the set. The median is not influenced by very large or small scores.

We say that the median is a more *robust* measure than the mean. It resists influences of individual scores in the set. The median is a *robust measure of central tendency*. It resists the influence of extreme (very large or very small) scores. The mean, on the other hand, can be strongly influenced by the presence of very large or very small scores.

Though not robust, the mean has many other properties that make it a very important measure in statistics. We will consider only one of these properties here.

THE "UNDOING PROPERTY" OF THE MEAN

It is sometimes important to regain information that may not have been given in a statistical report, for example. We can then use the undoing property of the mean.

■ **Example:**   Suppose that we are told that the mean value of scholarships given to ten people was $450. How much money was awarded for scholarships?

Solution:

We remember the rule for finding the mean of a set of numbers. Let $\overline{S}$ stand for the mean value of a scholarship:

$$\overline{S} = \frac{\text{sum of all scholarships}}{\text{total number}}$$

So we can say that

Sum of all scholarships = $\overline{S} \times$ total number
= $\$450 \times 10 = \$4500$

A total of $4500 was awarded in scholarships.

(If you did not follow this solution, remember this example:

$$\frac{12}{6} = 2 \text{ means } 12 = 2 \times 6)$$

Neither the median nor the mode has this undoing property. ∎

EXERCISES

1   Find the mean and median of these two sets of data:

(a)   10, 15, 20, 25
(b)   10, 15, 20, 2500

How have the mean and median been affected by replacing 25 in set (a) with 2500 in set (b)?

2   Find the mean and median of these two sets of data:

(a)   100, 110, 120, 130, 140
(b)   0, 110, 120, 130, 140

How have the mean and median been affected by reducing 100 in set (a) to 0 in set (b)?

3   Ron mowed lawns last summer. His weekly earnings for 7 weeks are given. Find the mean and median values of his earnings.

$22.00, $30.00, $5.50 $17.50,
$32.50, $28.00, $30.00

4   In Exercise 3 of this section, find the mode of Ron's weekly earnings. Which of the three averages—mean, median, or mode—presents a better picture of Ron's earning power last summer?

5   Refer to Exercise 6 of Section 2.1, which gives the response times of the McBride Fire Department. If the city council is attempting to demonstrate the effectiveness of the fire department, which center—mean or median—would it be likely to use in its report?

6   A dispute over salaries has arisen in a print shop between the workers and the owner. The owner believes she should include her $30,000 salary when reporting the mean (average) salary at the shop. The workers do not agree. Do you see why? Here are their salaries:

$5,000 $6,500 $7,000 $7,000 $10,000

If the owner knew a little more statistics, she could avoid this dispute. How? (How might the workers, if they also knew more statistics, respond?)

7   An outpatient clinic reported that the average (mean) charges per patient were $20.50 last month, and that a total of 110 patients were served by doctors during that month. What was the total of the doctors' charges for that month?

8   Mary got a 75 and an 80 on two tests this semester. If a grade of B requires an average (mean) of at least 80, what must she get on her next test in order to get a B average?

9   If Mary (in the above exercise) were able to persuade her teacher to average grades using medians instead of means, what minimum grade would she need on her next test to have a B average?

10   Two coins were tossed 12 times. Here are the outcomes (*HT* means one coin showed heads and one showed tails).

    HH TH HH TH HT HH
    HT HH TH HT HH TH

Make a table to show how many times these outcomes were obtained:

(a)   2 heads
(b)   1 head
(c)   0 heads

What is the modal number of heads obtained?

11   Here are the heights (in inches) of 25 ten-year-old girls:

54	55	54	52	54
53	56	56	57	56
58	56	55	54	60
57	58	55	54	57
53	50	59	56	55

Find the mean, median, and mode of these data. Compare them. Which measure of central tendency is more appropriate for reporting the "average" height of the 25 girls?

12   The number of issues of a monthly magazine read per year has been reported as follows:[1]

Number of issues read	0	1	2	3	4	5	6	7	8	9	10	11	12
% of people	40	15	5	2	1	1	1	0	1	1	3	10	20

(*Hint:* 40% of the people read no issues of the magazine, 15% read one issue, and so on.)

Find the mean, median, and mode for these data. What measure of central tendency would you use in reporting to the publisher of the magazine about how much the magazine is read?

13   **Key Problem Recap—*Help Wanted—Average Salary is $12,000:*** Reread the key problem for this chapter. What measure of central tendency was being used in the ad? In what measure do you think Jane would be more interested as she decides whether to take the job at XYZ?

2.3 SPREADS
(Measures of Variation)

In the first section of this chapter we learned that one important thing to know about a set of data is its center—mean, median, or mode. Now we will learn about another important property of a set of data, how much the data *vary*.

Tall people and short people, hot temperatures and cold temperatures, bright stars and dull stars—things that vary are a part of our world. Heights vary; temperatures vary; light intensity varies.

Table 2.7 shows the high temperature for each month in a town in northern Manitoba.

TABLE 2.7

MONTH	AVERAGE TEMPERATURE (°C)
Jan.	−22
Feb.	−18
Mar.	−10
Apr.	0
May	9
June	15
Jul.	18
Aug.	17
Sept.	10
Oct.	3
Nov.	− 8
Dec.	−17

These data show that the temperature varies a great deal in northern Manitoba. The winters are extremely cold and the summers are rather warm. The greatest variation in temperature is between the months of January and July, as illustrated by Figure 2.1.

As a second example, Figure 2.2 shows the prices for 1 pound of ground beef across the United States in June, 1979.[2] The price ranged from about $1.45 in Los Angeles to $1.88 in Dallas.

The *range* of a set of data is one way in which we can describe how much the data vary. From the table of mean temperatures, we see that the low temperature is −22° in January and the high is 18° in July. This is a *range* of 40° (from 22° below zero to 18° above zero).

The price of ground beef is about $1.88 in Dallas and $1.45 in Los Angeles. This is a range of $.43.

 The ***range of a set of data*** is $X_H - X_L$ where X_H is the highest value and X_L is the lowest value in the data.

If a stem-and-leaf table is used, the range of a set of data can be obtained rather easily.

FIGURE 2.1

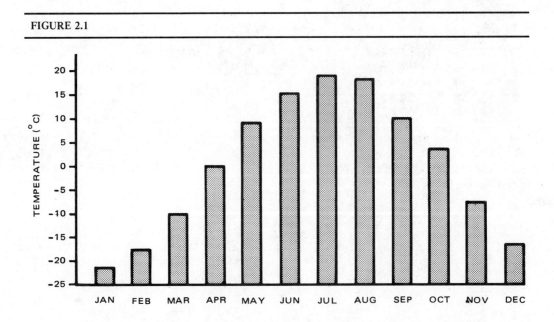

FIGURE 2.2

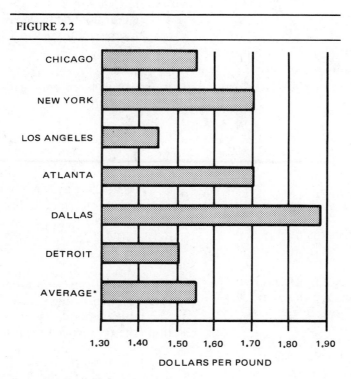

*FOR 19 CITIES SURVEYED

■ **Example:** What is the range of the weights of the 50 girls at Central (Table 2.8)?

TABLE 2.8

STEM	LEAF
8	8, 9, 9
9	2, 9, 8, 2
10	3, 5, 6, 8, 3, 0, 7, 1, 0
11	7, 4, 5, 4, 8, 4, 0, 9, 0, 2
12	2, 0, 9, 7, 4, 8, 1, 5
13	2, 8, 5, 7, 9, 2, 4, 1, 7, 5, 2
14	8, 6
15	3, 2

Solution:

We can see that the weights range from a low of 88 pounds to a high of 153 pounds. The range of these data is

Range = 153 − 88 = 65 pounds

While the range is easy to calculate and is often used, it has a serious drawback. It makes use of only two pieces of data, the highest and lowest, and ignores all other values. Consider this example:

Bill and Judy got these scores on the same 5 quizzes:

Bill: 4, 4, 4, 4, 9
         Range of quiz scores for Bill: 9 − 4 = 5

Judy: 4, 7, 7, 8, 9
         Range of quiz scores for Judy: 9 − 4 = 5

Notice that each set of scores has the same range, 5. But the sets of scores are rather different. Bill had mostly fours. His mean score was 5. Judy's scores were more evenly spread out between 4 and 9. Her mean score was 7.

EXERCISES

1 The following table shows the number of calories in yogurt and milk (values are for 1 cup of milk or 8 ounces of yogurt).[3]

Food	Calories
Whole milk	150
Low-fat milk	119
Strawberry low-fat yogurt	194
Plain low-fat yogurt	119
Strawberry whole-milk yogurt	211

Which food has the least number of calories? The greatest number? What is the range of number of calories in these foods?

2 The average price of a first-class hotel room in several cities around the world was reported in 1980 as follows.[4]

Cities	First-class Hotel Room (per night)
Zurich	$120
Chicago	98
New York	106
Stockholm	109
Tokyo	144
Paris	113
London	139

In what city is a first-class hotel room most expensive? Least expensive? What is the range of prices?

3 Below are the average (mean) hourly wages for workers in the indicated industries in 1947.[5]

Food Products	Tobacco Mfg.	Textile Mills	Printing	Chemicals	Oil
1.06	.91	1.04	1.48	1.22	1.50

Draw a graph of the data. Between which two industries is there greatest variation (difference) in wages?

4 The following table gives hourly wages for the same industries in 1977 (see Exercise 3).[5]

Food Products	Tobacco Mfg.	Textile Mills	Printing	Chemicals	Oil
5.34	5.50	3.97	6.09	6.39	7.27

Between which two industries is the variation now the greatest?

5 Comparing the data in Exercises 3 and 4 of this section, which industry has seen the greatest increase in the size of the hourly wage in the 30-year period 1947–1977? Which industry has seen the least increase?

6 Musical tones are produced by vibration, such as when strings on a guitar are plucked. The table below shows the amount of vibration (the frequency) for four musical notes.

Note	Frequency
Middle C	261
E	328
G_1	391
C	522

What is the range of frequencies?

7 Find the temperature range for the following cities, according to the information given in Figure 2.3.

City	High	Low	Range
Anchorage			
Denver			
Flagstaff			
Louisville			
Edmonton			

(*Note:* Canadian temperatures given in both Fahrenheit (F) and Celsius (C)).

8 Which city had the least (smallest) temperature range for the day reported? Which had the greatest?

FIGURE 2.3

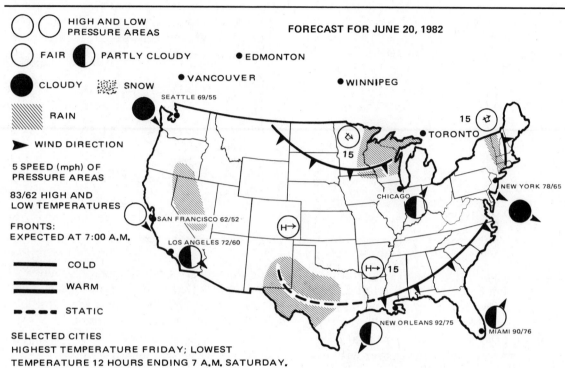

○○ HIGH AND LOW
PRESSURE AREAS

○ FAIR   ◑ PARTLY CLOUDY   ● EDMONTON

● CLOUDY   ⣿ SNOW   ● VANCOUVER   ● WINNIPEG

⣿ RAIN

▶ WIND DIRECTION

5 SPEED (mph) OF
PRESSURE AREAS

83/62 HIGH AND
LOW TEMPERATURES

FRONTS:
EXPECTED AT 7:00 A.M.

━━━━ COLD

═══ WARM

■━■━■ STATIC

FORECAST FOR JUNE 20, 1982

SEATTLE 69/55
15
TORONTO
CHICAGO
SAN FRANCISCO 62/52
NEW YORK 78/65
LOS ANGELES 72/60
H→
H→ 15
NEW ORLEANS 92/75
MIAMI 90/76

SELECTED CITIES
HIGHEST TEMPERATURE FRIDAY; LOWEST
TEMPERATURE 12 HOURS ENDING 7 A.M. SATURDAY.

CITY	HI	LO
Albany	75	54
Albuquerque	92	63
Amarillo	92	59
Anchorage	61	48
Asheville	78	65
Atlanta	81	65
Atlantic City	82	65
Austin	92	73
Baltimore	82	66
Billings	66	47
Birmingham	84	65
Bismark	60	48
Boise	88	59
Boston	79	61
Brownsville	95	77
Buffalo	67	55
Burlington	71	51
Casper	61	45
Charleston, SC	80	70
Charleston, W.VA	77	55
Charlotte, NC	83	67
Cheyenne	64	46
Cincinnati	75	53

CITY	HI	LO
Cleveland	73	51
Columbia, SC	81	68
Columbus	75	51
Dallas	89	70
Dayton	55	53
Denver	68	51
Duluth	60	38
El Paso	96	64
Fairbanks	62	44
Fargo	68	45
Flagstaff	78	39
Great Falls	66	46
Hartford	75	61
Helena	73	43
Honolulu	87	73
Houston	92	80
Jackson, Miss.	87	68
Jacksonville	82	70
Juneau	65	48
Kansas City	81	66
Las Vegas	102	73
Little Rock	82	63
Los Angeles	65	59

CITY	HI	LO
Louisville	78	55
Lubbock	93	62
Memphis	84	68
Miami	85	81
Minneapolis-St. Paul	73	53
Nashville	83	58

CANADIAN TEMPERATURES

CITY	F	F	C	C
Calgary	79	47	26	8
Edmonton	80	48	26	8
Montreal	73	58	22	14
Ottawa	71	55	21	12
Regina	73	49	22	9
Toronto	72	56	22	13
Vancouver	80	62	26	16
Winnipeg	65	48	18	8

9 The table below gives heights of basketball players on two teams. Which team has the least range in height? What is the range?

Position		Height (inches) Raiders	Bullets
Center		78	77
Forward	L	77	76
	R	77	76
Guard	L	72	74
	R	73	74

10 Randy is quarterback for the Maroons. His record for yards gained by passing for last year and this year are as shown in the table. For which year was his passing record more consistent? Explain.

Game Number	1	2	3	4	5	6	7	8	9	10
Yards Gained Last Year	43	89	54	88	54	58	72	77	88	67
Yards Gained This Year	54	32	84	121	87	43	38	53	73	63

PROJECT CORNER

1. Calculate the normal temperature (defined in Exercise 5 of Section 2.1) for your city. By going through old newspapers or other records, include temperatures on this day of the year for at least 20 years. Compare your answer with official records.

2. Get the Message?

Secret messages can be sent using coded messages called *cryptograms*. Here is a cryptogram:

XUTCO UEYET UETNU TYYVA BBLB

The letters actually making up the message have been substituted for other letters to make this coded message. The letters are in groups of five so that the spacing will not provide clues about short words and long words.

Cryptographers, people who study how to break codes, make use of the fact that letters of the alphabet appear with different frequencies in written English. The letter *E*, for example, has been found to appear frequently, while *Z* seldom appears. So cryptographers have made tables that show how often the letters of the alphabet appear in various kinds of writing. They then use this information to help them decide which letters in the coded message represent the letters in the original message.

The message above is too short and provides too small a sample to provide good estimates of the likelihood of various letters appearing in the coded message. It is given here just to give you an idea of what a cryptogram is. The message is:

USING STATISTICS IS A BREEZE

Notice, for example, that *B* appeared three times in the cryptogram. The letter *B* is one of the most frequently used letters in the cryptogram. (Not the *most* frequent though—there are five *U*s in the cryptogram.) On this basis we may think that *B* stands for *E*. (Or, we may think that *U* stands for *E*, based on our small sample of information.)

As a project, take a piece of writing of several hundred words, such as from a newspaper or magazine article. Make a table showing how often each letter of the alphabet is used in the article. Then make a cryptogram based on your table and see if a friend can decode your message.

This method of codebreaking is actually used. See, for example, the book *Secret and Urgent* by Fletcher Pratt, Dover Publications (reprinted).

2.4 BOXES AND WHISKERS

A picture is worth a thousand words. That's a good reason for using graphs in statistics. One way of showing some important information about a set of data, such as its center and spread, is to use a box-and-whisker diagram (or plot). We first show a stem-and-leaf table for the prices (in dollars) of 12 video games (Table 2.9).

TABLE 2.9

STEM	LEAF		BOX AND WHISKER
7	0, 0		
8	3	$\leftarrow Q_1$	
9	0, 1, 6, 6		Middle 50%
10	0, 5		of Data
11	4	$\leftarrow Q_3$	
12	1, 6		

A box is drawn in such a way that it encloses the middle half of the data. In order to draw in the box, we need to know where the middle half of the data is located. So we find the points which mark off the bottom one-fourth of the data and the top one-fourth of the data.

In our example, since there are twelve pieces of data, we need to know the lowest three prices (one fourth of 12 is 3) and the top three prices. The lowest 3 prices are:

$70, $70, $83

The middle 6 prices are:

$90, $91, $96, $96, $100, $105

The top 3 prices are:

$114, $121, $126

Since we have divided the data in two quarters, the points at which these divisions are made are sometimes called "quartiles."

The number dividing off the lower one fourth of the data is the "first quartile" (written Q_1). So,

$$Q_1 = \$90$$

The number dividing off the upper quarter of the data is the "third quartile" (Q_3). So,

$$Q_3 = \$105$$

As part of Table 2.9, we see the box drawn to include the middle range of prices—they are the six prices ranging from $90 to $105.

The "whiskers" are the vertical lines on either end of the box diagram, showing the range of the data. The range is from $70 to $126. The line dividing the box in two parts indicates, approximately, the median of the data. The median is $96.

■ **Example 1**: Do a box-and-whisker plot of the weights of the 50 girls. We use the data from before, but this time we order the "leaves" (Table 2.10). This makes it easier to find the middle 50% of the weights. Since one fourth of 50 is 12.5, we need to find the lowest (approximately) 12 weights and the highest 12 weights. Since the leaves are ordered, it's easy to count up for the lowest 12 values and down for the highest 12 values. The middle 50% of the girls weigh, approximately, from 103 lbs to 132 lbs. ■

TABLE 2.10

STEM	LEAF	BOX AND WHISKER
8	8, 9, 9	
9	2, 2, 8, 9 Q₁	
10	0, 0, 0, 1, 3, 3, 5, 6, 7, 8	
11	0, 2, 4, 4, 4, 5, 7, 8, 9	
12	0, 1, 2, 4, 5, 7, 8, 9	
13	1, 2, 2, 2, 4, 5, 5, 7, 7, 8, 9	
14	6, 8	
15	2, 3 Q₃	

■ **Example 2—A "Paneful" Problem**: Two companies that manufacture window glass have given bids for a contractor who is building a library. Since glass that varies in thickness can cause distortions, the contractor decided to measure the thickness of panes of glass from each factory, at several locations on each pane. Table 2.11 shows measurements for the two panes, one from each manufacturer.

TABLE 2.11

COMPANY A	COMPANY B
10.2	9.4
12.0	13.0
11.6	8.2
10.1	14.9
11.2	12.6
9.7	7.7
10.7	13.2
11.6	12.2
10.4	10.2
9.8	9.5
10.6	9.9
10.3	9.7
8.5	11.5
10.2	11.5
9.7	10.5
9.2	10.6
8.6	6.4
11.3	13.5

We now draw box-and-whisker plots of the measurements from each company.

First, let's prepare stem-and-leaf tables for each set of measurements (Table 2.12). (In the tables, the measurements are multiplied by 10.) Each leaf has been ordered in size from smallest to largest. In order to help us compare the two sets of data we use a common stem.

TABLE 2.12

COMPANY A LEAF	STEM	COMPANY B LEAF
	6	4
	7	7
5, 6	8	2
2, 7, 7, 8	9	4, 5, 7, 9
1, 2, 2, 3, 4, 6, 7	10	2, 5, 6
2, 3, 6, 6	11	5, 5
0	12	2, 6
	13	0, 2, 5
	14	9

Median = 10.3 (rounded)   Median = 10.6 (rounded)

Now we make some overall observations about the two sets of data. The range of measurements from Company B is greater than for Company A.

Company A: range = 12.0 − 8.5 = 3.5 millimeters

Company B: range = 14.9 − 6.4 = 8.5 millimeters

Notice that in Company A the data are bunched in the interval 10.1–10.9, while in Company B the data are spread out more evenly, particularly for the stems 9 to 13.

Box-and-whisker plots help to describe the two sets of data.

The first and third quartiles for the data are:

Company A	Company B
$Q_1 = 9.7$ mm	$Q_1 = 9.5$ mm
$Q_3 = 11.2$ mm	$Q_3 = 12.6$ mm

Table 2.13 shows the box-and-whisker plots.

TABLE 2.13

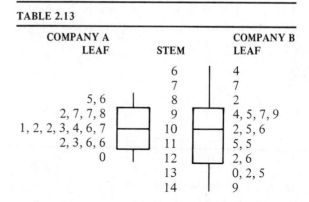

COMPANY A LEAF	STEM	COMPANY B LEAF
	6	4
	7	7
5, 6	8	2
2, 7, 7, 8	9	4, 5, 7, 9
1, 2, 2, 3, 4, 6, 7	10	2, 5, 6
2, 3, 6, 6	11	5, 5
0	12	2, 6
	13	0, 2, 5
	14	9

For Company A we see that the middle half of the data is between 9.7 and 11.2 millimeters and for Company B the middle half of the data is between 9.5 and 12.6 millimeters.

The window glass produced by Company A has less variation in thickness and should, therefore, have less distortion than that of Company B. ■

EXERCISES

1   Here are the number of home runs scored by the American League between 1951 and 1965 (from Table 1.8). Do a stem-and-leaf and box-and-whisker plot of the data.

33	52	43	32	37
52	42	42	42	40
61	48	45	49	32

2   Draw a box-and-whisker plot of the salary data from the key problem. What information does this plot give that will be useful to people thinking of working for Company XYZ?

3   Draw a box-and-whisker plot of the temperatures for the town in northern Manitoba (Section 2.3).

4 In a psychology experiment, each of 30 dogs learned to jump over a barrier. The data below tell the number of trials each dog took to learn to jump.[6]

```
 5   13   16   13    7
13   11   10   24   11
16   11    8    8    7
 6   13    7    9   13
 8   17    6   18   14
10    5   12   17   12
```

Draw a stem-and-leaf plot and a box-and-whisker plot of these data.

5 Penelope is looking for the perfect city, weatherwise, in which to live. She can't stand changes in temperature. She likes a climate that's pretty much the same year around. Which of the two cities listed below would she prefer? Use box-and-whisker plots in answering this question.

	Camelot	Eldorado
Jan.	29	15
Feb.	30	20
Mar.	35	22
Apr.	40	30
May	42	45
June	58	58
July	60	78
Aug.	59	77
Sept.	50	60
Oct.	42	58
Nov.	38	32
Dec.	30	20

6 The manufacturers of resistors for electronic circuits have put in bids to a television company. Their prices are comparable, so the television company purchases ten resistors of each brand. The resistors are marked 100 ohms. Each is measured for resistance and the following values are found.

Which company, Alpha or Beta, would you recommend as the supplier for the television company? Why?

Alpha	Beta
95	99
91	103
105	101
103	98
95	99
87	102
106	103
105	97
79	101
103	98

2.5 DEVIATION SCORES (The Snow of 1979)

In 1978–1979 Chicago had a very heavy snowfall, as is shown by Table 2.14. The average (mean) snowfall for Chicago is 42.6 inches. In 1978–79 the snowfall was 89.7 inches, or 47.1 inches above average ($89.7 - 42.6 = 47.1$).

TABLE 2.14

WINTER	SNOWFALL (INCHES)	DIFFERENCE (FROM AVERAGE SNOWFALL)
1967–68	28.4	$28.4 - 42.6 = -14.2$
1968–69	29.4	$29.4 - 42.6 = -13.2$
1969–70	77.0	$77.0 - 42.6 = +34.4$
1970–71	37.9	$37.9 - 42.6 = - 4.7$
1971–72	46.8	$46.8 - 42.6 = + 4.2$
1972–73	32.9	$32.9 - 42.6 = - 9.7$
1973–74	58.3	$58.3 - 42.6 = +15.7$
1974–75	52.2	$52.2 - 42.6 = + 9.6$
1975–76	43.3	$43.3 - 42.6 = + .7$
1976–77	54.1	$54.1 - 42.6 = +11.5$
1977–78	82.3	$82.3 - 42.6 = +39.7$
1978–79	89.7	$89.7 - 42.6 = +47.1$

Statistically, we can say that Chicago's record snowfall *deviated* from the average (mean) by +47.1 inches.

Using this same data, we can find a winter that had relatively little snow. For example, the 1967–68 winter had a snowfall of 28.4 inches, which is below the mean. This is a deviation of −14.2 inches from the mean. The negative sign tells us that the snowfall was 14.2 inches below the mean.

The winter of 1975–76, according to the table, had a normal snowfall. The deviation of this snowfall of 43.3 inches in only +.7 inches.

Deviation score: Suppose we have a set of scores and find their mean. The difference between a score and the mean is a *deviation score.*

■ **Example:**   Let's consider this set of quiz scores for five students (Table 2.15).

TABLE 2.15

STUDENT	SCORE	DEVIATION SCORE
Bill	9	$9 - 7 = 2$
Mary	6	$6 - 7 = -1$
Fred	7	$7 - 7 = 0$
Tom	4	$4 - 7 = -3$
Bea	9	$9 - 7 = 2$
Total	35	

$$\text{Average (mean) score} = \frac{35}{5} = 7.0$$

Deviation scores make it easy for the students to compare their performances on the quiz with the rest of the class.

Bill: Deviation score = 2.
         "I was 2 points above the average (mean)."

Mary: Deviation score = −1.
         "I was 1 point below average."

Fred: Deviation score = 0.
         "I got the average score."

Tom: Deviation score = −3.
         "I was 3 points below the average."

Bea: Deviation score = 2.
         "I was the same as Bill, 2 points above the average for the class."

A deviation score graph can sometimes be useful in showing how scores vary from one another.

The quiz scores are shown in the deviation score graph in Figure 2.4. ■

FIGURE 2.4

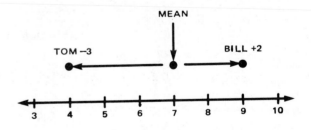

EXERCISES

1 The following were the high temperatures for June 20, 1982, for these cities:

City	°F
Atlanta	81
Bismarck	60
Great Falls	66
Miami	85
El Paso	96

What was the mean temperature for these cities? Find the deviation score temperature for each city. Which city was closest to the mean?

2 Express the heights of each of the players in the two basketball teams in Exercise 9 of Section 2.3 as deviation scores.

3 Table 2.16 gives average scores for collegiate football games during a season.[7]

TABLE 2.16

LEAGUE OR CONFERENCE	AVERAGE WINNER'S SCORE	AVERAGE LOSER'S SCORE
Ivy	30.6	11.0
Big Eight	22.2	7.6
Southeastern	25.7	9.4
Missouri Valley	25.8	10.6
Big Ten	23.9	12.0
Western Athletic	33.7	15.5
Pacific Eight	24.7	8.6

For the winners, express the average score for each league as a deviation from the mean winning score.

Now for the losers, express the average score for each league as a deviation from the mean losing score.

4 An important property of deviation scores for a set of scores is that they have a sum of zero, if founding has not occurred. Check that this is the case for the quiz scores in our example in Section 2.3. Now, take any set of, for instance, eight numbers. Keep them small in order to make the arithmetic simple. Find their mean (average). Change each number to a deviation score. Find the sum of these scores. It should be zero. Can you prove that this will always be the case?

PROJECT CORNER

Keep a record of the yards gained by your school football team during the past season. Draw a histogram of the data. Compute the mean, median, and mode of the data. Compute the deviation score for the yards gained for each game.

2.6 MEAN DEVIATION

The mean, or average, of the deviation scores can be useful in describing how much a set of scores varies. Let's consider this example of Pete's record in shooting baskets (Table 2.17). Early in the season, he was getting the number of baskets out of ten throws shown on the left. Late in the season, his record was as shown on the right.

TABLE 2.17
NUMBER OF BASKETS OUT OF 10 THROWS

GAME	EARLY IN SEASON	LATE IN SEASON
1	3	5
2	7	7
3	5	6
4	9	5
5	1	7
6	5	6
Totals	30	36
	Mean = $\dfrac{30}{6}$ = 5.0	Mean = $\dfrac{36}{6}$ = 6.0

We change each of these scores to deviation scores (Table 2.18).

TABLE 2.18

GAME	EARLY IN SEASON	LATE IN SEASON
1	3 − 5 = −2	5 − 6 = −1
2	7 − 5 = 2	7 − 6 = 1
3	5 − 5 = 0	6 − 6 = 0
4	9 − 5 = 4	5 − 6 = −1
5	1 − 5 = −4	7 − 6 = 1
6	5 − 5 = 0	6 − 6 = 0

In the early season, we see that Pete's shooting record was somewhat irregular. Sometimes he was as low as 4 points below his average of 5 baskets. Other times he was above average by the same amount. We can find the average distance he was from his average score by adding all the deviation scores (the vertical bars, which stand for *absolute value,* say to ignore the negative sign) and finding their mean.

Early in season:

$$\text{Mean deviation} = \frac{|-2|+|2|+|0|+|4|+|-4|+|0|}{6} = \frac{12}{6} = 2.0$$

Late in season:

$$\text{Mean deviation} = \frac{|-1|+|1|+|0|+|1|+|1|+|0|}{6} = \frac{4}{6} \doteq .66$$

We have described our observations statistically. Early in the season, the scores varied considerably. On the average, they were two points from the mean. Late in the season, Pete's shooting settled down. He consistently shot around his mean of six baskets and averaged only two-thirds of a point from his late season record of six baskets out of ten.

 Finding mean deviation:
1. Find the mean of the scores.
2. Change all the scores to deviation scores.
3. Find the mean (average) of all the deviation scores, ignoring the negative signs.

EXERCISES

1   Find the mean deviation of this set of quiz scores.

Bill	3
Jane	7
Mary	6
Pat	5
Phil	9

2   The wind speed for these cities, in miles per hour, is:[8]

Juneau	8.4
Chicago	10.3
Boston	12.5
Nashville	8.0
Miami	9.2

Find each: mean wind speed, median wind speed, range of wind speeds, and mean deviation of wind speeds. (Wind speeds are the yearly mean, for records kept through 1980.)

3   Recall that the normal temperature is defined as the mean temperature over the past 20 years. If the temperature at Anabru on February 30 is as given below, what is the normal temperature on February 30 in Anabru (rounded to the nearest whole number)?

Year	Temperature
1960	43
1961	48
1962	51
1963	30
1964	35
1965	38
1966	40
1967	41
1968	44
1969	48
1970	37
1971	42
1972	40
1973	41
1974	49
1975	50
1976	48
1977	47
1978	41
1979	43

4   A scientist recorded the earth vibrations produced by waterfalls. The measurements are given below.[9]

Waterfall	Vibrations (per second)
Lower Yellowstone	5
Yosemite	3
Canadian Niagara	6
American Niagara	8
Upper Yellowstone	9
Gullfoss (lower)	6
Firehole	19
Godafoss	21
Gullfoss (upper)	40
Fort Greeley	40

(a)   Find the mean of the earth vibrations.
(b)   Find the median of the earth vibrations.
(c)   Compare your results for (a) and (b). Why or why not do they differ?
(d)   Find the deviation score for each waterfall.
(e)   Find the mean deviation of these measurements.

5   Penelope is looking for the perfect city, weatherwise, in which to live. She can't stand changes in temperature. She likes a climate that's pretty much the same year around. Which of the two cities listed below would she prefer? Use mean deviation to help find your answer. Compare with your answer for Exercise 5 of Section 2.4.

	Camelot	Eldorado
Jan.	29	15
Feb.	30	20
Mar.	35	22
Apr.	40	30
May	42	45
June	58	58
July	60	78
Aug.	59	77
Sept.	50	60
Oct.	42	58
Nov.	38	32
Dec.	30	20

6   The manufacturers of resistors for electronic circuits have put in bids to a television company. Their prices are comparable, so the television company purchases ten resistors of each brand. The resistors are marked 100 ohms. Each is measured for resistance and the following values are found. Which company, Alpha or Beta, would you recommend as the supplier for the television company? Use mean deviation to help find your answer. Compare with your answer to Exercise 6 of Section 2.4.

Alpha	Beta
95	99
91	103
105	101
103	98
95	99
87	102
106	103
105	97
79	101
103	98

2.7 VARIANCE AND STANDARD DEVIATION

Range and mean deviation, we have seen, are statistics that describe how much a set of scores varies. Another statistic that is useful is the *variance* of a set of scores. We will show how the variance is found by using the data in Table 2.19.

We find the variance by squaring each deviation score and finding their mean (average).

Early in season:

$$\text{Variance} = \frac{(-2)^2 + 2^2 + 0^2 + 4^2 + (-4)^2 + 0^2}{6} = \frac{40}{6} \doteq 6.67$$

Late in season:

$$\text{Variance} = \frac{(-1)^2 + 1^2 + 0^2 + (-1)^2 + 1^2 + 0^2}{6} = \frac{4}{6} \doteq .67$$

The variance among the late-season scores is much smaller than that of the early-season scores. This shows, as does the mean deviation, that Pete's shooting was much more consistent late in the season.

The variance of the early season scores is 6.67. That is, the average (mean) square distance of the scores from their mean was 6.67 points.

— PROJECT CORNER —

Large bodies of water (such as the Great Lakes or the ocean) have a moderating effect on land temperatures. That is, the temperatures vary less because of the influence of the bodies of water. Obtain average monthly temperatures for a year for two cities at about the same latitude (distance from the equator): one located by a large body of water and another far away from such a body of water. For each set of data, calculate the statistic that would help show the moderating effect of large bodies of water on temperatures.

TABLE 2.19
NUMBER OF BASKETS OUT OF 10 FREE THROWS

EARLY IN SEASON	DEVIATION SCORE	SQUARE OF DEVIATION SCORE	LATE IN SEASON	DEVIATION SCORE	SQUARE OF DEVIATION SCORE
3	−2	4	5	−1	1
7	2	4	7	1	1
5	0	0	6	0	0
9	4	16	5	−1	1
1	−4	16	7	1	1
5	0	0	6	0	0
Totals 30	0	40	36	0	4

Similarly, the average (mean) square distance of the late-season scores from their mean was only .67 points.

Finding the variance of a set of scores;
1. Find the mean of the set.
2. Find each deviation score.
3. Square each deviation score.
4. Find the mean of these squared scores.

As a further example, consider this set of quiz scores.

Bill	8
Sue	6
Ted	6
Sally	7
Mickey	4
Jose	7
Fran	4
Sum =	42

$$\text{Mean} = \frac{42}{7} = 6.0$$

Table 2.20 shows how to calculate the variance.

TABLE 2.20

STUDENT	DEVIATION SCORE	(DEVIATION)²
Bill	8 − 6 = 2	2 × 2 = 4
Sue	6 − 6 = 0	0 × 0 = 0
Ted	6 − 6 = 0	0 × 0 = 0
Sally	7 − 6 = 1	1 × 1 = 1
Mickey	4 − 6 = −2	(−2) × (−2) = 4
Jose	7 − 6 = 1	1 × 1 = 1
Fran	4 − 6 = −2	(−2) × (−2) = 4

$$\text{Variance} = \frac{14}{7} = 2.0 \qquad \overline{14}$$

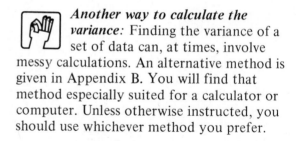

Another way to calculate the variance: Finding the variance of a set of data can, at times, involve messy calculations. An alternative method is given in Appendix B. You will find that method especially suited for a calculator or computer. Unless otherwise instructed, you should use whichever method you prefer.

STANDARD DEVIATION

Let's first review the ways in which we have described the *spread*, or *variation*, in a set of scores.

1. Range: The *distance* between the smallest and largest score in the set.

2. Mean deviation: The average (mean) *distance* of all the scores from the mean score of the set.

3. Variance: The average (mean) squared *distance* of all the scores from the mean score of the set.

There is another reasonable way of thinking about the amount of variation in a set of scores. This measure is called the *standard deviation* and is obtained by finding the square root of the variance. Recall that to get the variance of a set of scores, we squared each deviation score, added them up and found their mean. So, we can *undo* the effect of squaring by taking the square root of the variance. The result, the standard deviation, is really a sort of "average distance" of each score from the mean. The standard deviation has many important statistical properties. We will study some of these later in this book.

■ **Example 1:**   Consider Pete's basketball scoring record again.

Early in season:

$$\text{Variance} = 6.7$$
$$\text{Standard deviation} = \sqrt{6.7}$$
$$\doteq 2.6$$

Late in season:

$$\text{Variance} = .67$$
$$\text{Standard deviation} = \sqrt{.67}$$
$$\doteq .81$$

(If you do not remember how to find the square root of a number, please refer to the table of square roots, Appendix H.)

Just as the variance of Pete's late-season scores is less than the variance of his early-season scores, so is the standard deviation of his late-season scores less than the standard deviation of his early-season scores. The standard deviation of Pete's early-season scores, 2.6, is again a kind of distance. Taking the square root undoes the effect of squaring the deviation scores, which was done to find the variance. In this new way of thinking about average, we can say that, on the average, Pete's free throw record in ten throws was about 2.6 from his mean of five baskets. Late in the season, he was shooting much closer to his average score, or mean (which was now 6), since the standard deviation went down to .81. ■

■ **Example 2:**   Using Table 2.21, find the standard deviation of the temperatures of these Canadian cities. (Data from Figure 2.3).

TABLE 2.21

CITY	TEMPERATURE (°C)	DEVIATION SCORE	(DEVIATION)^2
Calgary	8	$8 - 11.8 = -3.8$	$(-3.8) \times (-3.8) = 14.44$
Montreal	14	$14 - 11.8 = 2.2$	$(2.2) \times (2.2) = 4.84$
Toronto	13	$13 - 11.8 = 1.2$	$(1.2) \times (1.2) = 1.44$
Vancouver	16	$16 - 11.8 = 4.2$	$(4.2) \times (4.2) = 17.64$
Winnipeg	8	$8 - 11.8 = -3.8$	$(-3.8) \times (-3.8) = 14.44$
	59		52.80

$$\text{Mean temperature} = \frac{59}{5} = 11.8$$

$$\text{Variance} = \frac{52.80}{5} = 10.56$$

$$\text{Standard deviation} = \sqrt{10.56} \doteq 3.2 \quad ■$$

EXERCISES

1   Find the variance of these quiz scores.

Bill	3
Jane	7
Mary	6
Pat	5
Phil	9

Compare the variance with the mean deviation calculated in Exercise 1 of Section 2.6.

2   Find the variance of the temperature of the cities given in Exercise 2 of Section 2.6.

3   What is the variance of this set of numbers? Explain.

$$5, 5, 5, 5.$$

4   Four students made the following number of spelling errors in an essay they wrote:

$$12, 7, 5, 4.$$

What is the variance of the number of errors made?

5   The four students in Exercise 4 were asked to rewrite their essays. The number of spelling errors then made by them were:

$$8, 3, 1, 0$$

Compare the mean and variance of the number of errors made first (Exercise 4) with the number of errors made in the rewrite.

6   Find the variance of the number of defective flash bulbs found in samples of 25 over a 5-week period:

$$4, 2, 0, 1, 3.$$

7   What is the variance of the normal temperatures at Camelot and Eldorado during the year? (See Exercise 5 of Section 2.6.)

8   Find the variance of the resistances for the two samples in Exercise 6 of Section 2.6. Answer the question asked in Exercise 6 of Section 2.6 using variance rather than mean deviation as the basis for comparison.

9   Find the variance and standard deviation of these data. They are the number of accidents occurring in a week on a busy freeway:

$$4, 0, 6, 10, 5.$$

10   Find the range, mean deviation, variance, and standard deviation of these data. They are the number of defective valves found in batches of 1000 in a machine shop:

$$2, 4, 0, 10.$$

11   Find the standard deviation of the quiz scores in Exercise 1 of Section 2.6.

12   Find the variance and standard deviation of the waterfall data in Exercise 4 of Section 2.6.

13   Calculate the standard deviation of the monthly temperatures at Camelot and Eldorado (see Exercise 7, above).

14   Find the variance and standard deviation of the fire alarm data, Exercise 6 of Section 2.1.

15   Find the variance and standard deviation of each of these two sets of data (from Exercise 1 of Section 2.2):

$$10, 15, 20, 25$$
$$10, 15, 20, 2500$$

16   Refer to Exercise 15 of Section 2.7. Find the mean deviation for each set of data. Which do you think is a more robust measure of variation, mean deviation, or standard deviation? ("Robust" was discussed in Section 2.2.)

Expected Value

KEY PROBLEM
How Much Gum?

The Tripl-Bubl Gum Company decides to promote its gum by enclosing a card bearing the flag of a country with each stick of gum. Suppose they use flags for six different countries and, when a person buys a stick of gum, the chances of getting any one of the six flags are equal. How many sticks of gum, on the average, would you expect to have to buy to get the complete set of all six flags?

We will solve this problem by doing an experiment. But before we do, make a guess as to how many sticks of gum you think you would have to buy. Do you think it would be a large number, such as 100 or more? Or would it be a smaller number? Many people choose 36, feeling, maybe, that a reasonable answer has something to do with 6, and 36 is, after all, 6 times 6.

UNITED STATES

CANADA

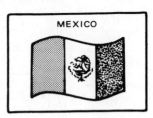

MEXICO

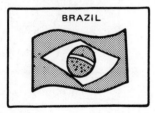

BRAZIL

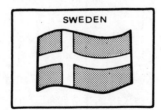

SWEDEN

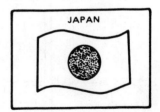

JAPAN

3.1 WHAT DO YOU EXPECT?

The weather forecast calls for a 50% chance of rain. Forecasts are based on looking at weather data (temperature, humidity, wind direction, and so on) and making judgments; for instance, in the past when we had weather conditions like this, it rained one-half, or 50%, of the time.

■ **Example:** Suppose a forecast of a 50% chance of rain is given in 12 cities around the country on a given day. In about how many of those cities, on the average, would you expect it to rain?

Solution:

There are several ways to go about answering this question. And you may already have an answer. But let's consider a method you may not have thought of using, one which will be useful later for solving more difficult problems.

One method to use would be to place a telephone call to each of the cities and ask if it had rained! That is a sensible approach, but it takes time and money. (Can you think of other objections?)

We propose, instead, that you conduct an experiment right in your classroom. Let's think of a model we could use for finding out what is likely to happen. One model would be a coin.

| Heads: | Rain |
| Tails: | No Rain |

A coin (let's assume it is fair, not bent or biased) has a 50–50 chance of falling heads. That is, it falls heads about one-half, or 50%, of the time when it is tossed. So a coin is a model of the chances of rain when the probability of rain is 50%, or one-half.

A PROBABILITY EXPERIMENT

In order to estimate how many of the 12 cities are likely to receive rain, on the average, toss the coin 12 times, once for each city. We say that the 12 tosses make up an *experimental trial.*

The outcomes of this trial are:

$$H\ H\ T\ T\ T\ H\ H\ T\ H\ H\ H\ T$$

We see that we have obtained 7 heads. So, we conclude from this experiment that 7 of the cities received rain.

THE RYATTS[1]

There are difficulties with this approach, though. Let's do another trial. Again, we toss the coin 12 times (once for each day). Would we get 7 heads again? Since this is an experiment involving chance, we cannot be sure how many heads we will get.

Here are the results:

$$T\ T\ T\ H\ T\ H\ H\ T\ T\ T\ H\ T$$

This time we get 4 heads. That is, 4 cities got rain. We could record our results in a table like Table 3.1.

TABLE 3.1

NUMBER OF HEADS	FREQUENCY (NUMBER OF TRIALS)
4	1
7	1

So, is our answer 4 days, 7 days, or some other number? It seems that the average (mean) of these two numbers would be a better answer.

$$\text{Mean number of heads} = \frac{4+7}{2} = 5.5$$

Still a better answer could be obtained by conducting more trials of tossing a coin 12 times, finding the number of heads obtained, and averaging our results.

We notice that the result of each trial can range from 0 heads (it *is* possible to toss a coin 12 times and get tails every time!) to 12 heads. So our data table will look like Table 3.2.

We now ask 48 more people each to toss a coin 12 times and record the results of that large experiment in our table (Table 3.3).

TABLE 3.2

NUMBER OF HEADS (CITIES GETTING RAIN)	TALLY	FREQUENCY
0		
1		
2		
3		
4	/	
5		
6		
7	/	
8		
9		
10		
11		
12		

TABLE 3.3

NUMBER OF HEADS	TALLY	FREQUENCY
0		0
1		0
2	/	1
3	//	2
4	ЖЖ /	6
5	ЖЖ ЖЖ //	12
6	ЖЖ ЖЖ	10
7	ЖЖ //	7
8	ЖЖ	5
9	ЖЖ /	6
10	/	1
11		0
12		0
		Total = 50

Based on these 50 trials, we find the mean number of heads out of 12 tosses to be

$$\frac{\text{total number of heads obtained}}{\text{50 trials}}$$

$$= \frac{0\times0 + 0\times1 + 1\times2 + 2\times3 + 6\times4 + 12\times5 + 10\times6 + 7\times7 + 5\times8 + 6\times9 + 1\times10}{50}$$

$$= \frac{0 + 0 + 2 + 6 + 24 + 60 + 60 + 49 + 40 + 54 + 10}{50} = \frac{305}{50} = 6.1 \ \blacksquare$$

So, based upon these 50 trials, we estimate the number of cities likely to get rain as about 6.1. If we use a fair coin to conduct a very large number of trials, such as 10,000, we would find the number of cities to be very close to 6.0. We call the value 6.0 the (theoretical) *expected value* of the outcomes of this large number of trials.

EXERCISES

1   The following table gives the number of heads obtained in 100 trials. A trial consists of tossing a coin 12 times. Find the average (mean) number of heads obtained per trial.

N Heads	Frequency (f)
0	0
1	0
2	0
3	4
4	11
5	16
6	25
7	21
8	18
9	4
10	0
11	1
12	0

2   The following table gives the number of tails obtained in 50 trials. A trial consists of tossing a coin 12 times. Find the average (mean) number of tails obtained per trial.

N Tails	Frequency (f)
0	0
1	0
2	0
3	4
4	6
5	10
6	12
7	8
8	7
9	2
10	1
11	0
12	0

3   Suppose, in Exercise 1 of this section, the experiment was carried out to estimate how many of 12 cities were likely to receive rain when the forecast for rain was 50% for each city. Interpret your answer to Exercise 1 in terms of the estimated number of cities receiving rain.

4   In Exercise 2 of this section, suppose the experiment was carried out to estimate the number of cities *not* getting rain when the forecast for no rain was 50%. Interpret the answer in terms of the number of cities not getting rain.

5   The weather forecast is for a 50% chance of rain in Vancouver, where Brian lives. Brian hears the forecast but decides to take a chance and not take his umbrella. Assuming the forecast for rain stays at 50% for each day, how many days in a row can he expect to get by doing this before he's rained on? Conduct trials to estimate the number of days. You should do 20 trials (you may want to get friends to help you).

3.2   MODELS OF CHANCE

The idea of a model is a useful one in mathematics and science. For example, an engineer who is designing a new type of aircraft might first build a model aircraft that incorporates the new features and fly it in a wind tunnel. Similarly, the chemists who discovered the structure of the DNA molecule built a model to help them in their investigation.

IS FOOTBALL COIN TOSS FOR REAL?

I've heard that the coin toss on the football field, just before the game, to determine who kicks off is really just for show and that the actual deciding toss takes place earlier. Is that true?–T.S., Chicago

Yes and no. From 1946 to 1976 the National Football League had the team captains and officials meet 30 minutes before the game for the deciding toss, which was then re-enacted on-field for show. This allowed coaches to plan strategy and prepare either offensive or defensive lineups. This was changed last season and the coin toss on-field three minutes before kick-off is the real thing.

The National Collegiate Athletic Assn. and the Illinois High School Assn. report that the coin toss is handled either way, depending on the preference of the coaches.

In this book we will use models to help us solve problems. Our models will usually be probabilistic models, such as coins or dice. For example, a toss of a coin could be used to represent whether a child is a boy or a girl, as in the following problem.

■ **Example:**   What is the expected number of girls in a three-child family?

Solution:

One way in which to solve this problem would be simply to go out and conduct interviews of many three-child families, keeping records of how many girls there are in each family. The results of such a survey might be recorded in a table like Table 3.4.

TABLE 3.4

NUMBER OF GIRLS	NUMBER OF FAMILIES
0	13
1	32
2	38
3	17
	Total 100

Our survey showed that the average (mean) number of girls in our sample of three-child families was

$$\frac{0 \times 13 + 1 \times 32 + 2 \times 38 + 3 \times 17}{100}$$

$$= \frac{0 + 32 + 76 + 51}{100} = \frac{159}{100} = 1.59$$

However, instead of actually interviewing families, we could use the approach suggested in Section 3.1. We can think of a model of a three-child family. Instead of actually interviewing families, then, we can toss one coin three times, once for each child, where we let heads stand for a girl and tails stand for a boy.

Heads: Girl
Tails: Boy

Therefore, a coin is a model for a child, and the outcome of tossing a coin represents the child's being a boy or girl. We then conduct an experiment in which we wish to find the average number of heads (or tails) obtained.

Such an experimental approach to solving problems is called the *Monte Carlo method*, named in honor of the well-known casinos on the Mediterranean Sea. Monte Carlo methods are very important in research and have been used a great deal in mathematics, physics, and economics to solve problems that are extremely difficult—if not impossible—to solve by the usual theoretical approaches. Hence, the methods we use in this chapter and elsewhere in this book have importance in other fields of study as well.

We summarize the steps to follow in using a Monte Carlo solution to a problem, relating them to our example of the three-child family:

Step 1—*Identify the model:*   In our example of the three-child family, an appropriate model was a coin. (You will see later that other models might be equally appropriate. For example, you might roll a die and use this rule: even number = boy, odd number = girl.)

Step 2—*Define a trial:*   A trial consists of tossing the coin three times, once for each child in the family, and counting the number of heads obtained.

Step 3—*Record the statistic of interest:*   In this case, we are interested in the number of heads (number of girls) obtained in each trial.

Step 4—*Repeat the experiment until a sufficient number of trials is obtained:*
When you do a probabilistic experiment like tossing coins, your answers will tend to be different from one experiment to another. This happens because the results of the experiment are estimates of a theoretical value. Each experimental result will probably be different from the theoretical value. But, the average (mean) value will get closer and closer to the theoretical value.

In this case, the number of heads when a fair coin is tossed three times and repeated for a great number of trials, has a *theoretical expected value* of one-half times three

$$\frac{1}{2} \times 3$$

or 1.5.

You will see later in this book that the more trials you do, the closer the results of probability experiments will be to the theoretical value being estimated. For now, we assume that about 100 trials are sufficient to give us enough accuracy in our estimates.

The experiment of tossing a coin three times was repeated 100 times with these results.

Heads	Frequency
0	11
1	29
2	44
3	16
Total	100

Step 5—*Find the average (mean):*   Add the values of the statistic of interest found in Step 3 and divide by the total number of trials

given in Step 4. For our example, the mean number of heads obtained was

$$\frac{0\times11 + 1\times29 + 2\times44 + 3\times16}{100} = \frac{165}{100} = 1.65$$

We now repeat these five steps in the coin-tossing experiment, and compare the results obtained in 100 more trials.

Step 1—*Model:* Use a coin.

Heads: Girl
Tails: Boy

Step 2—*Trial:* Toss a coin three times, once for each child, and count the number of heads obtained.

Step 3—*Statistic of interest:* Of interest is the number of heads obtained in each trial.

Step 4—*Repeat trials:* Repeat Steps 2 and 3 for a total of 100 trials. The results of the tosses are in Table 3.5.

TABLE 3.5

N HEADS (GIRLS)	FREQUENCY
0	14
1	34
2	36
3	16
	Total = 100

Step 5—*Find mean:*
$$\frac{0\times14 + 1\times34 + 2\times36 + 3\times16}{100} = \frac{154}{100} = 1.54$$

For this experiment, the mean number of girls expected in a three-child family is estimated to be 1.54. ∎

This is close to the theoretical expected value of $1.5 \times 3 = 1.5$. In other problems, the theoretical value may not be so easy to find. So we will estimate it using our five-step method.

EXERCISES

Use the five-step Monte Carlo method to solve these exercises. Unless otherwise indicated, use only 50 trials.

1   Solve the three-child family problem by doing your own coin tosses. That is, toss a coin three times and repeat for 100 trials. Keep a record of the number of heads on each trial. Write out the five steps. Compare your estimated number of heads with those obtained in this section. How did the results of other class members compare with yours?

2   In a four-child family, what average (mean) number of boys would be expected?

3   Repeat Exercise 2 of this section, but use a six-sided die instead of a coin as your model. (*Hint:* Let odd number = boy and even number = girl.)

4   In a two-child family, what is the average (mean) number of girls expected? Use a deck of playing cards to do the trials. Let red card = girl and black card = boy.

5   A baseball player has a batting average of .250. This means that on the average, the player has been getting one hit out of every four trips to the plate. What is the number of hits the player can expect out of 20 times at bat? (*Hint:* A probability of .250 can be obtained by using only four sides of a regular six-sided die. For example, you could let 1 = hit; 2, 3, and 4 = no hit; 5 and 6 = ignore.)

6   What is the theoretical number of hits expected for the batter in Exercise 5.

7   Suppose another baseball player has a batting average of .300. What is the player's expected number of hits out of 20 times at bat? First you must find a model for a probability of .3.

8   On the island of Bulama, a new chief has introduced the law that, for every family in his tribe, the family inheritance must go only to the firstborn daughter. If there are no daughters, the family wealth goes to the chief. As a result, all couples continue to have children until they have a girl. Under this law and the assumption that the probability of a child being a girl is .5, what is the expected average family size on Bulama?

9   Sue doesn't like to be bothered with an umbrella. So if the forecast for rain is .25 (25% chance of rain), she doesn't carry an umbrella. Using this rule, how many days in a row can she expect to keep dry (not be rained on) when the forecast for rain is .25 day after day?

10   On the average, how many hits in a row can be expected from a baseball player with a batting average of .250? With a batting average of .300?

11   If the chances of rain are .75 (75%), how many rainy days in a row can be expected?

12   It is claimed that .50 (50%) of people connected with a telephone answering machine when they make a call will hang up. Under this assumption, how many people in a row could be expected to hang up when they are answered by a telephone answering machine?

3.3   INDEPENDENCE

The idea of independence is very important in statistical thinking. We will introduce it in terms of examples now and make it a little more exact in the next chapter.

Roughly speaking, we say that two outcomes are *independent* if the occurrence of one has no influence upon the chance of the occurrence of the other. For example, if you toss a coin and roll a die, the outcomes of the two activities are independent. (You could, of course, make them dependent by actually fastening the die to the coin. But we assume that you have given the coin a good, honest toss and the die a fair roll.)

It may not be quite so obvious that individual tosses of the same coin are, likewise, independent. When you toss a coin, the outcome of the next toss is not at all influenced by the outcome of the preceding toss. Hence, tosses of coins are independent.

Also, because of independence, the model of one coin tossed three times is equivalent to the model of three coins tossed once. Similarly, the model of one coin tossed four times is equivalent to that of four coins tossed once. And so on. The choice of model is just a matter of which one you find to be easier or more sensible to use in solving a particular problem.

Now, back to the independence of outcomes having to do with the weather. We can think of examples where the likelihood of rain in one city is closely linked to the likelihood of rain in another, and therefore the outcomes are to some extent dependent. Suppose, for example, the cities are twin cities, like Minneapolis and St. Paul, and are in the same storm system!

Let's, instead, think of two cities in which the chances of rain are independent. For example, two cities such as Chicago and San Juan, Puerto Rico, may both have a forecast of rain of 50%. But whether it rains in Chicago would seem to be independent of rain in San Juan. We will make the assumption of independence when we carry out the next experiment.

■ **Example:**   On 12 days, the forecast for rain in both Chicago and San Juan has been 50%. On the average, on how many of these days would we expect it to rain *in both cities*?

Solution:

We follow our five steps to solve this problem.

Step 1—*Model*:   As a model for this problem, we now need two coins, one for each city. The outcomes of the two tossed coins are independent of each other, as are the outcomes of rain (we assume) in Chicago and San Juan. For example,

> Penny: Chicago
> Nickel: San Juan

In each case, heads means *rain* and tails means *no rain*.

Step 2—*Trial*:   A trial consists of tossing the penny (for Chicago) and nickel (for San Juan) a total of 12 times each (once for each of the 12 days as stated in the problem).

Step 3—*Statistic of interest*:   We are asked to estimate the mean number of days when it rained in *both cities*. So we want to know how many times we obtained *two heads* when the two coins are each tossed 12 times.

For example, here are the results we obtained by actually tossing a penny and a nickel 12 times each (the outcome of the penny is shown first).

Two heads were obtained 5 times. So, for this trial, the statistic of interest is 5.

The data may be recorded in a table like Table 3.6.

TABLE 3.6

NUMBER OF DAYS	TALLY	FREQUENCY
0		
1		
2		
3		
4		
5		
6		
7		
8		
9		
10		
11		
12		

Step 4—*Repeat trials*:   We repeat the trial of tossing the penny and nickel 12 times for a total of 50 trials (Table 3.7).

TABLE 3.7

NUMBER OF DAYS	TALLY	FREQUENCY
0		0
1	卌 /	6
2	卌 ////	9
3	卌 卌 卌 ///	18
4	卌 卌 /	11
5	////	4
6	/	1
7	/	1
8		0
9		0
10		0
11		0
12		0
		50

Step 5—*Find mean:*   The mean number of days (out of 12) when it rained in both cities is the mean number of trials resulting in two heads.

$$\text{Mean} = \frac{6\times1 + 9\times2 + 18\times3 + 11\times4 + 4\times5 + 1\times6 + 1\times7}{50}$$

$$= \frac{6 + 18 + 54 + 44 + 20 + 6 + 7}{50} = \frac{155}{50} = 3.1$$

So we expect rain in both cities an average of 3.1 out of 12 days. ■

EXERCISES

1   Two coins are tossed for a total of ten tosses and the number of times two heads are obtained is recorded. This is repeated 50 times. In this experiment, what is a trial? What are the possible outcomes of a trial? How many tosses make up a trial? How many trials are there?

2   Three coins are tossed for a total of 20 tosses. The number of times *two or more* heads were obtained in the 20 tosses was recorded. This is repeated 100 times. In this experiment, what is a trial? How many trials are there?

3   Two coins were tossed 12 times for a total of 100 trials. The results in the table show how many times two heads were obtained in a trial. What is the average (mean) number of times out of 12 that two heads were obtained?

N Occasions for 2 Heads	Frequency
0	4
1	12
2	21
3	30
4	22
5	8
6	2
7	1
8	0
9	0
10	0
11	0
12	0

4   The table below gives the results of tossing two coins 12 times for a total of 1,000 times. Compare the average (mean) number of occasions two heads were obtained with the answer for Exercise 3 which was based on 100 trials.

N Occasions	Frequency
0	26
1	127
2	250
3	253
4	184
5	104
6	46
7	7
8	2
9	1
10	0
11	0
12	0
Total	1000

5   Suppose the forecast for rain was 50% in both San Juan and Chicago on 20 occasions. Conduct an experiment consisting of 60 trials to estimate on how many occasions it could be expected to rain in both cities.

6 There are two traffic lights on Sally's route to class. Each light is green for 30 seconds and both lights have a 60-second cycle. That is, in 60 seconds the lights go through green, amber, and red. Assuming the lights operate independently of each other, how many mornings out of 10 would you expect her to get a green light on both lights?

7 In Exercise 6, on how many mornings out of 10 can Sally expect to get a green light on *neither* light?

8 In three cities—Anabru, Burbank, and Casbah—the forecast is for a .5 chance of rain. Assume rain in all three cities to be independent of each other. If the same forecast is given on 12 days, in how many days would rain be expected, on the average, in all three cities?

9 Robert insists on taking true-false quizzes by guessing. He says that it's not worth studying for them, and he doesn't even bother to read the questions. Using this method, how many questions out of ten can he expect to get right (just by guessing) in the long run?

3.4 ABOUT HOW MANY . . . ?

We are now ready to solve somewhat more complicated problems. Let's start with a problem from psychology. People may be divided into three personality types: introvert, extrovert, or ambivert. Introverts are shy, withdrawn people. Extroverts are outgoing, social types. Ambiverts are those who find a balance between extroversion and introversion. A well-known English psychologist, Dr. H. J. Eysenck, has estimated that

one-third of the population are introverts, one-third are extroverts, and one-third are ambiverts.

■ **Example:** In a research project on group dynamics 18 couples are needed, where each couple consists of one introvert and one extrovert. If the couples are formed by picking people by chance, about how many couples would we expect to consist of one introvert and one extrovert?

Solution:

Step 1—*Model:* We need a model for picking a couple—two people—by chance and for determining whether each is an introvert or extrovert. Let's use a pair of dice, one die for each person, where

1 or 2 Dots:	Introvert
3 or 4 Dots:	Ambivert
5 or 6 Dots:	Extrovert

(*Note:* We are assuming that the two people in a couple are chosen *independently* and that we are using Eysenck's theory of personality.)

Step 2—*Trial:* A trial consists of rolling two dice 18 times (once for each couple).

Step 3—*Statistic of interest:* We are interested in whether we have obtained an introvert and an extrovert in our chance selection of two people. We could record the outcomes in a table like Table 3.8. We want to count outcomes where we get both an extrovert and an introvert.

TABLE 3.8

TRIAL	INTROVERT (1 OR 2 DOTS ON EITHER DIE)	EXTROVERT (5 OR 6 DOTS ON EITHER DIE)	OTHER OUTCOMES

Step 4—*Repeat trials:*   At least 50 trials should be carried out.

Step 5—*Find mean:*   Find the mean value of the statistic of interest by finding the total number of couples consisting of one introvert and one extrovert (Step 3) and dividing by the total number of trials in Step 4. In a sample trial using rolls of a die, suppose that we get for couple one, 5 dots and then 4 dots. This is written simply as:

$$5 \quad 4$$

In order to save space, the results of the dice rolls may be written close together, with a space after every five rolls to help you keep your place in the list. For example, a list of outcomes of 36 rolls of a die might be:

54 62 (61) 36 46   46 43 43 36 36
64 63 (51) (25) 36   21 35 44

We can read off the digits in twos, as they are marked in this list, and think of each pair of digits as a pair of persons picked at random for the psychological research. We see from this list that there are three couples made up, according to our model, of an introvert (digits 1 or 2) and an extrovert (digits 5 or 6). The pairs of digits giving us those couples are circled in the above list. They are:

$$\begin{array}{cc} 6 & 1 \\ 5 & 1 \\ 2 & 5 \end{array}$$

So, our first trial gave us three couples with whom to do our research. Notice that we need 36 digits (18 pairs of digits) for one trial.

The results of our 50 trials are summarized in a frequency table (Table 3.9).

TABLE 3.9

NUMBER OF COUPLES CONSISTING OF ONE INTROVERT AND ONE EXTROVERT	f
0	6
1	9
2	18
3	10
4	3
5	0
6	2
7	1
8	0
9	1
10 or more	0
Total	50

The mean number of couples consisting of one introvert and one extrovert is

$$\frac{9 + 36 + 30 + 12 + 12 + 7 + 9}{50} = \frac{115}{50} = 2.3 \quad \blacksquare$$

Another example for us to consider is our key problem for this chapter.

KEY PROBLEM RECAP—*How Much Gum?*
About how many packages of bubble gum, on the average, would we expect to have to buy to obtain a complete set of six different flags? Assume that on each purchase of a stick of bubble gum, the chances of obtaining each of the six flags are equal.

Solution:

We solve this problem as we did others in this chapter—by conducting a probability experiment. We could, of course, go out and purchase lots of bubble gum. However, that is not practical (or even desirable), even for the most avid bubble gum enthusiast. Instead, we will conduct an experiment to represent, or simulate, the purchasing of bubble gum and keep records of the kinds of cards obtained.

The problem is a probabilistic one, since on any particular purchase we are not sure of the outcome. It could result in any of six different ways (each of the six cards) with, we assume, equal chances for each card. Hence we can use a familiar probabilistic model to help us solve the problem—a die. See Figure 3.1.

Step 1—*Model*:   We will use a six-sided die, each side corresponding to a different one of the six possible outcomes (six different

bubble gum cards). One roll of the die corresponds to the purchase of one stick of bubble gum, thus getting one of the six cards.)

Step 2—*Trial*:   A trial consists of rolling the die until all six sides of the die are obtained.

Step 3—*Statistic of interest*:   In this problem, we are interested in estimating the number of rolls of a die (sticks of bubble gum) required to obtain all outcomes of the die (all six flags). We record the number of outcomes in a

FIGURE 3.1

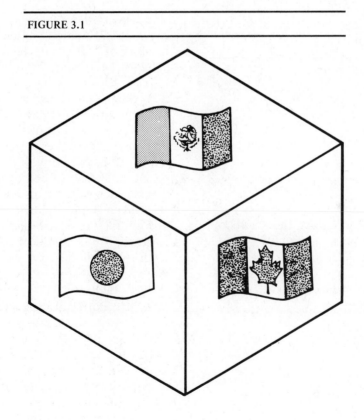

trial, as shown in the table below. The circled tally mark indicates the last roll of the die for that trial. For example, in the first trial, the last outcome rolled was a 2. (Notice that the *last* outcome obtained is *always* a single tally since you roll until you get every outcome at least once.)

Step 4—*Repeat trials:*   About 100 trials should be obtained.

Step 5—*Find mean:*   Compute the average number of rolls of the die from Table 3.10.

TABLE 3.10

			OUTCOMES				N
TRIAL	1	2	3	4	5	6	ROLLS
1	///	①	＃＃	//	//	///	16
2	//	//	//	//	//	①	11
3	//	///	①	//	//	//	12
4	①	////	//	////	///	////	18
5	//	//	①	///	///	＃＃ ////	20

In our table, we have recorded the results of only five trials in order to show how the problem is solved.

The mean number of rolls of the die (sticks of bubble gum purchased) is

$$\frac{16 + 11 + 12 + 18 + 20}{5} = \frac{77}{5} = 15.4$$

You should obtain at least 50, and preferably 100, trials before reporting results of an experiment. In advanced mathematics courses it can be shown that if a very large number of trials were obtained, the average number of rolls of the die needed to obtain all six outcomes would be about 14.7.

EXERCISES

Solve each of these exercises by first writing the five steps for doing the required probability experiment. Unless otherwise stated, do 50 trials.

1   Using the method in this section for finding couples consisting of an introvert and an extrovert, do 30 trials and estimate the average number of such couples out of 18 that you would expect to find.

2   Do another 50 trials of the bubble gum problem. Compare your average (mean) number of rolls of the die with the theoretical value of 14.7.

3   Snap-Krak Cereal advertises that you get a ball point pen with every box of cereal you buy. There are eight different colored pens in all. How many boxes of cereal would you expect to have to buy, on the average, to get the complete set of pens if the chances of getting each of the pens are the same? (*Note:* If you can get an eight-sided die, called an octahedral die, you should use it as the model for this problem. Otherwise, you could use an eight-sided spinner or a deck of eight different playing cards for your model.)

4   Mrs. Potts always keeps her classroom door locked, and the key is kept in her purse along with five other loose keys. When she arrives for class, she pulls out a key without looking in her purse and tries it. If the key does not fit, she returns it to her purse and repeats the process. (Maybe she will get that same key again!) On the average, how many tries will she have to make before she succeeds in opening her door?

5   It was stated in this section that the average number of rolls of a die needed to get all six outcomes gets closer and closer to 14.7

as the number of trials gets larger and larger. The formula giving this result is

$$\frac{6}{1}+\frac{6}{2}+\frac{6}{3}+\frac{6}{4}+\frac{6}{5}+\frac{6}{6}=14.7$$

Similarly, it can be shown that for an eight-sided die, the theoretical number (expected value) of rolls needed to get all eight sides is

$$\frac{8}{1}+\frac{8}{2}+\frac{8}{3}+\frac{8}{4}+\frac{8}{5}+\frac{8}{6}+\frac{8}{7}+\frac{8}{8}=21.7$$

Use this pattern to find the expected number of rolls of a 12-sided die to get all 12 sides. Then use a 12-sided die, a spinner, or some other appropriate model and conduct 50 trials. Compare your estimated number of rolls of the die with the theoretical value given by this formula.

6   How Many Games to Win the World Series?[2]   Suppose for the World Series the National League team is believed to be better than the American League team. Let's say the National League team won two-thirds of its games during the season. What is the expected number of games to be played to win the World Series? (Remember that in order to win the series, a team must win four games and the series goes to a maximum of seven games.) What assumptions are being made when we take the probability of the National League winning a game as being two-thirds (or .67)?

3.5   PASS IT ON!

Experiments like the ones we have been doing are actually used in many fields of study—for example, in the study of the spread of diseases. Here we will look at a very simple example of how epidemics might be studied.

■ **Example—The Hermit's Epidemic:**   Six hermits live on an otherwise deserted island. An infectious disease strikes the island. The disease has a 1-day infectious period and after that the person is immune (cannot get the disease again). Assume one of the hermits gets the disease (maybe from a piece of Skylab). He randomly visits one of the other hermits during his infectious period. If the visited hermit has not had the disease, he gets it and is infectious the following day. The visited hermit then visits another hermit. The disease is transmitted until an infectious hermit visits an immune hermit, and the disease dies out. There is one hermit visit per day. Assuming this pattern of behavior, how many hermits can be expected, on the average, to get the disease?

Solution:

Step 1—*Model:*   Use a six-sided die, where a side represents each of the six hermits.

Step 2—*Trial:*   Roll the die to see which hermit gets the disease first. Then roll the die to see which hermit is visited and gets the disease. We continue rolling until an immune hermit is visited. Then the disease dies out. (If the same side is obtained one after the other, we ignore the second roll, since even a hermit does not visit himself!)

Step 3—*Statistic of interest:*   We want to know how many sides of the die were obtained in a trial (how many hermits got the disease).

Step 4—*Repeat trials:*   About 100 trials should be obtained.

Step 5—*Find mean:*   Find the average number of sides obtained. One trial (epidemic) is demonstrated in detail. We use these die rolls:

1st Epidemic	2nd Epidemic	3rd Epidemic	4th Epidemic	5th Epidemic
26454	111356445	6124441	455522636	24462

Since (in the first epidemic) a 2 is rolled first, hermit 2 has the disease first. This is marked in Table 3.11. The next roll gives a 6. That is, the next day, hermit 2 visits hermit 6, who has not had the disease and so gets it. The next roll is a 4, so hermit 4 gets the disease from hermit 6. Then hermit 5 contracts the disease. Hermit 5 visits hermit 4. But hermit 4 is immune (he had the disease two days ago). Thus the epidemic ends after four people were infected. Four more such trials (epidemics) are reported in the table. Recall that if the same outcome of a die is repeated in successive rolls, the repeats are ignored. For example, in trial (epidemic) 2, hermit 1 gets the disease. Ignore the next two 1s. Hermit 1 visits hermit 3, and so on. See Table 3.11.

TABLE 3.11

TRIAL (EPIDEMIC NUMBER)	|	HERMIT				NUMBER OF HERMITS INFECTED
	1	2	3	4	5	6	
1		†		†	†	†	4
2	†		†	†	†	†	5
3	†	†		†		†	4
4		†	†	†	†	†	5
5		†		†		†	3

Total = 21

Mean number of hermits infected per epidemic

$$= \frac{21}{5} = 4.2 \ \blacksquare$$

EXERCISES

1   Do 50 epidemics on your own. Compare your result with ours.

2   Suppose there were eight hermits on the island instead of six. How many would be expected to be infected under the same assumptions as given in the problem above? (Refer to Exercise 3 of Section 3.4 for a suggestion for a model for this problem.)

3   Suppose there were 12 hermits on the island. Conduct an experiment to estimate how many would get the disease under the same assumption as given in the example in this section. (*Note:* For a model for 12 random outcomes, see the following Project Corner.)

─── **PROJECT CORNER** ───

A Homemade 12-Sided Die

In this book you will find problems requiring a 12-sided die as a model. Rather than build such a die, you can do something that will give the same results as rolling a 12-sided die: You just toss a coin and a 6-sided die together. Why this works is explained in Chapter Four.

3.6   TAKE A WALK, BY CHANCE

Random outcomes, such as the results of tossing coins or rolling dice, can be used by scientists to solve complex problems: For example, the movement of molecules of gas can be thought of as random.

Physicists sometimes use probability experiments to study the paths of molecules. Or, the behavior of a nuclear reactor can be studied with the help of chance or random outcomes to estimate the effectiveness of radiation shields, the extent of radiation pollution in the atmosphere, and so on. The path of a particle moving at random is called a *random walk*.

A simplified version of a random walk is shown in Figure 3.2.

A woman takes a walk, always starting from zero on the line shown in Figure 3.2. In order to decide whether to go to the left or right, she tosses a coin. If the coin falls heads, she walks one block to the right. If the coin falls tails, she walks one block to the left. For example, if she obtained *H, H, T,* for three tosses, she would then be located at 1. On the average, how far from zero would she expect to be after 10 tosses of a coin?

Table 3.12 gives the results of 12 trials (random walks) consisting of 10 tosses each. If you prefer, you could use a die to solve this problem, instead of tossing coins. One rule to use would be:

Odd Digit: Walk one block to the right
Even Digit: Walk one block to the left

TABLE 3.12

RANDOM WALK NUMBER (10 TOSSES OF A COIN)	1	2	3	4	5	6	7	8	9	10	11	12
LOCATION AT END OF WALK	2	0	0	−2	2	−4	−4	−2	−4	2	6	−2

FIGURE 3.2

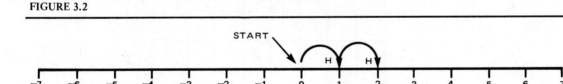

The average distance of stopping point from starting point is

$$\frac{|2| + |0| + |0| + |-2| + |2| + |-4| + |-4| + |-2| + |-4| + |2| + |6| + |-2|}{12}$$

$$= \frac{2 + 0 + 0 + 2 + 2 + 4 + 4 + 2 + 4 + 2 + 6 + 2}{12} = \frac{30}{12} = 2.5$$

In two dimensions, the random walk problem becomes more interesting.

Think of Figure 3.3 as being a city map. A person takes a walk, starting at (0, 0). The person can walk either north, south, east, or west. In order to decide, the person tosses two coins and uses this rule:

HH: One block north
HT: One block south
TH: One block west
TT: One block east

FIGURE 3.3

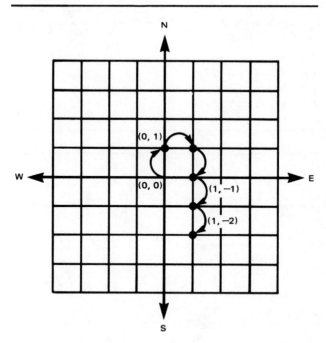

Suppose that two coins are tossed five times, with these results:

HH, TT, HT, HT, HT

Satisfy yourself that after these five tosses, the person will be at the location $(1, -2)$.

If we need to know how far this person is from the starting point, we can use the Theorem of Pythagoras, which says that the square of the hypotenuse (of a right triangle) is equal to the sum of the squares of the other two sides. This theorem enables us to find the length of the hypotenuse of a right triangle if we know the lengths of the other sides. For example, in the triangle in Figure 3.4, the length C of the hypotenuse is 5 units.

FIGURE 3.4

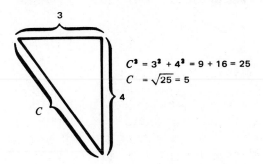

$$C^2 = 3^2 + 4^2 = 9 + 16 = 25$$
$$C = \sqrt{25} = 5$$

Applying this relationship to our city map, we have the triangle in Figure 3.5.

FIGURE 3.5

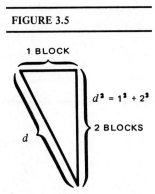

1 BLOCK

$$d^2 = 1^2 + 2^2$$

2 BLOCKS

d

The actual distance of the person from the starting point is

$$d = \sqrt{1^2 + 2^2} = \sqrt{5} \doteq 2.2 \text{ blocks}$$

■ **Example—The Temperature of a Point on a Steel Plate:**   A steel plate is heated so that the temperature of each edge is maintained as indicated in Figure 3.6. What is the temperature of an arbitrary point P on the plate?

FIGURE 3.6

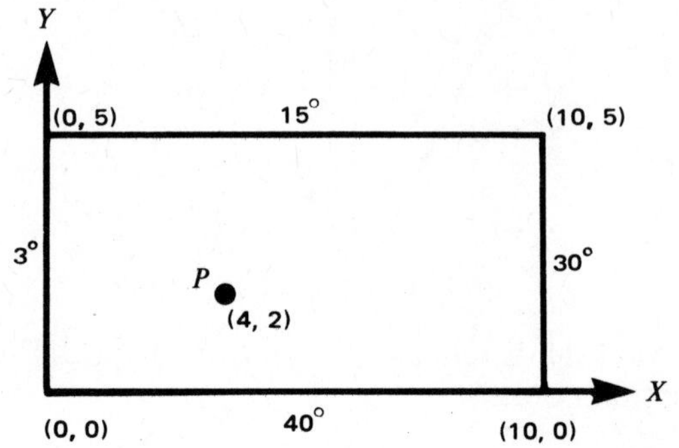

Solution:

This problem can be solved by using a two-dimensional random walk starting at P. But in this case, the walk continues until one of the edges is reached. When an edge is reached, the temperature of that edge is recorded. The walk then resumes at P and is repeated. After sufficient trials (random walks), the average of all the temperatures obtained is taken to be an estimate of the temperature at P.

Table 3.13 gives the result of 100 random walks, or trials, estimating the temperature of the point P at (4, 2).

TABLE 3.13

TEMPERATURE (T)	30°	3°	15°	40°
NUMBER OF WALKS	2	17	34	47

$$T = \frac{(2 \times 30°) + (17 \times 3°) + (34 \times 15°) + (47 \times 40°)}{100}$$

$$= 25.01°$$

For 500 random walks, done by a computer, the result in Table 3.14 was obtained for the temperature at the point (4, 2).

TABLE 3.14

TEMPERATURE (T)	30°	3°	15°	40°
NUMBER OF WALKS	13	42	164	281

$$T = \frac{(13 \times 30°) + (42 \times 3°) + (164 \times 15°) + (281 \times 40°)}{500}$$

$$= 28.43°$$

It can be shown by advanced mathematics that this method produces the theoretical solution as the number of trials gets larger (and the size of the steps taken in the walk gets smaller). ■

EXERCISES

1   Continue the simple, one-dimensional random walk of Table 3.12 for another 18 trials. What average distance is the stopping point from the starting point after 30 trials of 10 coin tosses each?

2   Describe how a die could be used to do the random walk of Exercise 1.

3   Suppose the random walk of Exercise 1 is changed as follows. A wall is put up at the point −5 and the point 5. When the person reaches either wall, the walk ends. What is the average number of blocks the person could expect to walk before reaching one of the walls? Base your estimate on at least 30 trials.

4   In the two-dimensional random walk involving the street map (Figure 3.3) take a walk from $(0, 0)$ based on 10 tosses of two coins. What is the average distance of the point of termination from $(0, 0)$?

5   Repeat Exercise 4 for a total of 10 trials (random walks), each based on 20 tosses of two coins. What is the average distance of the point of termination from the point of origin?

6   In the temperature problem, estimate the temperature at each of the given points, based on 30 random walks for each point. (Use Figure 3.6.)

(a) $(1, 3)$
(b) $(2, 4)$
(c) $(3, 1)$

CHAPTER

4

Probability

FIGURE 4.1

20% CHANCE A PIECE OF SKYLAB HITS A CITY[1]

by James Coates
CHICAGO TRIBUNE PRESS SERVICE

WASHINGTON—The National Aeronautics and Space Administration (NASA) has been assuring the public that the odds that the plummeting Skylab space station will hit any given person as it crashes to earth are 600 billion to 1.

However, studies conducted by a private research firm under commission from NASA say the odds are less astronomical.

AMONG PREVIOUSLY unpublicized aspects of the Skylab fall—expected sometime in mid-July—are the following findings of the private study:

• Despite NASA statements that the earth is three-quarters water and that a strike thus is most likely to come in the ocean, scientists have determined that there are 7 chances in 10 that at least one piece of the spacecraft will strike land.

• The chances of a piece weighing more than 250 pounds hitting land is slightly better than 50-50.

• There are 2 chances in 10 that a heavy piece, such as a 2-ton lead safe aboard the space station, will strike land somewhere.

• Odds are just slightly below 1 in 5 that some piece of Skylab—probably a small one—will strike a city over 100,000 in population. On the other hand, there is only 1 chance in 40 that a piece weighing 250 pounds or more will strike a city of that size.

KEY PROBLEM

Probability and the Law

A police officer had noted the position of the valves on the front and rear tires on one side of a parked car. He used a technique like that used by pilots in noting directions: one valve was pointing to 1:00 and the other value to 6:00. The nearest hour is used in both cases.

After the allowed time had run out, the police officer found the car in the same place with the two valves still pointing toward 1:00 and 6:00. The driver, however, denied that she had overparked. She claimed that she had driven away, come back, and parked with the wheels in the very same position as before!

The court had an expert compute the probability of such an event occurring, and the judge acquitted the driver.[2]

We live in a probabilistic world. An editorial in a magazine speaks of increased chances of nuclear war. We are told that a vaccination will reduce our chances of getting the flu. We send out several applications believing that this will increase our likelihood of getting a job. We pay more for life insurance as we grow older because the risk to the company is increasing.

Postscript: In acquitting the driver, the judge remarked that if all *four* wheels had been checked and found to point in the same direction as before, the driver's claim would have been rejected as too improbable. In such a case, the four wheels, each with 12 positions, could combine in a total of

$$12 \times 12 \times 12 \times 12 = 20,736 \text{ ways}$$

So, the probability of a chance repetition would be only

$$\frac{1}{20,736}$$

(It's assumed here that the wheels of an automobile turn independently of each other. In reality, they do not.)

4.1   RELATIVE FREQUENCY

While a captive during World War II, J. E. Kerrich of Denmark conducted an experiment in which he tossed a coin 10,000 times and kept a record of the outcomes.[3] The results are given in Table 4.1. The entries in the right-hand column, the relative frequencies, are calculated by dividing the number of heads obtained by the number of tosses of the coin. For example, in the first row, the relative frequency is

$$\frac{4}{10} = .400$$

We can interpret the right column as giving estimates, based on an experiment of tossing a coin, of the probability or likelihood of getting heads when tossing a coin. Do the results of Kerrich's experiment bear out what you would expect?

We notice that Kerrich obtained heads very close to half the time, especially as the number of tosses got larger. Sometimes, he obtained heads a little less than half the time (for example, in the first 100 tosses he got only 44 heads). Sometimes he obtained heads a little more than half the time (for example, in the first 1,000 tosses he got 502 heads). But the results do support what we would expect from a *fair*, or unbiased, coin. The coin can fall either heads or tails, and it falls heads about half the time, as the number of trials gets larger.

As we noted in Chapter One, a relative frequency can be thought of as a probability. It is an *experimental* probability because it is calculated involving data gathered from an experiment. A relative frequency is a fraction. The denominator tells us how many observations have been made. The numerator tells us how many observations are of interest. For

TABLE 4.1
10,000 TOSSES OF A COIN

NUMBER OF TOSSES	NUMBER OF HEADS	RELATIVE FREQUENCY (EXPERIMENTAL PROBABILITY)
10	4	.400
20	10	.500
30	17	.567
40	21	.525
50	25	.500
100	44	.440
150	71	.473
200	98	.490
300	146	.487
400	199	.498
500	255	.510
600	312	.520
700	368	.526
800	413	.516
900	458	.509
1,000	502	.502
2,000	1,013	.507
3,000	1,510	.503
4,000	2,029	.507
5,000	2,533	.507
6,000	3,009	.502
7,000	3,516	.502
8,000	4,034	.504
9,000	4,538	.504
10,000	5,067	.507

example, Kerrich tossed a coin 10,000 times and got 5,067 heads. The relative frequency is

$$\frac{5,067}{10,000}$$

We often state

$$P(\text{outcome}) = \frac{\text{Number of outcomes of interest}}{\text{Total number of outcomes}}$$

Note: Outcomes of interest are also called "successful outcomes."

In this book, we will conduct many experiments and calculate probabilities using the data from these experiments. We will often think of these experimental probabilities as estimating theoretical probabilities.

As another example of experimental probability as a relative frequency, suppose that an automobile club is investigating the chances of being able to use roadside telephones to call for emergency help. The club members try 30 telephones and find that 24 are in working order. So, the club reports the probability of being able to use a telephone to request emergency aid is

$$P(\text{telephone works}) = \frac{24}{30} = .80$$

On the basis of the information, we can expect to get through on a phone in case of a future emergency about 80% of the time.

We should note here that certain assumptions are made about how the survey was carried out. We assume, for example, that the sample of telephones is a fair one and that the service club members did not bias their selection by trying, for example, only telephones that had a battered appearance!

EXERCISES

1   In Table 4.1, what estimated probability of getting heads was obtained for Kerrich's data, based on

(a)   100 tosses of the coin?
(b)   500 tosses of the coin?

Do you think that Kerrich was using a well-balanced coin? Why or why not?

2   Suppose a building-supplies producer wishes to estimate the probability that a certain type of carpet tack will fall in the point up position (position A) when dropped on a smooth wood floor. So a researcher drops one of the tacks 500 times and finds that it falls point up 27 times. What is the estimated probability that the tack will fall point up? What assumptions are made in order for the experiment to be valid?

3   A certain baseball player has made 24 hits out of the last 92 times at bat. What is the estimated probability that the player will get a hit at the next time at bat? (*Note:* You have calculated the player's batting average.)

4   Look up the batting average of three baseball players, either in a sports magazine or the sports section of a newspaper. Interpret these figures as probability estimates.

5   Joe wants to estimate his chances of getting an ace when he draws one card from a well-shuffled deck. So he performs the following experiment 100 times. He takes an ordinary deck of cards, shuffles it well, deals a card, examines it, and records whether or not it is an ace. The table below shows the number of times an ace was obtained.

Card	f
Ace	14
Not an ace	86

What is $P(\text{ace})$?

6   Try Exercise 5 yourself. What estimated value of $P(\text{ace})$ do you get?

7   The birthday problem is well known in the study of probability. We will consider a related problem, the birth*month* problem. Simply stated it is:

> What is the probability that in a group of, for instance, four people at least two share a birth-month (were born in the same month).

Solution: In order to solve this problem experimentally, a 12-sided die is needed with each side labeled one of the 12 months. Roll the die once. Roll three more times for the birthmonths of the remaining three in the group of four. Suppose the four numbers obtained on four rolls of a 12-sided die are:

3, 7, 1, 10

1	2	3	4	5	6	7	8	9	10	11	12
x		x				x			x		

Note: 1 = January, 2 = February, etc.

In this first set of four rolls, no month was shared as a birthmonth. Continue getting sets of four rolls, recording the months obtained. The table shows results for 9 more sets of four rolls (trials). Find the estimated value of P(shared birthmonth).

Trial	Outcome				Shared Birthmonth?
2	12	7	6	10	no
3	4	10	4	4	yes
4	2	11	10	1	no
5	3	10	7	3	yes
6	3	12	12	6	yes
7	5	7	5	11	yes
8	9	11	5	11	yes
9	5	5	2	10	yes
10	2	1	1	2	yes

Based on 10 trials

$$P(\text{shared birthmonth}) = \frac{7}{10} = .7$$

Now do 40 more trials and find P(shared birthmonth).

8   Find these probabilities for the birth-month problem. Use 50 sets of rolls of the 12-sided die for each estimate.

(a) P(shared birthmonth in a group of 3 persons)

(b) P(shared birthmonth in a group of 6 persons)

(Or you could use the data you generated in the project in Section 3.5.)

9   From your everyday life, name three events with which a probability is associated.

10   The Illinois Water Survey reports that in the Champaign-Urbana, Illinois, area only 34 out of the last 90 Christmases were white (that is, there was measurable snow on the ground on Christmas Day). Use this data to estimate the probability of a white Christmas in Champaign-Urbana.

11   The weather forecaster says, "The chances of rain today are 25%."* Interpret this statement from a relative frequency point of view. That is, on the average, how often would we expect it to rain?

12   Casey Jones' batting average is .250. Interpret this from a relative frequency point of view.

---

*Statisticians prefer that probability be expressed as a decimal fraction (.25) rather than as a percentage (25%). However, the use of percentage is common in magazines, on television and so on.

13 XYZ Airlines finds from its records that out of 2,000 persons who made reservations last month, only 1,850 persons actually showed up for the flights. From these data, what is the estimated probability that a person who makes a reservation will show up for the flight? What is the estimated probability of a no-show?

14 In the 1969 World Series the New York Mets played the Baltimore Orioles. Before the series was played, a mathematics class made up probabilities for each team to win each game, up to the maximum of seven games (see Table 4.2). These estimates were based on information such as where the game was to be played, who were the opposing pitchers, and other factors.

For each game, what influence were game location and pitchers judged to have on the class estimates of the probabilities of winning for each team?

TABLE 4.2
LOCATION OF GAME, PITCHERS, AND PROBABILITIES[4]

GAME	LOCATION	SCHEDULED PITCHERS	P(METS)	P(ORIOLES)
1	Baltimore	Seaver vs. Cuellar	0.4	0.6
2	Baltimore	Koosman vs. McNally	0.5	0.5
3	New York	Gentry vs. Palmer	0.4	0.6
4	New York	Seaver vs. Cuellar	0.7	0.3
5	New York	Koosman vs. McNally	0.5	0.5
6	Baltimore	Gentry vs. Palmer	0.3	0.7
7	Baltimore	Seaver vs. Cuellar	0.5	0.5

SPECIAL INTEREST FEATURE

Population experts say that there are generally 5% more male babies born than females, but the mortality rate for male babies is higher than for females. By marrying age, the ratio of males to females is about even.[5]

4.2   WHAT ARE THE CHANCES?

We can now use the idea of relative frequency, together with the method discussed in the last chapter, to solve probability problems.

■ **Example—Three-Girl Families:**   What is the probability that in a family of three children, all are girls?

Solution:

One approach to this problem, as was mentioned in Section 3.2, would be simply to survey a large number of three-child families and record the number of girls in each. Table 4.3 might result from such a survey of 100 families having three children.

TABLE 4.3

N GIRLS	f
0	11
1	40
2	32
3	17
	Total = 100

From this table, we see that 17 of the 100 families had three girls. So, we estimate

$$P(3 \text{ girls}) = .17 \quad ■$$

As before, another approach to estimating the probability of three girls in a three-child family would be to use a coin and toss it three times, once for each child, where

Heads:	Girl
Tails:	Boy

Each three tosses of the coins would represent surveying one three-child family. We would then find how many of the tosses resulted in three heads (three girls).

We use our five-step method, but change Step 3 and Step 5. In Step 3, we define a successful trial. In Step 5, we find the estimated probability of a successful trial.

Step 1—*Model*:   Since the probability that a child is a girl is assumed to be .5, we use a coin.

Heads:	Girl
Tails:	Boy

Step 2—*Trial*:   A trial consists of tossing a coin three times and counting the number of heads (girls) obtained.

Step 3—*Successful trial*:   In our problem, we are asked to estimate the probability that a family consists of three girls. So, a successful trial results when three heads are obtained.

Step 4—*Repeat steps 2 and 3*:   We should obtain about 100 trials for a good estimate of $P(3 \text{ girls})$.

Step 5—*P(success)*:   The probability of a successful trial is estimated by dividing the number of successful trials (found in Step 3) by the total number of trials (found in Step 4).

In our example, we tossed the three coins 50 times. You will be asked to do more trials in the exercises. We obtained the results in Table 4.4.

TABLE 4.4

N HEADS	f
0	6
1	20
2	16
3	8
	Total = 50

Our estimated probability is

$$P(3 \text{ girls}) = \frac{8}{50} = .16$$

■ Example—A Slightly Different Problem:
Suppose, instead the problem were: What is the probability that in a family of three children, *two or more* are girls?

Solution:

We need to make very little change in our approach.

Step 1—*Model*:   Use a coin.

Heads: Girl
Tails: Boy

Step 2—*Trial*:   A trial consists of tossing a coin three times (one for each child) and counting the number of heads.

Step 3—*Successful trial*:   A trial is a success, in terms of our problem, if *two or more* heads are obtained.

Step 4—*Repeat steps 2 and 3*:   You should obtain at least 50 trials.

Step 5—*P(success)*: $P(2 \text{ or more girls}) =$

$$\frac{\text{number of successful trials}}{\text{total number of trials}}$$

We will use the results of the 50 tosses of the three coins reported in the last table, repeated as Table 4.5.

TABLE 4.5

N HEADS	f
0	6
1	20
2	16 } 2 or more heads obtained
3	8

Now the number of successes in 50 trials is $16 + 8 = 24$.

$$P(2 \text{ or more girls}) = \frac{24}{50} = .48$$

The theoretical probability $p(2 \text{ or more girls})$ is .50. So, this estimate is a rather good one. ■

■ Example—Batter Up!!
Roscoe has a batting average of .250. That is, on the average, he has been getting one hit out of four times at bat (since $.250 =$ one-fourth). What are the chances that, out of the next ten times at bat, he will get three or more hits?

Solution:

We now require a model for a probability of one-fourth. In Section 3.6 we used two coins, since, for example, we expect to get two heads one time out of four when we toss two coins. Instead, let's use a six-sided die and ignore a five or a six if rolled.

Step 1—*Model*:   We use four sides of a six-sided die (there are four-sided dice, called tetrahedral dice, which would be very good to use, too).

1: get a hit
2, 3, 4: no hit
5, 6: ignore, roll again

Step 2—*Trial*:   A trial consists of rolling the die ten times, once for each time at bat. (If you get a five or six, do not count that roll.)

Step 3—*Successful trial*:   A successful trial occurs when we get "*one*" for a total of three times or more (a "*one*" means "*a hit*").

Step 4—*Repeat steps 2 and 3*:   Conduct at least 50 trials.

Step 5—*P(success):* $P(3$ or more hits in 10 times at bat$) =$

$$\frac{\text{number of successful trials}}{\text{total number of trials}} \blacksquare$$

We will conduct two trials to demonstrate. The following outcomes of the rolls of a die were obtained and recorded in Table 4.6.

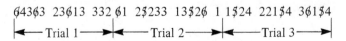

6́436́3 236́13 332 6́1 26́233 136́26́ 1 16́24 2216́4 36́16́4

|←———— Trial 1 ————→|←———— Trial 2 ————→|←———— Trial 3 ————→|

TABLE 4.6	
TRIAL	**NUMBER OF HITS**
1	1
2	3
3	3

We see in our first three trials that we obtained three or more hits twice. Table 4.7 gives the results of 50 trials.

TABLE 4.7	
***N* HITS**	**TRIALS**
0	5
1	10
2	17
3	11 ⎫
4	2 ⎪
5	1 ⎪
6	1 ⎪
7	2 ⎬ 18
8	1 ⎪
9	0 ⎪
10	0 ⎭
	—
Total	50

For our 50 trials, we have

$$P(3 \text{ or more hits}) = \frac{18}{50} = .36 \blacksquare$$

EXERCISES

Give the five steps for solving each problem. Do 50 trials unless otherwise told.

1   Toss three coins 50 times to make the total number of tosses obtained in Table 4.4 equal to 100. Now find $P(3$ girls$)$ based on 100 trials and compare with the answer for 50 trials found earlier.

2   Using the 100 trials resulting from Exercise 1 of this section, find $P(2$ or more girls$)$ and compare with $P(2$ or more girls$)$ $= .48$, which was found earlier based on 50 trials.

3   In a family of four children, what is the probability of two or more boys?

4   In a family of three children, what is the probability of at least one girl?

5   The probability of rain is given as .25 in each of three cities in widely separated parts of the world. What is the probability that it will rain in two or more cities?

6   Casey Jones has a batting average of .333. What are his chances of getting a hit five or more times in the next ten times at bat? (*Hint:* You can use a die to get a probability of one-third.)

7   One of the parents of a four-child family is infected by strep throat. The family physician informs them that the chances of others in the family getting strep is about .15 (for each person). What is the estimated probability that at least one of the four children gets strep? (The value of .15 was actually provided by the physician of one of the authors.) (*Hint:* approximate .15 by one-sixth.)

8   Ron didn't study for his history quiz, which was a true-false test. He had no idea about the answers for eight of the questions, so he guessed. What are his chances of getting six or more of them correct? (*Hint:* Use a coin where, for instance, heads correct and tails wrong.)

9   In January 1983, it was estimated that one out of three persons in the United States had seen the movie "ET (The Extra-Terrestrial)."[6] What is the probability that in a group of four people at least two of them had, at that time, seen the movie "ET"?

10   A certain marksman shoots at a target with a rather small bull's eye. At first the probability that any one of his shots hits the bull's eye is .5. In other words, his hit-miss sequence is just like the heads-tails sequence of a fair coin. But, unlike a coin, our marksman has a *memory* and is trying to make a good score. Therefore, if he misses the bull's eye *three times in a row*, he aims more carefully, and then his probability of hitting the bull's eye is .75. He continues to aim carefully until he finally hits the bull's eye. Then he relaxes and goes back to the old pattern. In 40 shots, how many bull's eyes will he hit on the average? (*Hint:* This is a *two-model* problem. What model, involving coins, can be used when the marksman is taking careful aim?)[7]

4.3 RANDOM DIGITS

So far, we have been able to solve a variety of problems by using only a coin or a die. Often, though, it is useful and convenient to have other models to help solve problems. Random digits are often used in this way. Let's illustrate with an example.

■ **Example:**   A camera manufacturer finds that 10% (.1) of the springs used in the shutters are defective. Of the next five springs the manufacturer tests, what is the probability of finding that *one or more* is defective?

Solution:

We need some way of representing a probability of .1 for a defective spring. A spinner marked off in ten equal regions (called a 10-spinner) like the one shown in Figure 4.2 would do.

FIGURE 4.2

Each of the ten sectors of the spinner is labeled with one of the digits 1, 2, 3, 4, 5, 6, 7, 8, 9, 0. Let's suppose that if we give this device a spin and the arrow stops, then the chances of getting a four are one out of ten or .1. Similarly, each other digit will occur with a probability of .1. We now have a model for producing probabilities of .1.

So, we can do lots of spins and write down the results. For example, we did eight spins and got:

$$3, \ 5, \ 3, \ 6, \ 8, \ 9, \ 6, \ 6$$

In order to solve the problem, many such random digits are needed. Statisticians have found ways to produce such long lists, usually by using a computer. One of the well-known lists, consisting of 1,000,000 random digits, was produced by the Rand Corporation and published in 1955.[8] (It is reported that this book had very little plot!)

A list of random digits is often called a *random digit table.* Table 4.8 is an example of such a table. Each row is numbered. We can use the table reading across the rows, taking each digit we find. It may help you to understand how this table is produced by imagining that a 10-spinner was used—one spin for each digit. So for example, in Row 1, five spins resulted in the digits 7, 0, 7, 4, 0.

In this table (Table 4.8), there are 45 digits in a row (9 groups with 5 digits each). In order to try and see how often each of the ten digits appears, we made a frequency table (Table 4.9) of the first 50 digits in Table 4.8 (up to the digits 51537 in Row 2).

TABLE 4.9

DIGIT	TALLY	f
1	7卌 //	7
2	////	4
3	////	4
4	////	4
5	7卌 ///	8
6	////	4
7	7卌 //	7
8	/	1
9	//	2
0	7卌 ////	9
	Total	50

So, for these 50 "spins" we see that a 9 was obtained 2 out of 50 times. Therefore,

$$P(9) = \frac{2}{50} = .04$$

Also, since 1 was obtained 7 times, we have

$$P(1) = \frac{7}{50} = .14$$

However, each of these probabilities, based on 50 trials, estimates a *theoretical* probability of .1. Since we have only 50 trials, our estimates of the theoretical probability vary quite a bit. ∎

TABLE 4.8

ROW									
1	70740	34910	59535	40227	07000	12166	21557	01147	85636
2	51537	46599	54789	84716	38925	18099	24548	65938	54301
3	95200	71041	95802	49728	55936	74399	60006	50279	46380

Let's use Table 4.8 to solve our problem about the camera springs.

Step 1—*Model*: Use random digits with this rule:

> 0: defective spring
>
> 1 through 9: not defective

Step 2—*Trial*: A trial consists of doing five spins, one for each of the five springs inspected.

Step 3—*Successful trial*: In terms of our problem, a trial is successful if *at least one* zero (defective spring) is found. (We do not suggest that the factory regards this as a success!)

Step 4—*Repeat steps 2 and 3*: At least 50 trials should be conducted.

Step 5—*P(success)*: The estimated probability of finding at least one defective spring is

$$\frac{\text{number of successful trials}}{\text{total number of trials}}$$

We will now conduct ten trials using Rows 2 and 3 of Table 4.8. We simply read off the digits, row by row, as shown in Table 4.10.

TABLE 4.10

TRIAL	DIGITS FROM TABLE 4.8	SUCCESS?
1	5, 1, 5, 3, 7	No
2	4, 6, 5, 9, 9	No
3	5, 4, 7, 8, 9	No
4	8, 4, 7, 1, 6	No
5	3, 8, 9, 2, 5	No
6	1, 8, 0, 9, 9	Yes
7	2, 4, 5, 4, 8	No
8	6, 5, 9, 3, 8	No
9	5, 4, 3, 0, 1	Yes
10	9, 5, 2, 0, 0	Yes

Based on only ten trials, which is entirely too few to give a reliable estimate of the probability, we found three successes. That is,

$$P(\text{at least one defective spring}) = \frac{3}{10} = .30 \ \blacksquare$$

OTHER MODELS OF CHANCE

In Exercise 7 of Section 4.2, we needed a model for obtaining a probability of .15, which was a physician's estimate of the probability that a person would get strep throat from someone else in the family who has the disease. It was suggested there that a die could be used for approximating that probability, since the theoretical probability of getting a 1, for example, when rolling a die is 1 out of 6, or .1666 (which is close to .15). Now we will see how a table of random digits can be used for the required model when a probability of .15 is required.

We will use pairs of digits from Table 4.11. The 100 numbers that can be produced using two digits are: 01, 02, 03, 04, 05, 06, 07, 08, 09, 10, 11, . . . , 98, 99, 00. We use 00 rather than 100 since we are using *pairs of digits* made up from Table 4.11.

TABLE 4.11

FIRST DIGIT	SECOND DIGIT									
	0	1	2	3	4	5	6	7	8	9
0	00	01	02	03	04	05	06	07	08	09
1	10	11	12	13	14	15	16	17	18	19
2	20	21	22	23	24	25	26	27	28	29
3	30	31	32	33	34	35	36	37	38	39
4	40	41	42	43	44	45	46	47	48	49
5	50	51	52	53	54	55	56	57	58	59
6	60	61	62	63	64	65	66	67	68	69
7	70	71	72	73	74	75	76	77	78	79
8	80	81	82	83	84	85	86	87	88	89
9	90	91	92	93	94	95	96	97	98	99

We see from Table 4.11 that 100 numbers (01 to 99, plus 00) can be formed using the ten digits 0 through 9. As we will see later in this chapter, it can be shown that if each of the ten digits is equally likely to be chosen (that is, with a probability of .1, or one-tenth), then pairs of such random digits are chosen with a probability of .01, or one-one-hundredth. So, if we want a model for producing an event with a probability of .15, we can use this rule:

> Digits 01 to 15:   event occurs
> Digits 16 to 99, plus 00:   event does not occur

Now, let's look at another example.

■ **Example:**   For strep throat, the chances of contracting the disease by coming into contact with an infected person have been estimated as .15. What are the chances of at least one of four children in a family getting the disease, assuming each has come into contact with an infected person?

Solution:

Step 1—*Model:*   Use pairs of random digits, with the rule:

> 01 through 15:   gets strep throat
> 16 through 99, plus 00:   does not get strep throat

Step 2—*Trial:*   Read four pairs of random digits, one for each child in the family.

Step 3—*Successful trial:*   A trial is a success if at least one of the four pairs of digits is in the range 01–15.

Step 4—*Total trials:*   Repeat for a total of 100 trials (six are shown in Table 4.12, using the random digits provided and reading across the table, row by row).

Summary of 100 trials:

number of successes = 44

TABLE 4.12

ROW									
1	69531	54637	06640	35956	26693	27891	06397	70132	29186
2	56905	10986	53970	31729	18700	91782	65398	63865	81835
3	04905	82632	60729	55072	48677	48613	26881	16880	19349
4	00285	86659	06887	53425	24104	81243	46620	39849	95102

SAMPLE TRIAL	DIGITS	SUCCESS?
1	69, 53, (15,) 46	Yes
2	37, (06,) 64, (03)	Yes
3	59, 56, 26, 69	No
4	32, 78, 91, (06)	Yes
5	39, 77, (01,) 32	Yes
6	29, 18, 65, 69	No

Step 5—*Probability of success:* P(success) =

P(at least one child gets strep) = $\dfrac{44}{100}$ = .44 ∎

EXERCISES

Unless otherwise indicated, do 50 trials for each exercise. If needed, additional random digits are in Appendix F.

1 Conduct another 20 trials of the experiment done earlier to find the estimated probability of at least one defective spring in five.

2 Either conduct a total of 100 trials for Exercise 1 of this section, or share your results with others in your class to provide a reliable estimate of the probability of at least one defective spring in five. Theoretical mathematics gives a probability of .410 if an infinite number of trials were to be conducted. How does your answer compare to this value?

3 A certain type of plastic bag is found to burst 20% of the time under moderate pressure. If a prospective buyer tests one-half dozen bags, what is the estimated probability that at least one will burst? What is the estimated probability that two or fewer will burst? Do about 60 trials.

4 In Exercise 3 of this section, what is the estimated probability that none of the one-half dozen bags will burst? You may use the same trials as in Exercise 3, but add another 40 trials to them to improve your estimate.

5 Wilma has a batting average of .330. What is her estimated probability of getting four or more hits out of the next ten times at bat?

6 Find the reported batting average of your favorite baseball player from the newspaper. Calculate the estimated probability that this player will get four or more hits out of the next ten times at bat. (*Hint:* If necessary, use sets of *three* random digits.)

7 A film manufacturer advertises that 95 out of 100 prints will develop. Ms. Jones buys a roll of film and finds that 2 prints do not develop. If the manufacturer's claim is true, what is the estimated probability that in a roll of 12 prints, 2 or more will not develop?

8 A pharmaceutical company knows from previous testing that a certain antibiotic capsule falls below prescribed strength 6% (that is, .06) of the time, making the capsule ineffective. What is the estimated probability that a prescription of 20 such capsules will contain 2 or more with ineffective dosages?

9 Margo is taking a multiple-choice test that has three choices for each question. She doesn't know the answers to 12 of the questions, so she guesses. Find the estimated probability of getting 6 or more of the questions correct by guessing.

10 XYZ Airline has a no-show rate of 18% (.18) according to past records. Suppose they decide to overbook a flight by accepting 42 reservations for a 40-seat airplane. On the average, how many persons could they expect to show for such a flight? What is the estimated probability that everyone will get a seat on a flight with this policy of overbooking?

11   A geologist is studying the number of quartzite pebbles in a stream. Existing records give reason to believe that for this stream the probability of picking up a pebble that contains quartzite is .7. In a sample of 10 pebbles, what is the probability of getting 8 or more pebbles containing quartzite? Of getting 6 or fewer pebbles containing quartzite? [9]

4.4   INDEPENDENT EVENTS

Let's return to our example of rain in San Juan and Chicago. We assume that whether it rains in one city is unrelated to whether it rains in the other.

■ **Example:**   The weather forecast for a certain day is a 60% chance of rain (.60) for San Juan and a 50% chance of rain (.50) for Chicago. What is the estimated probability of rain in both cities?

Solution:

We'll use a table of random numbers to solve this problem.

Step 1—*Model:*   A model of a probability of .60 for San Juan:

> 1, 2, 3, 4, 5, 6:   rain in San Juan
> 7, 8, 9, 0:   no rain in San Juan

A model of probability of .50 for Chicago:

> 1, 2, 3, 4, 5:   rain in Chicago
> 6, 7, 8, 9, 0:   no rain in Chicago

Step 2—*Trial:*   A trial consists of reading two digits (one for each city) from the table and recording them. We use Table 4.13.

Step 3—*Successful trial:*   In terms of our problem, a successful trial occurs when it rains in both cities. That is, we get one of the digits 1–6 for San Juan and one of the digits 1–5 for Chicago.

Step 4—*Number of trials:*   Conduct at least 50 trials.

Step 5—*P(success):*   P(rain in both San Juan and Chicago) =

$$\frac{\text{number of successes}}{\text{number of trials}}$$

For our source of random digits, we will use Row 1 of Table 4.15 (page 90) showing *rolls of a 10-sided die*, in order to conduct four trials. The digits are 47914 16071 and are written as follows:

$$4,7 \quad 9,1 \quad 4,1 \quad 6,0 \quad 7,1$$

TABLE 4.13

TRIAL	RAIN IN SAN JUAN (1–6)	RAIN IN CHICAGO (1–5)	SUCCESS?
1	4 = rain	7 = no rain	No
2	9 = no rain	1 = rain	No
3	4 = rain	1 = rain	Yes
4	6 = rain	0 = no rain	No
5	7 = no rain	1 = rain	No

Therefore, based on our very few number of trials—only five—we have the estimated probability

$$P(\text{rain in both San Juan and Chicago}) = \frac{1}{5} = .20 \quad ■$$

In order to get a more reliable estimate of the probability, we conducted a total of 100 trials, with the results in Table 4.14.

TABLE 4.14

| NUMBER OF TRIALS | NUMBER OF TIMES IT RAINED | | NUMBER OF SUCCESSES* |
	IN SAN JUAN	IN CHICAGO	
100	42	59	26

*Rain in both cities.

From Table 4.14, we find

$$P(\text{rain in both San Juan and Chicago}) = \frac{26}{100} = .26$$

But also note from Table 4.14 that:

$$P(\text{rain in San Juan}) = \frac{42}{100} = .42$$

$$P(\text{rain in Chicago}) = \frac{59}{100} = .59$$

These estimates bring us to an important property or law about independent events.

The law of experimental independence: If two events are independent, then the estimated probability that they both occur is approximately equal to the product of the two individual probabilities (for a reasonably large number of trials).

Suppose the two events are A and B. Then, this law states

$$P(A \text{ and } B) \doteq P(A) \times P(B) \ \blacksquare$$

Let

$$A = \text{rain in San Juan}$$
$$B = \text{rain in Chicago}$$

Then

$P(A \text{ and } B) = P(\text{rain in both San Juan and Chicago})$
$\qquad \doteq P(A) \times P(B)$
$\qquad = P(\text{rain in San Juan}) \times P(\text{rain in Chicago})$

Now, from our experiment, we found

$$P(\text{rain in both San Juan and Chicago}) = \frac{26}{100} = .26$$

$$P(\text{rain in San Juan}) = .42$$

$$P(\text{rain in Chicago}) = .59$$

$P(\text{rain in San Juan})$ times $P(\text{rain in Chicago}) =$

$$.42 \times .59 \doteq .25$$

which is approximately .26.

You will be asked in the exercises to conduct your own experiment to obtain more trials and show that this law provides a good estimate for the probability of two independent events.

Later in this chapter, you will learn how the theoretical probability of two independent events taking place can be calculated exactly, rather than just estimated, if the theoretical probability of each event is known. (You probably can guess the rule.)

EXERCISES

Use the random digits provided in Table 4.15. Do 50 trials unless otherwise told.

1   Conduct another ten trials for the rain experiment, as done above. Give all the details required. To illustrate, we have done one trial, Trial 6, using the next two digits, 3,6, in Table 4.15.

Trial	San Juan	Chicago	Success?
6	3 = rain	6 = no rain	No

You should do Trials 7–15.

TABLE 4.15

ROW	COIN TOSSES (OUTCOMES ARE H AND T. *H STANDS FOR HEADS, T FOR TAILS.*)									
1	TTHHH	TTTTT	THTTT	TTHHH	HHHTT	TTHHT	HTHTT	HTTTT	HTHTH	HHTTT
2	TTTHT	HTTTH	HHHHH	TTHTT	TTTTH	THHHT	TTTTH	TTHTH	TTTHH	HHTHH
3	TTHHH	HHTTT	HTHHH	HHHTT	THTTT	HHTHT	TTHTT	THHTH	THTHT	HTTHT
4	TTTHT	HHHHT	HHTHH	THTTT	HHTTT	HTTTT	HHTTT	THHHH	THHTT	HHTTH
5	THHHT	HTTTH	TTHHH	HTHTH	HHTTT	HHHTH	HHTHH	HTTTH	TTTHT	HHTTT

ROW	ROLLS OF A 4-SIDED DIE (OUTCOMES ARE 1, 2, 3, 4.)									
1	34431	42314	21413	11332	41244	42134	24223	32143	14411	13312
2	42223	22224	13413	12323	31323	24141	43244	43143	43221	22221
3	14323	22224	24411	11113	32232	22132	42123	14412	13222	44243
4	13442	13244	13132	13213	21133	31244	21141	43421	11324	44141
5	42431	22144	14222	13232	32423	34223	33314	44323	41332	33233

ROW	ROLLS OF A 6-SIDED DIE (OUTCOMES ARE 1, 2, 3, 4, 5, 6.)									
1	51316	53462	52315	12145	42135	46352	34241	12561	65115	51151
2	26641	15643	46336	55636	41456	42664	64135	26446	34311	21622
3	66544	66324	42661	23343	46423	23244	36632	46112	33231	41453
4	31462	43516	61523	33561	51163	51415	56443	35635	24311	45214
5	23155	64515	51143	22416	41564	61451	14113	64524	54414	65516

ROW	ROLLS OF A 10-SIDED DIE (OUTCOMES ARE 1, 2, 3, 4, 5, 6, 7, 8, 9, 0.)									
1	47914	16071	36031	59991	06587	20041	72258	43406	89002	57786
2	87803	01790	94569	84915	35267	19243	18401	49493	51687	55104
3	73195	84474	62039	01874	30812	08586	81047	11382	49653	72071
4	70782	99421	57972	49804	45883	27459	51242	48024	01517	43432
5	95902	41240	99773	47194	23315	49300	12459	03162	06584	68944

ROW	ROLLS OF A 12-SIDED DIE (OUTCOMES ARE 1, 2, 3, 4, 5, 6, 7, 8, 9, 0, E, T. *E STANDS FOR 11, T STANDS FOR 12.*)									
1	51EE4	11672	7E891	05707	25405	84E83	21170	69T85	45947	53T77
2	T87T1	55379	588T4	T1E24	60920	73111	33985	389E9	82506	37992
3	05978	2TT54	95511	E478T	94575	T1T49	0E6T6	T4002	7601E	50T54
4	9T3E3	89673	191E3	4E2E5	55077	58517	869ET	29619	61070	41792
5	TE4E8	346TE	34E75	9T680	173T0	79592	24275	58039	819T8	483T5

2 Suppose the rain experiment was conducted for 80 trials and 26 successes were obtained. What is P(rain in both cities)?

3 Suppose in the 80 trials for Exercise 2 of this section, you get rain in Chicago on 42 trials and rain in San Juan on 50 trials. Compute P(rain in Chicago) $\times$ P(rain in San Juan) and compare with P(rain in both cities) found in Exercise 2 of this section.

4   There are two traffic lights on your route to class. Each has a 60-second cycle. Out of those 60 seconds, each light is green for 20 seconds. Assuming the lights work independently of each other, conduct an experiment to estimate the probability of getting a green on both lights. Then apply the law for the probability of independent events and see how well your results hold for that law.

5   Suppose the first traffic light (Exercise 4 of this section) is green for 30 seconds out of 60 and the second is green for 20 seconds out of 60. Find the estimated probability of getting a green light on both. Then apply the law for independent events and see how well your results hold for that law.

6   A two-stage rocket must fire in both stages before it will lift off. The probability that each stage will fire is .75. The stages fire independently. Conduct an experiment to estimate the probability of a lift-off. Verify the Law of Experimental Independence.

7   Assume that an airplane can make a safe flight if at least one-half of its engines are in operation. What is the probability of a safe flight of a two-engine plane if each engine operates independently and the probability of failure of an engine is .10.

8   Repeat Exercise 7 for a four-engine plane. Then decide which is safer, a four-engine or two-engine plane?

9   Using Table 4.15, solve this problem: What is the probability that in a family of three, at least one is a girl? (*Hint:* You could use this rule. Odd digits 1, 3, 5, 7, 9 for a girl. Even digits 0, 2, 4, 6, 8 for a boy.)

10   A ten-question, multiple-choice test has three choices for each question. If a student guesses each answer, what are his or her chances of passing (obtaining more than five correct responses)?

(*Hint:* There are three choices for the answer, and if the student guesses, each choice has an equal chance to be picked. You could use the rule

> Digits 1, 2, 3:   Choice A
> Digits 4, 5, 6:   Choice B
> Digits 7, 8, 9:   Choice C

Ignore 0.)

11   Suppose that automobile accidents are equally likely on each day of the week. What is the probability that if two accidents occur, they will both be on the weekend (that is, Saturday or Sunday)?

12   Refer to the key problem for this chapter. Devise a probability experiment to estimate the probability that the driver actually had not overparked. (Recall that she claimed that she had driven away, returned, and parked with the wheels in the same position as before.)

13   Records show that during a certain season, National Football League teams won 65% of their home games. Do an experiment to estimate the probability of winning three home games in a row.

4.5   THEORETICALLY SPEAKING

So far, we have solved probability problems by estimating the required probability after conducting some sort of experiment and collecting data. But probability may be approached from another point of view. We have already used this viewpoint, for example, as we have tried to decide which model to use in a given experiment. We first need to discuss the idea of equally likely outcomes.

EQUALLY LIKELY OUTCOMES

In many situations involving probability, all outcomes are equally likely. That is, if the situation were to be repeated a large number of times, each of the outcomes would occur with the same relative frequency. A coin toss, for example, has two equally likely outcomes: heads and tails (see Figure 4.3a). A roll of an ordinary die has six equally likely outcomes: 1, 2, 3, 4, 5, 6 (see Figure 4.3b).

We can use the idea of equally likely outcomes to define theoretical probability. For the toss of a coin, we can say

$$p(\text{heads}) = \frac{\text{number of successful outcomes}}{\substack{\text{total number of} \\ \text{equally likely outcomes}}} = \frac{1}{2} = .5$$

For the roll of a die, we can say

$$p(1 \text{ or } 2) = \frac{2}{6} = \frac{1}{3}$$

It is important to note that we did not do an experiment to find this probability. Instead, we used some basic ideas about equally likely outcomes and successful outcomes. We are using a definition based upon an understanding of the meaning of probability and an analysis of the situation involved. Such probabilities are theoretical and may not ever be observed exactly in the real world. Recall that we use a lowercase p, as in

$$p(\text{heads}) = \frac{1}{2} = .5$$

to indicate a theoretical probability. Earlier in this chapter, we referred to the coin tossing

---

FIGURE 4.3

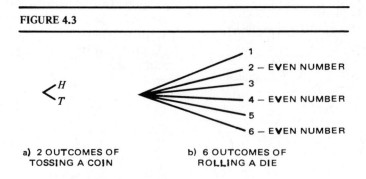

a) 2 OUTCOMES OF
   TOSSING A COIN

b) 6 OUTCOMES OF
   ROLLING A DIE

experiment of Kerrich and found, based on 10,000 tosses of the coin, that

$$P(\text{heads}) = .507$$

The capital P shows that we are talking about an *experimental* probability based on data obtained from a certain number of trials. So, $P(\text{heads})$ is an estimate of $p(\text{heads})$. That is, the *experimental* probability of getting heads is an estimate of the *theoretical* probability of getting heads on the toss of an honest coin.

 Theoretical probability (for equally likely outcomes: Suppose we have a set of N equally likely outcomes, of which S of them are successes. Then the theoretical probability of a successful outcome is

$$p(\text{success}) = \frac{\text{number of successful outcomes}}{\text{total number of outcomes}} = \frac{S}{N}$$

For example, for a roll of a die, the probability of getting a 3 is

$$p(3) = \frac{1}{6}$$

As a slightly more complicated example, suppose we want to find the probability of getting an even number when rolling a die.

According to Figure 4.3b, three of the outcomes give even numbers. So, $p(\text{even number when rolling a die}) =$

$$\frac{3}{6} = \frac{1}{2} = .5$$

Now let's see how to find the theoretical probability of getting two heads in the toss of two coins. Figure 4.4, called a *tree diagram*, shows the possible outcomes.

The tree diagram helps us keep track of the possible outcomes when tossing two coins. Along the top branch of the tree, we find the outcome *HH*. Along the bottom branch, we have the outcome *TT*, and so on.

From this diagram, we count a total of four equally likely outcomes; only one of these is two heads. So the theoretical probability is

$$p(2 \text{ heads}) = \frac{1}{4} = .25$$

If we want the theoretical probability of getting one or more heads when tossing two coins, we can count three outcomes from the tree diagram (Figure 4.5).

FIGURE 4.4

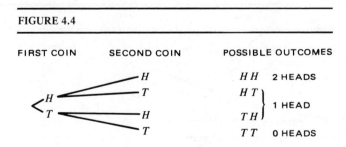

FIGURE 4.5

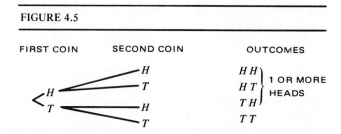

FIRST COIN   SECOND COIN   OUTCOMES

We find that three of the outcomes give one or more heads. So,

$$p(1 \text{ or more heads}) = \frac{3}{4} = .75$$

If we want to find the theoretical probability of one or more girls in a two-child family, for example, we can draw a family tree that is just like the diagram for the two coins (see Figure 4.6).

Again, counting the number of families with one or more girls, we find three. So,

$$p(1 \text{ or more girls}) = \frac{3}{4} = .75$$

Finally, this process can be continued. For example, for a three-child family, the number of possible family combinations is found as shown in Figure 4.7.

FIGURE 4.6

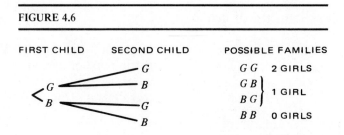

FIRST CHILD   SECOND CHILD   POSSIBLE FAMILIES

FIGURE 4.7

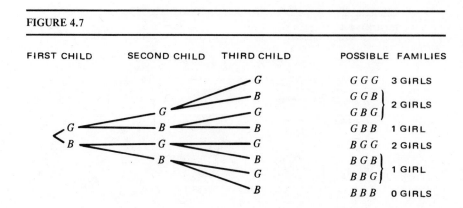

FIRST CHILD   SECOND CHILD   THIRD CHILD   POSSIBLE FAMILIES

Notice that we can use a rule to find the total number of outcomes (families, in this case). To find the total, we find the product of the outcomes for each of the individual events. Since each child can be either a boy or a girl and there are three children, we have

$$2 \times 2 \times 2 = 8$$

possible family combinations. For example, we find that three girls in a three-child family happens in one of the eight possible family combinations. So,

$$p(3 \text{ girls}) = \frac{1}{8} = .125$$

When we solved this problem in Section 4.2 by tossing coins (Table 4.4), we obtained an experimental probability of

$$P(3 \text{ girls}) = .16$$

It is also important to note that our procedures for finding probabilities apply both to coins and to children in families.

The law of theoretical independence: If two events are independent, then the theoretical probability that they both occur is (exactly) equal to the product of the two individual probabilities.

If we have the two independent events, A and B, then, by this law,

$$p(A \text{ and } B) = p(A) \times p(B)$$

Suppose, for example, that we toss a coin and roll a six-sided die. What is the theoretical probability of heads on the coin and a 6 on the die?

Since

$$p(\text{heads}) = \frac{1}{2}$$

and

$$p(6 \text{ on die}) = \frac{1}{6}$$

and since the tossing of a coin and the rolling of a die are independent, we have

$$p(\text{heads and } 6) = \frac{1}{2} \times \frac{1}{6} = \frac{1}{12} \doteq .08$$

■ **Example 1:**   Suppose we choose a pair of random digits, each taken independently from 0 to 9. What is the probability of getting the pair 2, 3?

Solution:

We have

$$p(2) = \frac{1}{10}.$$

and

$$p(3) = \frac{1}{10}$$

So,

$$p(2 \text{ and } 3) = p(2) \times p(3) = \frac{1}{10} \times \frac{1}{10} = \frac{1}{100} = .01$$

This property also applies to more than two events for both estimated and theoretical probabilities. ■

■ **Example 2:** (Three independent events involving estimated probability.) Suppose we collect weather data on three cities—San Juan, Chicago, and Calcutta—and find these estimated probabilities of rain on a certain day:

$$P(\text{rain in San Juan}) = .42$$
$$P(\text{rain in Chicago}) = .59$$
$$P(\text{rain in Calcutta}) = .34$$

We assume that rainfall in each city is independent of rainfall in the other cities. So we estimate the probability of rain in all three cities to be $P(\text{rain in San Juan}) \times P(\text{rain in Chicago}) \times P(\text{rain in Calcutta}) =$

$$.42 \times .59 \times .34 \doteq .08 \ ■$$

■ **Example 3:** (Three independent events involving theoretical probability.) What is the theoretical probability of three heads in three tosses of a fair coin?

Solution:

For each toss,

$$p(\text{heads}) = \frac{1}{2}$$

Therefore,

$$p(3 \text{ heads}) = p(\text{heads on first toss})$$
$$\times\ p(\text{heads on second toss})$$
$$\times\ p(\text{heads on third toss})$$
$$= \frac{1}{2} \times \frac{1}{2} \times \frac{1}{2} = \frac{1}{8} = .125$$

Compare this answer with the one obtained earlier in this section for $p(3 \text{ girls})$ using a tree diagram. ■

EXERCISES

Unless otherwise noted, remember that the probabilities required in these exercises are theoretical, not experimental.

1   Name three probability experiments having equally likely outcomes. Be prepared to say why the outcomes are equally likely.

2   A four-sided die is called a *tetrahedral* die. Assume the outcomes of a four-sided die are 1, 2, 3, 4. Under the assumption of equally likely outcomes, find the following:

(a) $p(1)$
(b) $p(2)$
(c) $p(3)$
(d) $p(4)$
(e) $p(\text{even number})$
(f) $p(\text{number less than 4})$

3   A ten-spinner is shown.

What is the theoretical probability of obtaining each of the digits 0–9? What is the theoretical probability of obtaining an odd digit?

4   Suppose you have 19 classmates. One of you is to be chosen to take part in a field trip to Washington, D.C. What is the theoretical probability you will be chosen, assuming each class member has an equal chance of being selected?

5   A regular deck of playing cards has 13 hearts (red), 13 diamonds (red), 13 clubs (black), and 13 spades (black. Suppose that you deal the top card from a well-shuffled deck, which is face down. Find:

(a) p(black card)
(b) p(heart)
(c) p(ace)
(d) p(king of diamonds)

6   From the records of the U. S. Department of Health, find the number of boys and girls born during the year of the most recent census. Calculate P(boy), the estimated probability of a child being a boy, using the relative frequency approach to probability. Compare this with our assumed theoretical value of p(boy) = .5.

7   In the key problem at the beginning of this chapter, suppose that the valve of a tire can be in one of 12 positions, corresponding to the numerals on a clock. What is the probability that a tire will by chance stop so that the valve is in a given position, such as straight down (6:00)? What is the probability that the tire valve will by chance stop in the same position two times in a row?

8   Suppose a 6-sided die is rolled. What is the probability of getting an odd number? That is, find p(odd number).

9   A prime number is a number (greater than one) whose only divisors are itself and one. Suppose a 6-sided die is rolled. What is p(prime number)?

10   Suppose a pair of dice is rolled. Find:

(a) p(sum of dots is less than 5)
(b) p(sum of dots is 7)
(c) P(sum of dots is greater than 11)

(*Note:* When a pair of dice are involved, and we are interested in sums, Table 4.16, rather than a tree diagram, is more useful in counting the various outcomes. The sums of the two dice are shown in the table. For example, we see that 5 can be obtained in four ways from the sum of two rolls of the dice.)

TABLE 4.16

FIRST DIE	SECOND DIE					
	1	2	3	4	5	6
1	2	3	4	5	6	7
2	3	4	5	6	7	8
3	4	5	6	7	8	9
4	5	6	7	8	9	10
5	6	7	8	9	10	11
6	7	8	9	10	11	12

11   Roll the two dice 50 times and find the experimental probabilities asked for in Exercise 10 of this section. That is, find:

(a) P(sum of dots is less than 5)
(b) P(sum of dots is 7)
(c) P(sum of dots is greater than 11)

12   Suppose two coins are tossed. Find:

(a) p(1 head)
(b) p(at least one head)
(c) p(0 heads)

13   Find the experimental probabilities (use 40 tosses) for each of the outcomes requested in Exercise 12 of this section.

14   Suppose three coins are tossed. Find:

(a) p(1 head)
(b) p(2 or more heads)
(c) p(3 heads)

15   Find the experimental estimates of the probabilities required in Exercise 14 of this section. Do 50 trials for each experiment.

```
┌─────────────── PROJECT CORNER ───────────────┐
│                                               │
│ Spin a coin on its edge and let it come to    │
│ rest. See whether it falls heads or tails.    │
│ Repeat many times (at least 50) and get       │
│ an estimate of $P$(heads). For some newly     │
│ minted coins, $p$(heads) can deviate con-     │
│ siderably from .5.                            │
│                                               │
└───────────────────────────────────────────────┘
```

4.6   THANKS FOR THE COMPLEMENT

Suppose two coins are tossed. E is the event that two heads are obtained. If event E does *not* take place, then two heads are *not* obtained. (For example, we may get 1 head and 1 tail.)

These two events are called *complements* of each other. (This is different from compliments, which are nice things to say about people.)

Here is an event: F is the event that a person makes a reservation for dinner at a restaurant.

The complement of F is that a person does not make a reservation for dinner. (If the person shows up for dinner, of course, the head waiter may make uncomplimentary remarks as well!)

Complementary events: We can define complementary events in this way. If E is some event, then *not E* is the complementary event.

■ **Example**:   When three coins are tossed 50 times, the number of times each outcome is obtained is recorded in Table 4.17.

TABLE 4.17

HEADS	f
0	7
1	16
2	18
3	9
Total	50

From this table, we see that three heads were obtained 9 times out of 50. If E is the event 3 heads, then not E happened

$$50 - 9 = 41 \text{ times}$$

More simply, we can say that in the 50 tosses of the three coins, 3 heads were *not* obtained 41 times.

In terms of experimental probability, we can say that

$$P(3 \text{ heads}) = \frac{9}{50} = .18$$

and

$$P(\text{not } 3 \text{ heads}) = \frac{41}{50} = .82$$

Notice, though, that we can also say that

$$P(\text{not } 3 \text{ heads}) = 1 - P(3 \text{ heads}) = 1 - .18 = .82$$

The property of complementary events always holds.

***Property of complementary events
for experimental probability:*** For
some event E, the probability that E
will not happen is 1 − (the probability that E
happens). That is,

$$P(\text{not } E) = 1 - P(E)$$

In terms of theoretical probability, we have
the same rule. Look at the tree diagram for
three coins in Figure 4.8. The event, E, three
heads, occurs in one out of the eight possible
outcomes. Therefore, not E occurs in seven
out of the eight outcomes.

So, we have

$$p(3 \text{ heads}) = \frac{1}{8}$$

and

$$p(\text{not } 3 \text{ heads}) = 1 - \frac{1}{8} = \frac{7}{8} \blacksquare$$

***Property of complementary events
for theoretical probability:*** For
some event E, the theoretical probability that E will not happen is

$$p(\text{not } E) = 1 - p(E)$$

EXERCISES

1   Table 4.18 gives the results of tossing
two coins 100 times. Let E be the event no
heads. What, in words, is not E? From this
table, find $P(E)$ and $P(\text{not } E)$.

TABLE 4.18	
HEADS	f
0	20
1	52
2	28
	Total = 100

2   Let E be the event no heads when two
coins are tossed. What are the theoretical
probabilities $p(E)$ and $p(\text{not } E)$?

3   A deck of playing cards are shuffled. Let
event A = getting an ace. Write in words the
event not A.

4   Describe an experiment to find $P(\text{not } A)$
in Exercise 3 of this section.

5   A survey of 60 emergency telephones
shows that 6 do not work. Let event W =
phone does not work. What are $P(W)$ and
$P(\text{not } W)$? Write out not W in words.

FIGURE 4.8

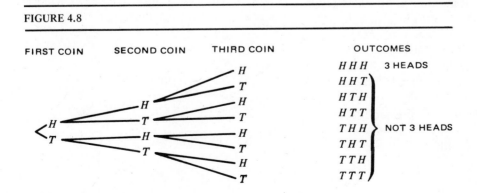

FIRST COIN	SECOND COIN	THIRD COIN	OUTCOMES	
		H	HHH	3 HEADS
		T	HHT	
	H	H	HTH	
	T	T	HTT	
H		H	THH	NOT 3 HEADS
T	H	T	THT	
	T	H	TTH	
		T	TTT	

6   In a family of four children, what is the estimated probability that at least one is a girl? Let event G = at least one is a girl. Find $P(\text{not } G)$. Do 50 trials.

7   Use a tree diagram to find $p(G)$ and $p(\text{not } G)$ in Exercise 6 of this section. Compare your experimental value obtained with the theoretical value of $p(\text{not } G)$.

8   A six-sided die is rolled twice. Find:

(a) $p(12)$
(b) $p(6)$

9   A student says that her probability of passing mathematics is .8 and of passing English is .6. Assuming these two events to be independent, what is the probability that she passes both? What is the probability that she fails both?

10   A rocket has three stages, A, B, and C. The probability that each fires is $p(A) = .9$; $p(B) = .8$; $p(C) = .7$. If their firings are independent of each other, what is the probability that all fire?

11   What is the probability that *none* of the stages of the rocket in Exercise 10 of this section fires?

12   A manufacturer of transistors submits a bid on each of four government contracts. A firm will receive the contract if it submits the lowest bid. In the past, this manufacturer's bid has been the low bid 15% of the time. Assuming this continues to be the case and assuming independence from one bid to another, what is the probability that the manufacturer will *not* receive any of the four contracts? Conduct an experiment of about 50 trials.

13   Each of ten persons randomly chooses two others in the group as friends. A person who ends up with no friends by this random choice is called an *isolate*. Use random digits to estimate:

(a)   the mean number of isolates produced by this process of choosing friends; and
(b)   the probability that *no one* is an isolate.

Do 50 trials. A table like Table 4.19, which shows one trial, will help you keep track of the friend-choosing process.

TABLE 4.19

PERSON CHOOSING	PERSONS CHOSEN AS A FRIEND									
	1	2	3	4	5	6	7	8	9	0
1			X						X	
2					X			X		
3				X				X		
4		X								X
5	X		X							
6		X					X			
7	X	X								
8			X				X			
9						X		X		
0		X				X				

14   Records show that during a certain season, NFL teams won .68 of their home games. Let G be the event winning a home game. What is the complement of G? What is the probability of not G? What is the estimated probability of losing two home games in a row?

Making Statistical Decisions

KEY PROBLEM

Do People Prefer Fizzle Pop?

Statistical ideas can help us understand the meaning of data that we read in the newspaper or hear on the radio or television news.

Fizzle Pop soda has been conducting a poll at shopping centers. They have filled two identical glasses with pop, one with Fizzle Pop and the other with a well-known soft drink. By a secret mark on the bottom of the glass, the polltaker tells which was preferred.

Fizzle Pop finds that out of ten people interviewed during the first hour of the poll, seven prefer Fizzle Pop. Can we conclude that more people prefer Fizzle Pop than the well-known drink?

In this chapter we will see how statistics can be used to help answer this question.

CHAPTER 5

5.1   SAMPLES AND POPULATIONS

When we take samples, like a bite of newly baked cake, or a spray from a new brand of perfume, we are not usually interested in that sample alone. We want to see whether the entire cake is good or whether we should buy a whole bottle of perfume.

Samples are believed to represent something larger, which we call a *population*.

Here are some examples:

1.   A national poll of 1,500 people is used to determine the mood of the country.

> Sample: 1500 responses
> Population: entire country

2.   A swimming-pool attendant takes a small bottle of water from the pool and tests it for chlorine content.

> Sample: the bottle of water
> Population: the water in the swimming pool

3.   A grain-elevator operator takes a container full of corn from an unloading truck and tests it for moisture content.

> Sample: the container of corn
> Population: the truckload (or perhaps several truckloads) of corn from a certain farm

So far in this course we have calculated such values as the mean, range, variance, and standard deviation for a set of data (that is, for a sample).

■ **Example 1:**   Five persons are asked their opinion of the taste of a new food product. They rate it from 5 (yummy) to 0 (terrible). Their responses are:

$$3 \quad 2 \quad 3 \quad 4 \quad 3$$

The mean response is

$$\frac{3 + 2 + 3 + 4 + 3}{5} = \frac{15}{5} = 3.0$$

The variance is

$$\frac{(3-3)^2 + (2-3)^2 + (3-3)^2 + (4-3)^2 + (3-3)^2}{5}$$

$$= \frac{0 + 1 + 0 + 1 + 0}{5} = \frac{2}{5} = .40$$

The standard deviation is

$$\sqrt{.40} \doteq .63$$

Now suppose we want to use this information to give the food producer some idea not about how these five people like the new product (in practice, we would never use such a small sample, but let's keep it simple here), but about how the public, or people in general, will like the product.

Solution:

We could use these sample values to *estimate* the opinion of the population that the sample is believed to represent. There is an important idea here. To illustrate, let's suppose that there are 10,000 potential buyers of this new food product. These 10,000 people make up the population in which we are interested. In terms of the opinions of these 10,000 people concerning the new product, we can think of a population mean and variance. To get the population mean, we add up all 10,000 responses (from yummy to terrible) and divide by 10,000. This is a lot of work, but it could certainly be done. Similarly, we could—if we had to—find the variance of these 10,000 ratings. ■

In statistics, we learn how to use samples to estimate properties of populations. We can use the mean of a sample to estimate the mean of the population from which it was taken and the variance of a sample to estimate the variance of the population.

■ **Example 2:**   In Exercise 11 of Chapter One we were given the number of home runs scored by each year's leaders in the American League for the 60 seasons from 1921 through 1980. Table 5.1 shows a stem-and-leaf table of those runs.

TABLE 5.1

STEM	LEAF
1	
2	2, 4
3	2, 2, 2, 2, 2, 2, 3, 3, 3, 4, 5, 6, 6, 6,
*	7, 7, 7, 7, 9, 9, 9
4	0, 1, 1, 1, 2, 2, 2, 3, 3, 4, 4, 4, 4, 5, 5,
*	6, 6, 6, 6, 6, 7, 8, 8, 9, 9, 9, 9, 9, 9
5	2, 4, 8, 8, 9
6	0, 1
7	
8	2

Solution:

We can find the mean and variance of these 60 numbers in the usual way.

$$\text{Mean number of runs} = 42.85$$
$$\text{Variance} = 95.26$$

Now let's think of this example from a sampling point of view. The information in Table 5.1 is a population. We estimate the average number of runs by a random sample of data. For instance, the data might be in yearbooks, and we have available only a random sample of, say, 10 yearbooks between 1921 and 1980.

Let's say our random sample of size 10 is in Table 5.2.

TABLE 5.2

STEM	LEAF
2	
3	2, 6, 7
4	1, 3, 6, 7, 9
5	8
6	1

The mean of these 10 numbers is 45.0. This appears to be a reasonably good estimate of the *true* mean of the 60 numbers (that is, 42.85).

As we will see later in this book (Chapter Twelve), the mean of a random sample serves very well as an estimator of the mean of the population from which the sample was drawn.

The variance of these 10 numbers is 78.0. This is not a very good estimate of the population variance of 95.26. As we will also see later, this way of finding the variance does not usually serve very well as an estimator of the population variance (we call it a *biased* estimator of the population variance, as we shall see later). This is particularly a problem when the random sample is small (of size less than 20). We will see later how to remove the bias in the sample variance when used to estimate the variance of the population from which the sample was taken. ■

An important part of statistics has to do with using information from samples to make decisions about the populations that the samples are assumed to represent. In this chapter we will learn a few of these methods. More advanced methods are discussed in Chapters Twelve and Thirteen.

EXERCISES

1 For each of the following, identify the sample and the population.

(a) A poll of 40 seniors is taken to find out the views of the senior class on raising the drinking age from 18 to 21.

(b) A few small pieces of rock from a river valley region are analyzed at a laboratory to see how much gold-bearing mineral they contain.

(c) Julia's child, Maurice, is taking a course in French cooking. He has made a pot of onion soup and tastes a spoonful to see if he has added enough seasoning.

(d) Mork and Mindy are at the beach. They want to go swimming, so Mork puts his toe in the water to help decide whether they should go in.

(e) A building inspector wants to see whether the concrete being used is of high-enough quality. She takes a small sample of mix from a truckload every morning for 5 days and sends it to the laboratory for analysis.

(f) Lisa has symptoms of appendicitis. Her doctor does a blood test to see if her white blood cell count is normal.

2 From Exercise 1 above, find:

(a) An example of a sample mean (which we label $\bar{X}$) and describe the population mean being estimated.

(b) An example of a sample variance (S^2) that might be obtained and describe the population variance being estimated.

3 Make up your own example of a population about which you might want to get information and the sample you would obtain to bet the information needed.

4 Bring in two examples from newspapers or magazines that involve samples and population. Tell which is the sample and which is the population

5 The following is an article from *Education Week*.[1] What is the sample and what is the population?

> Most American students today have positive feelings about music, but many teenagers consider art less important than did their counterparts five years ago, . . . according to recently released reports of the National Assessment of Educational Progress.

6 Figure 5.1 shows an article from Weekend Magazine. What is the sample and what is the population to which the article refers?

7 Describe the sample and population referred to in the following article.

> College students take more than twice as many courses in the fields of science, mathematics and engineering as they do in the humanities, a government-sponsored survey indicates. The statistics were based on a survey of course selections at 760 colleges and universities in the United States.[3]

8 In Figure 5.2, identify the sample and the population.

9 Identify the sample or samples and population in the following.

> An environmental group is trying to estimate the number of fish in a lake. A spot in the lake is chosen using a table of random digits. A catch of fish is made there. The fish in the catch are counted, tagged, and then quickly released back into the lake. In a few days' time, another spot in the lake is chosen (again, using a table of random digits). A second catch of fish is made. The total number of fish caught and the number of tagged fish is noted. This information is used to estimate the total number of fish in the lake.[5]

FIGURE 5.1

ARMCHAIR ATHLETES DO MORE THAN WATCH[2]

Tomorrow is Grey Cup day and many of us will spend it glued to our television sets. Does the enthusiasm of Canadians for sports extend beyond the armchair? It certainly does. Almost two-thirds of the people surveyed in the *Weekend* Poll say they have participated actively in sports during the last 12 months—and those who watch sports on television are more likely to be active than those who don't. Swimming turns out to be the most popular sport; more than 30 percent of those active in sports have engaged in swimming over the last year. Next in order of

Have you actually participated in sports over the last year?		Those who watch TV sports	Those who don't watch TV sports
Yes	63%	69%	·57%
No	35	31	43

preference are cycling (20 percent) and jogging (18 percent).

About 64 percent watch sports on television: hockey is by far the most popular game, followed by football and baseball. Thirty-nine percent have attended a paid-admission sporting event within the last year: again, hockey is the most popular event.

On the average, Canadians spent 6.4 hours per week last year on sports, half of it in active participation. They spend 2.1 hours watching TV sports and 1.1 hours attending sports events. A slight majority of people say the time they devote to active participation has increased more than the time they spend watching sports.

FIGURE 5.2

THE CHANGING MOOD OF THE NATION'S VOTERS[4]

What America Is Thinking

More than 130,000 households responded to FAMILY WEEKLY's recent "Timely Issues" survey. Here's a summary of their opinions.

Should the government provide more health services, to be paid for ultimately by taxes?

Yes **23.9%**   No **67.6%**
Not sure **8.1%**

Will marriage fade in importance?

Yes **27%**   No **59.7%**
Not sure **12.9%**

Are today's schools adequately preparing children for the future?

Yes **8.1%**   No **82.3%**
Not sure **8.5%**

5.2 USING INFORMATION FROM SAMPLES

Opinion polls are an example of how information from a sample is used to make statements about the population from which the sample was taken.

■ **Example:**   A social studies class wants to know whether their community is in favor of raising the minimum age for getting a driver's license. They ask 25 people chosen at random from the telephone book.

The results are:

Number of persons in favor of raising driving age = 15
Number of persons not in favor (opposed)   = 10

Total persons in poll   = 25

From these results, should the class conclude that the community is in favor of raising the driving age?

We will solve this problem by taking an approach that will at first seem a little unusual. Later, as you see how the approach is used, you will see how powerful it is.

Solution:

In the example we are asked whether the community is in favor of raising the drinking age. We will solve this problem by *assuming* that the community is evenly split on the issue. That is, half of the people are in favor of raising the driving age and half are opposed. So,

$$p(\text{in favor}) = 0.5 .$$

We will estimate, using five steps, how likely it is, in a random sample of 25 persons, to get 15 who are in favor of raising the drinking age, assuming that the community being sampled is evenly split.

There is something else we must keep in mind, as we estimate our probability. Even though we are told that 15 of the 25 persons in the sample favored raising the drinking age, we are also interested in those trials in which *more than 15* people who favor raising the drinking age were chosen by chance. So our statistic of interest is not only 15 in favor, but any values greater than 15.

Step 1—*Model:*   We use a (fair) coin:

Heads: In favor
Tails: Opposed

Step 2—*Trial:*   A trial consists of tossing the coin 25 times, once for each person in the poll.

Step 3—*Outcome of trial:*   We record the number of heads—the number of persons in favor of raising the driving age.

Step 4—*Repeat trials:*   We should do about 100 trials in order to get a good estimate of $p(15$ or more in favor). (*Note:* Since *more than 15* in favor would also be of interest, we estimate $p(15$ or more in favor) in our sample of 25).

Step 5—*Probability of statistic of interest:*
From our table of outcomes (Table 5.3), we see that 15 or more heads were obtained (in 25 tosses of a fair coin) 23 times in 100 trials.

So,

$$P(15 \text{ or more in favor}) = \frac{23}{100} = .23 \ \blacksquare$$

raising the driving age *even though the community is evenly split on the issue.* On the basis of our experiment, we conclude that getting 15 or more people in a sample of 25, happens fairly often (in our 100 trials, it happened 23 times).

From these results, the class should *not* conclude that the community favors raising the driving age.

Let's notice that small numbers of heads occurred by chance, as well. For example, *10 or fewer* heads occurred in 19 of the 100 trials.

TABLE 5.3

N HEADS	FREQUENCY	
0	0	
1	0	
2	0	
3	0	
4	0	
5	0	
6	0	
7	1	
8	2	
9	7	
10	9	
11	14	
12	18	
13	15	
14	11	
15	9	
16	7	
17	5	
18	2	
19	0	
20	0	15 or more in favor
21	0	
22	0	
23	0	
24	0	
25	0	

EXERCISES

1   Using the table of outcomes (Table 5.3), find these probabilities tossing a fair coin 25 times.

(a) $P(16 \text{ or more heads})$
       $P(18 \text{ or more heads})$
       $P(20 \text{ or more heads})$

(b) $P(12 \text{ or fewer heads})$
       $P(9 \text{ or fewer heads})$
       $P(5 \text{ or fewer heads})$

2   Answer Exercise 1 by doing your own experiment. Toss a coin 25 times. Record the number of heads. Repeat for 20 trials. Compare your estimates with the ones obtained in Exercise 1 using 100 trials.

3   Suppose that the social studies class mentioned earlier in this section obtained 18 persons in a random sample of 25 voters who were in favor of raising the driving age. What should the class conclude now? Use the data for 100 trials from Table 5.3.

Let's think about our result. Just by chance, we obtained 15 or more heads in 25 tosses of a fair coin 23 times out of 100. In terms of our problem, this is like getting 15 more people in a sample of 25 people who favor

4   From 1963 to 1982, the National League beat the American League in the All Star Game 19 times out of 20. Are the Leagues equally matched, does it seem? Here is a table of 100 trials where a trial consists of tossing a fair coin 20 times. Use Table 5.4 to answer the question.

TABLE 5.4

N HEADS	FREQUENCY
0	0
1	0
2	0
3	0
4	0
5	2
6	3
7	8
8	13
9	12
10	16
11	19
12	9
13	12
14	5
15	1
16	0
17	0
18	0
19	0
20	0

5   Voters are believed to be evenly split on a constitutional amendment. A telephone poll of 20 voters results in 14 in favor of the amendment. Should it be concluded, based on this poll, that the community favors the amendment? Use the data in Table 5.4. (*Hint:* Does getting 14 or more voters, out of 20, happen often, by chance? Let's say that by *often*, we mean that it happens more than .10 of the time.)

6   Suppose in Exercise 5, above, that the poll had shown 7 out of 20 persons in favor

of the amendment. Should it be concluded, based on this poll, that the community is *opposed* to the amendment? (*Hint:* The hint for Exercise 5 may be of help, too.)

7   An English lady claims she prefers her tea served with cream put in the cup before the tea. (The usual way is to pour the tea then add the cream.) Suppose she cannot tell which was done first and just guesses. What is the probability that for 12 cups of tea with cream, she correctly identifies 8 or more simply by guessing which of the cups had the cream put in first?

For this problem, use Table 5.5. It shows that results of 100 trials, where each trial consists of tossing a fair coin 12 times.

TABLE 5.5

N HEADS	FREQUENCY
0	0
1	0
2	3
3	3
4	20
5	22
6	28
7	12
8	8
9	3
10	1
11	0
12	0

8   A jury consisting of 3 women and 9 men is selected. The lawyer for the defense believes there was sex bias (in favor of men) in selecting the jury. What is the probability of choosing 9 or more men (out of 12 people) by chance (that is, equal likelihood of choosing a man or woman on each selection)? What is the probability of choosing 8 or more men for a jury by chance? Use Table 5.5.

SPECIAL INTEREST FEATURE

Statistics and the Stars

People have studied the stars for untold centuries. But only within the last one hundred years have the relatively modern concepts of statistics and probability been applied to astronomy. As we look at the stars, it may seem that they are scattered at random in the heavens. Closer observation, however, reveals groups of stars—double stars and sometimes chains of four or five which form a glittering necklace. Have such arrangements occurred by chance?

An astronomer in the eighteenth century, John Michell, calculated the probability that the pairings he found did occur by chance. He obtained such small probabilities that he concluded that the stars must be physically linked together. This conclusion was later verified by the astronomer William Herschel.

Such calculations of unlikehood, or improbability, have been applied to other configurations of stars and even to galaxies as a part of man's continuing quest for more knowledge about the origins of the universe.[6]

5.3 YES-NO POPULATIONS* (Optional)

So far in this book, we have used models like coins, dice, or random digits to produce data for solving problems. In this section, we will consider another method of producing random outcomes, that of sampling boxes.

A *sampling box* can be used to obtain random samples from a population in a quick and easy manner. The population is a large number of beads or BB shot of the same size, which are available in two colors. (For example, BB shot is manufactured in two metals—brass and copper.)

A random sample of these beads is taken by means of a paddle, which is drilled with holes just large enough to hold one bead or BB (see Figure 5.3). The number of holes drilled in the paddle determines the size of the sample of beads that are drawn. For example, in Figure 5.3, the 20 holes in the paddle will give a sample of size 20.

In this section, we will deal with a 50-50 population. That is, we will produce a population of beads in which exactly one-half are black and one-half are white. For example, let's assume we place a large number of beads, say 300 black beads and 300 white beads in the sample box. Then, if we draw a bead at random from the box, it has a probability of .5 of being black.

FIGURE 5.3
MAKING A SAMPLING BOX

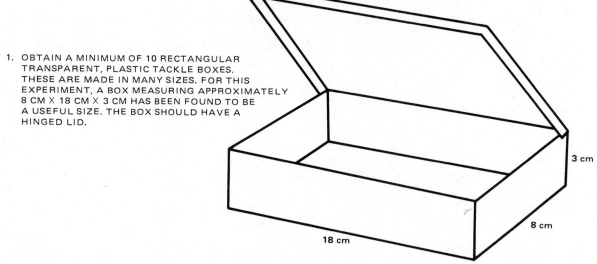

1. OBTAIN A MINIMUM OF 10 RECTANGULAR
   TRANSPARENT, PLASTIC TACKLE BOXES.
   THESE ARE MADE IN MANY SIZES. FOR THIS
   EXPERIMENT, A BOX MEASURING APPROXIMATELY
   8 CM X 18 CM X 3 CM HAS BEEN FOUND TO BE
   A USEFUL SIZE. THE BOX SHOULD HAVE A
   HINGED LID.

2. CUT AN 8 CM X 18 CM PIECE OF PLEXIGLASS (4 MM THICK) THAT
   FITS EXACTLY IN THE BOTTOM OF THE BOX.

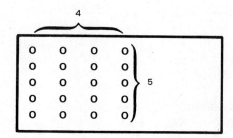

SAMPLING "PADDLE" FOR SAMPLE OF SIZE 20

Since a bead can be either black or white, the following correspondence could be produced:

Black: Yes
White: No

Because of this two-valued property of the populations, it is sometimes called a *yes-no* population.

■ **Example 1:**   Central Academy, a large high school, has an equal number of boys and girls. A random sample of 20 students is chosen to discuss a proposal to raise the driving age. What is the probability that 8 *or fewer* of the students chosen by chance will be girls?

Solution:

Step 1—*Model:*   Use the bead box, with half black and half white. The number of beads in the box should be large, corresponding to the number of students in Central Academy.

Black: Girls
White: Boys

Step 2—*Trial:*   A trial consists of using a 20-paddle and drawing a sample of 20 beads.

Step 3—*Outcome of trial:*   Record the number of black beads in the sample. Shake the box to return the beads to the population.

Step 4—*Repeat trials:*   Obtain 50 samples, recording the outcome of each (Table 5.6).

TABLE 5.6

N	FREQUENCY	
0	0	
1	0	
2	0	
3	0	
4	0	8 or fewer
5	1	
6	3	
7	7	
8	6	
9	6	
10	7	
11	11	
12	6	
13	2	
14	1	
15	0	
16	0	
17	0	
18	0	
19	0	
20	0	

Step 5—*Probability of statistic of interest:*   We estimate from Table 5.6 that

$$P(8 \text{ or fewer girls}) = \frac{17}{50} = .34 \;\blacksquare$$

■ **Example 2:**   A social studies class wants to find out whether their town is in favor of lowering the minimum age for getting a driver's license by 1 year. They take a poll of 30 people picked at random from the telephone book. The results of the poll are 19 yeses and 11 noes. How likely is it to get 19 or more persons in favor of lowering the driving age if the town is, in fact, evenly split on this issue?

Solution:

Step 1—*Model:*   Use a bead box, with half copper shot and half lead shot.

Copper: Yes
Lead: No.

(The total number of shot in the box should be large, representing the number of persons in the town.)

Step 2—*Trial:*   Use a 30-paddle, since sample of size is 30 involved.

Step 3—*Outcome of trial:*   Record number of copper shot obtained in the sample.

Step 4—*Repeat trials:*   Shake box in order to obtain a new random sample of shot. Obtain 100 samples.

Our results are shown in Table 5.7.

TABLE 5.7

N	FREQUENCY	
0	0	
1	0	
2	0	
3	0	
4	0	
5	0	
6	0	
7	0	
8	0	
9	1	
10	1	
11	7	
12	8	
13	7	
14	14	
15	15	
16	16	
17	8	
18	10	
19	5	
20	5	
21	2	
22	1	
23	0	19 or
24	0	more
25	0	
26	0	
27	0	
28	0	
29	0	
30	0	

Step 5—*Probability of statistic of interest*:
We find from our 100 trials that

$$P(19 \text{ or more yes}) = \frac{13}{100} = .13 \quad \blacksquare$$

EXERCISES

Do 50 trials unless otherwise stated.

1   What is the estimated probability of choosing 6 or fewer girls by chance if a random sample of 20 students is required? Use data in Table 5.6.

2   What is the estimated probability of choosing 13 *or more* girls by chance if a random sample of 20 students is required? Use data in Table 5.6.

3   Answer Exercise 1 of this section by using your own bead box. Compare your estimated value of $P(6 \text{ or fewer girls})$ with the value provided by the data in Table 5.6.

4.   Answer Exercise 2 of this section by using your own bead box. Compare your answer with the one obtained for Exercise 2.

5   Suppose that in the social studies poll discussed in Example 2 of this section, 17 yeses had been obtained. How likely is this to happen by chance, if the town is evenly split on the issue? Use data in Table 5.7 to find the estimated probability $P(17 \text{ or more yeses})$.

6   Use a bead box to generate your own data to find $P(19 \text{ or more yeses})$ for the social studies poll (Example 2 of this section).

7   Use data obtained in Exercise 6 of this section to find $P(17 \text{ or more yeses})$ in a sample of 30 persons.

8   Long-range records show that in a certain community 50% of the people own their homes. A recent survey of 40 families shows that 30 own their homes. Do you believe that the records are still accurate? (*Hint:* How likely would it be to find 30 or more persons in a survey of 40 persons who own their own homes if it is the case that only 50% own their homes?)

9   In Exercise 8, suppose it was found that 25 owned their homes. Use the data of Exercise 8 to find $P(25$ or more own homes).

10   A jury is selected to decide a case involving sex discrimination. It consists of 9 women and 3 men. How likely is it to get 9 or more women on a 12-person jury by chance? What paddle would you use? (Assume the jury is made up from a long list of equal numbers of males and females.)

5.4   ACCEPTING A MODEL

The first step in doing a probability experiment, as we all know, is to decide on a model to produce the outcomes with the appropriate probabilities. The model may be a coin, a die, random digits from a table or a computer, beads drawn from a box, or some other device to give outcomes with the desired probabilities.

In statistical decision making, we are concerned with the choice of a model that is likely to have produced the outcomes. Consider this example.

■ **Example**:   Rob claims he has extra-sensory perception (ESP). So, he doesn't have to study for tests! He just concentrates and the answers flash before him. Mary is doubtful and does an experiment. She gives Rob a 10-question, true-false test on classical Greek, written in Greek. Rob takes the test, concentrating on each question, and gets 7 out of 10 of the questions correct.

Mary says, "I don't think you have ESP. You only got 7 correct answers." Rob says, "Well, I didn't say ESP would enable me to get a perfect score. But it helps me do better than I would by *just guessing*."

We need to make a decision. Does Rob have ESP or not, on the basis of our data (Rob got 7 out of 10 questions right)?

Solution:

We make our decision in a way that may seem strange, at first, but is really a powerful form of argument.

We *assume* he *is* guessing (that is, he doesn't have ESP) and then estimate how likely it is that he would get 7 or more out of 10 correct *under this assumption* (or model).

If such an event (a score of 7 or more correct) occurs fairly often by chance, we conclude that he may only be guessing and does not necessarily have ESP. That is, although the data *hint* that Rob has ESP, his performance on the test is not good enough to be conclusive evidence that he has ESP.

Had we found, however, that such a score (that is, 7 or more correct) was *unusual* to get by guessing, we would have concluded that he was *not* guessing and very likely does have ESP.

This sort of argument is not simple. But it is powerful and is used in various life situations. A common example is that of a trial by jury. A person charged with a crime is assumed innocent until proven guilty. That is, under the jury system, the model assumed is that of *innocent*. If the jury is not presented with enough evidence to persuade them of the defendant's guilt, they may pronounce the person not guilty. Other individuals looking at the evidence may suspect or conclude otherwise (and indeed, the defendant may indeed be guilty). But the jury's decision means that they have not had sufficient evidence for guilt. So, under benefit of doubt, innocence is presumed. Similarly, we are not willing to conclude that Rob has ESP until we have convincing evidence. (Is 7 out of 10 convincing evidence?)

We use our familiar five steps to help us make the decision. The names of the steps are changed only slightly from Chapter 4 to show we are now using the steps to help make decisions. Let's see how we can decide if Rob has ESP.

Step 1—*Choose a model:* We assume Rob *is* guessing. So, a good model to use is a fair coin.

Heads: Correct
Tails: Wrong

We could, of course, use a die, a deck of cards, or any other model that would give us results of guessing between two answers with equal likelihood (a 50-50 chance of being correct).

Step 2—*Trial:* A trial consists of tossing the coin ten times, once for each item on the test.

Step 3—*Outcome of trial:* We record the number of heads obtained (number of correct answers) by guessing.

Step 4—*Repeat trials:* We do about 100 trials.

Step 5—*Probability of obtained (or greater) value of statistic:* The value of *obtained statistic* comes from the data in our problem. We are told that Rob got 7 questions correct. Since Rob claims he is guessing, any score greater than 7 would also support his claim. So, we find

$$P(7 \text{ or more correct})$$

We do 100 trials and get the results in Table 5.8.

TABLE 5.8		
OUTCOME HEADS	*f*	
0	0	
1	0	
2	2	
3	14	
4	17	
5	32	
6	14	
7	15	}
8	5	} 7 or more
9	0	} correct
10	1	}
	100	

We see from this table that the probability of getting 7 or more correct by guessing is estimated as

$$\frac{21}{100} = .21.$$

That is, 7 or more answers out of 10 are correct by chance 21 times out of 100.

These 100 trials suggest such a result is hardly unusual. Rob's score (7 out of 10 correct) is *not* convincing evidence of anything other than chance behavior.

Suppose, instead, that Rob had gotten 9 out of 10 questions correct. Without knowing anything about statistics, we have the feeling that this is a very good score to get by just guessing. But we now have a way of testing our hunch. We see from our frequency table that getting 9 (or more) correct by guessing happened only 1 out of 100 times. Therefore, we would conclude Rob is *not* guessing. The data convince us that he is getting the right answers by some method other than guessing (maybe by ESP!). So, a coin is *not* a suitable model for producing outcomes of interest.

This decision-making procedure involves a sixth step.

Step 6–*Make a decision about the model:* If P(obtained statistic or greater) is *large,* then decide the model may be acceptable. (That is, the obtained statistic may be regarded as a chance outcome produced by the model.) ■

Large and *small* are defined differently in different circumstances. For now, we simply use this rule:

 A large probability is greater than or equal to .10. *A small probability* is less than .10.

Unless otherwise stated, you can assume this definition to hold. In practice, a value of .05 or .01 (or even smaller) is used.

EXERCISES

Do 50 trials unless otherwise stated.

1   Suppose in the example in this section, Rob got 8 questions correct. Use the data of the frequency table to decide whether such a score is evidence that he has ESP. Explain.

2   Answer the sample exercise (about whether Rob has ESP) by following the six-step procedure in this section. Does the decision you make about whether Rob has ESP agree with the decision we made in the example?

3   Refer to the problem about the tea taster in Exercise 7 of Section 5.2. Follow the six steps of this section to decide whether or not she can tell if the cream is put in the cup first.

4   Follow the six steps of this section to decide whether, according to their record between 1963 and 1982, the National League and the American League teams are equally matched. (Exercise 4 of Section 5.2.)

5   A newspaper editorial claims that a certain community is split evenly on an upcoming referendum on raising the drinking age from 18 to 21. A sociology class takes a telephone poll of 30 persons picked at random from a telephone book and asks, "Do you favor raising the drinking age from 18 to 21?" They find that 19 say Yes and 11 say No. Based on these 30 responses, do you think that the community is, in fact, evenly divdied on the issue? Follow the six steps, using 100 trials, to find your answer.

6   A jury is selected to decide a criminal case involving a female defendant. The jury consists of eight males and four females. The defense lawyer claims bias in selecting the jury in that males are favored over females in the selection process. Follow the six steps to decide whether males and females were chosen with equal liklehood. Do 100 trials.

7   A biologist believes she has developed a method for irradiating the genes of fruit flies so that the ratio of males to females in the

offspring will get smaller (there will be more females than males). In the first 24 of the resulting offspring after the treatment, there are 15 females and 9 males. Follow the six steps to decide whether the radiation treatment works.

8 Solve the key problem for this chapter.

9 Suppose in the key problem that 25 shoppers are asked for their preference and 15 prefer Fizzle Pop. Make a decision whether Fizzle Pop is preferred by shoppers, based on this random sample.

10 A certain die is believed to be loaded in favor of even numbers. You roll it 50 times and get an even number 35 times and an odd number 15 times. Decide whether or not the die is loaded by following the six steps.

11 Let's look again at Exercise 5 of Section 5.2 and focus on the no responses (11 out of the 30 people said No about raising the drinking age). We could ask the probability of getting 11 *or fewer* noes under the assumption that the community is evenly split. Use the data you generated in Exercise 5 of Section 5.2 to answer this question. Use the six steps.

12 Suppose in the jury problem (Exercise 6 of this section) we ask whether females are discriminated against. Therefore, the problem is to estimate the probability that four or fewer females are selected under the assumption of no discrimination. Redo Exercise 6 asking this question.

13 In Exercise 7, above, suppose that in 24 offsprings there were 16 females and 8 males. Estimate the probability of 8 or fewer males under the assumption of an equal ratio of females to males.

5.5 DIFFERENT MODELS FOR DIFFERENT PROBLEMS

All of the problems we have solved so far using the six-step method have involved models for outcomes with a probability of .5. Now we see how to solve problems using models for probabilities different from .5.

■ **Example 1:** Fred has been shooting foul shots with a season score of .6 (that is, he gets a basket 60% of the time that he shoots). His parents send him to basketball camp for 2 weeks. When he gets back home he goes to the basketball court and gets 10 baskets out of 12 shots. Has his shooting improved?

Solution:

We will assume that Fred's shooting has not improved. So we need a model that will produce outcomes of interest (getting a basket) with a probability of .6.

Step 1—*Model:* Recall what we did in Chapters Three and Four. We have several ways of getting outcomes with a probability of .6.

We could use random digits 0, . . ., 9, where

1-6:   Basket
7-9, 0:   No basket

Or, we could use a spinner marked off in ten equal sectors (Figure 5.4).

FIGURE 5.4

Or, a 12-sided die would serve our purpose.

> 1-6:   Basket
> 7-9, 0: No basket
> 11, 12: Ignore, roll die again

Step 2—*Trial*:   A trial consists of reading 12 digits (one for each shot) from a random digit table (or from a computer).

Step 3—*Outcome of trial*:   Record the number of baskets (digits 1, 2, 3, 4, 5, and 6) obtained in each trial.

Step 4—*Repeat trials*:   Do about 100 trials.

Step 5—*Probability of obtained value (or greater) of statistic*:   Fred got 10 baskets in 12 shots. We want to estimate the probability of getting 10 *or more* baskets from our frequency table. The outcomes of the 100 trials are in Table 5.9.

TABLE 5.9

N BASKETS	FREQUENCY	
0	0	
1	0	
2	2	
3	2	
4	16	
5	20	
6	25	
7	16	
8	10	
9	5	
10	3	⎫
11	1	⎬ 10 or more baskets
12	0	⎭
Total =	100	

Step 6—*Decision*:   The probability of getting 10 or more baskets is estimated from the table of outcomes as

$$P(10 \text{ or more}) = \frac{4}{100} = .04.$$

Since this event is unusual (it happens less than .10 of the time by chance), we conclude Fred did *not* get the baskets just by chance. ■

In some problems, the statistic of interest includes values *less than* the sample (obtained) value given.

■ **Example 2:**   A certain disease is contagious for 30% (.30) of the people exposed to the disease. A new vaccine is given to 20 persons exposed to the disease and only 4 of them become infected. Is the new vaccine effective?

Solution:

We assume that the vaccine is not effective. So we need a probability model for .3. Then, we are interested in those cases in which out of 20 persons, 4 persons *or less* were infected.

Step 1—*Model*:   We will use a ten-sided spinner where:

> 1, 2, 3:   Infected
> 4, 5, 6, 7, 8, 9, 0:   Not infected

Step 2—*Trial*:   A trial consists of doing 20 spins, one for each person given the vaccine.

Step 3—*Outcome of trial*:   Record the number of persons infected in each trial.

Step 4—*Repeat trials*:   Do 100 trials (see results in Table 5.10).

TABLE 5.10

N PERSONS OUT OF 20 INFECTED	FREQUENCY	
0	0	
1	1	
2	6	4 or less
3	7	
4	7	
5	14	
6	24	
7	15	
8	13	
9	6	
10	5	
11	2	
12	0	
13	0	
14	0	
15	0	
16	0	
17	0	
18	0	
19	0	
20	0	
	Total 100	

Step 5—*Probability of obtained statistic (or less):* We see from Table 5.10 that 4 or less persons out of 20 became infected in 21 out of the 100 trials. So, we have

$$P(4 \text{ or less persons infected}) = .21.$$

Step 6—*Decision:* It is not very unusual to have 4 or less persons out of 20 infected by this disease. We have *no* conclusive evidence that the vaccine is effective. ■

EXERCISES

1   Suppose Fred had gotten 8 out of 12 baskets. Follow the six steps to decide whether his shooting has improved. Use the data in Table 5.9.

2   Answer Exercise 1, above. But this time, generate your own data for 50 trials. Compare your decision with the one you made in Exercise 1.

3   A certain medication cures a skin rash .8 of the time, according to medical records. A new medication is tried on 20 patients and cures 17 of them. Follow the six steps to decide whether the new medication is more effective than the old. Use data in Table 5.11.

TABLE 5.11

N PERSONS OUT OF 20 CURED	FREQUENCY
0	0
1	0
2	0
3	0
4	0
5	0
6	0
7	0
8	0
9	0
10	1
11	1
12	0
13	3
14	7
15	11
16	8
17	10
18	5
19	3
20	1

4   Suppose in Exercise 3 of this section, the new medication is tried on 20 patients and is

effective for 19 of them. What is your decision about the medication now?

5   Factory records show that when a lathe is properly adjusted, it will produce valves at a 5% defective rate. That is, 5 out of 100 of the valves produced are defective. As a check on the lathe, the first 25 valves produced on the shift are checked. Suppose 3 defectives are found. Should production be stopped to adjust the lathe?

6   A producer of television sets believes that the transistors they are getting from a supplier are not as reliable as they used to be. In the past, 2% of them have been defective. In the batch of 100 they just received, they find 9 defectives. Is this conclusive evidence of the television manufacturer's claim?

7   Company X claims that 30% of its cars made in 1970 last 10 years. A consumer group finds that in a random sample of 20 cars made in 1970, only 3 were still running in 1980. Do you believe the claim of company X? (*Hint:* You can use Table 5.10 to solve this problem. Reread Example 2 of this section to see why.)

8   A politician believes she has the support of 55% of the people. A poll of 100 randomly chosen voters shows that 45 support her. Do you believe the politician's claim, based on this poll?

9   **The Case of the Vanishing Women Jurors:** A jury (which consists of twelve persons) is selected from a jury list. The list consists of persons eligible for jury duty. In the U.S. District Court for the State of Massachusetts, the jury for a particular trial consisted of 11 men and 1 woman. Since 30% (.30) of the persons on the jury list were women, do you believe that persons for the jury were drawn at random (that is, so that p (female) = .3)?[7]

5.6   THE MEDIAN TEST* (Optional)

The median can be used to help make decisions about two sets of data.

■ **Example:**   Suppose there is a new way of teaching spelling. A teacher has two classes of students. She teaches one of the classes using the new method, A. She teaches the other class using her usual method, B. Then she gives both classes the same spelling test. These scores are obtained:

Method A: 10 10 10 12 15 17 17 19 20 22 25 26
Method B: 6 7 8 8 12 16 19 19 22

We want to decide whether method A is better than method B. That is, did the students get a higher average (median) test score with method A than with method B?

Solution:

It is assumed here that the classes have comparable spelling ability before being taught by the two methods. When actual research is done, getting comparable groups is often carried out by assigning persons (or objects) to the groups at random.

From a statistical point of view, we recognize that two groups of students could get different scores on the same test just by chance. We will follow our six steps to see how likely it is that the scores differ by chance. Under the assumption that the groups do *not* differ, we put them both together and get one set of 21 scores. The median of this set of scores is 16.

Now we write a + for each score that is above the median score of 16. Since the given scores are in order, we see that we have:

Method A: + + + + + + +
Method B: + + +

It appears that method A is better for teaching spelling than method B. But again, is this only a chance happening? Let's follow our six steps to make our decision.

Step 1—*Choose a model*:   We assume that the results obtained by method A and method B are due to chance. So, we choose a model that we can use to see how often the results can be expected to happen by chance.

We will form two groups—method A and method B—as in our problem, but we make assignment to each group by chance. One way to do this is to use a deck of 21 playing cards (one for each of the scores in the combined group). Write one of the scores on each of the cards. Shuffle the deck well.

Step 2—*Trial*:   Deal 12 cards for group A. The remaining cards are the new group B. The median of the combined group is still 16, since we still have the same 21 scores.

Step 3—*Outcome of trial*:   Find the number of scores in group A that are greater than 16. Record this number.

Step 4—*Repeat trials*:   Do 50 trials.

Step 5—*Probability of obtained value (or greater) of statistic*:   Table 5.12 shows the results of 50 trials. We find that 7 or more scores in group A were greater than the median of the combined group in 15 of the 50 trials.

TABLE 5.12

N SCORES GREATER THAN 16	f
0	0
1	0
2	0
3	1
4	8
5	10
6	16
7	10 ⎫ 7 or
8	5 ⎭ more
	50

Step 6—*Decision*:   Getting 7 (or more) scores in group A that are larger than the median for the combined group happens often by chance. In our 50 trials, it happened 15 times.

$$P(7 \text{ or more}) = \frac{15}{50} = .30$$

Because getting 7 or more scores in group A larger than the median is *not* a rare event, we decide that the model chosen in Step 1 is appropriate. The differences between groups would be expected by chance. ∎

We do not have evidence that the new way of teaching spelling is more effective than the old way.

The problem of deciding whether two groups are the same or different is a very important one in statistics. It is called a *two-sample problem*.

EXERCISES

1   Suppose that in the example (page 119), students in group A obtained 8 scores which were above the median (instead of 7 scores as in the example). Use Table 5.12 to help decide whether the new method of teaching spelling appears to be more effective than the old.

2   A new feed for cattle is being developed. It is tried out in the laboratory with a group of cattle (group I). Group II continues to use the old feed. The following weight gains are recorded:

Group I: 104 109 127 143 187 204 209 266 277
Group II: 62   83   90 101 106 109 109 205

Do you conclude that the new feed is effective in producing weight gain?

3   Two groups of rats run a maze under the influence of two different drugs. Do the drugs have different effects on the error scores made by the rats?

Group A: 14 12   8 9 14 17 14
Group B: 12   8 11 7 14   8   8

4   Two groups, one consisting of farmers and the other made up of union members, were given an attitude test on the use of food stamps. Below are their scores.[8] Do the groups differ in their attitudes?

Farmers	Union Members
12	28
10	26
10	25
9	24
8	24
7	19
7	18
6	10
4	9
2	6

CHAPTER 6

Chi-Square

One of the most commonly used statistics is that of χ^2, or chi-square. The symbol χ is the Greek letter chi, pronounced to rhyme with *sky*. This chapter's key problem illustrates a use of the chi-square statistic.

KEY PROBLEM

Accidents at Irongate

The Irongate Foundry, Ltd., has kept records of on-the-job accidents for many years. Accidents are reported according to which hour of an 8-hour shift they happen. Table 6.1 shows their accident report:

TABLE 6.1

HOUR OF SHIFT	NUMBER OF ACCIDENTS
1	19
2	17
3	15
4	24
5	20
6	26
7	22
8	25
	Total 168

The union at the foundry wants to know whether accidents are more likely to take place during one hour of the shift rather than another. For example, do more accidents tend to be toward the end of the shift? The chi-square statistic can help us answer this question.

6.1   IS THE DIE FAIR?

For thousands of years, persons have used dice to help them make decisions. Some mathematics teachers have been known to use a die to decide those who are to turn in their homework! In a military draft, the selection of candidates often involves methods depending upon chance outcomes, like a lottery or a roulette wheel. For example, during World War II, each man in each selective service district was assigned a number between 1 and 9,000. (The number 9,000 was greater than the number of men in any selective service district.) The numbers were placed in identical capsules, which concealed their contents, and the capsules were drawn one at a time from a larger container until all were removed.

Senior officials used radio to announce the order in which the numbers were drawn. The order determined the order in which men were drafted. Table 6.2 is a frequency table showing the first 50 numbers drawn.

TABLE 6.2

INTERVAL	f
1-1000	5
1001-2000	0
2001-3000	3
3001-4000	1
4001-5000	7
5001-6000	8
6001-7000	11
7001-8000	7
8001-9000	8
	50

What do you think of these data? Do you think the draft was fair? Why or why not?

A natural question to raise in a chance experiment is that of fairness. If by rolling a die, I lose a game 5 times in a row, is the die fair? Is the state lottery fair? Was the lottery used in the military call-up fair?

One interpretation of fairness has to do with our expectations. Did the results meet our expectations? If the outcomes of a probabilistic experiment depart greatly from our expectations, we might question the fairness of that experiment. Our thinking here is similar to that in Chapter Five. But we will use a different statistic to help us make our decisions.

Let's consider the fairness of a die. Suppose we roll a six-sided die 60 times and obtain the outcomes in Table 6.3.

TABLE 6.3

OUTCOME OF DIE	TALLY	f	P
1	卌 ////	9	.15
2	卌 卌 ////	14	.23
3	卌 卌	10	.17
4	卌 ///	8	.13
5	卌 //	7	.12
6	卌 卌 //	12	.20
	Total	60	1.00

Do we think this die is fair? In order to help us decide, let's think about outcomes we would expect from a fair die. Recall that for a fair die, we have

$$p(1) = \frac{1}{6}$$

$$p(2) = \frac{1}{6}$$

$$p(3) = \frac{1}{6}$$

$$p(4) = \frac{1}{6}$$

$$p(5) = \frac{1}{6}$$

$$p(6) = \frac{1}{6}$$

That is, the *theoretical* probability of obtaining each of the six sides equals one-sixth. So, as we toss a fair die thousands and thousands of times, we would expect to get each side the same proportion of the time. For example, in 12,000 rolls, we would expect about one-sixth of them (or 2000) to be ones, the same proportion to be twos, and so on.

Another way of stating this is that we should find, for a fair die, that the experimental probability of getting a one will get closer and closer to the theoretical probability of getting a one. From the table, we got a one 9 times in 60 rolls of the die. So,

$$P(1) = \frac{9}{60} = .15$$

As the number of rolls of the die (if it is fair!!) gets larger, $P(1)$ gets closer and closer to one-sixth ($\doteq .167$.) Similarly, $P(2)$ gets closer and closer to .167, and so on.

Figure 6.1 shows a graph for the 60 rolls of the die in Table 6.3 and (in dotted lines) the expected proportion of outcomes for a fair die.

FIGURE 6.1

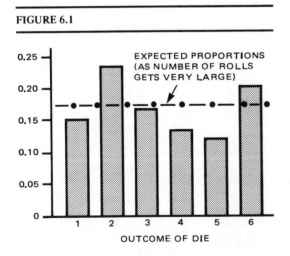

EXERCISES

1   Suppose we roll a six-sided die each number of times given. If the die is fair, how many times would we expect *each side* to occur?

(a)   150 times
(b)   300 times
(c)   600 times

2   Roll a six-sided die 90 times and record the outcomes you obtain. Draw a graph of your outcomes, following the example of the graph in Figure 6.1. On this graph, show the expected proportion of times each of the six sides of the die would be obtained, assuming the die to be fair.

3   Draw a graph (like Figure 6.1) of the following die-rolling data. Then draw a smooth curve to show the expected results, in the long run, for a true die.

Outcome	f
1	4
2	7
3	4
4	6
5	8
6	1
Total	30

4   Polly and Pete go through one page from a telephone book and write down the last digit of 50 telephone numbers. Here are their data. Draw a graph of these data, following the example of Figure 6.1. If the telephone numbers were random, show on the graph the expected proportion of each of the 50 digits.

Digit	f
0	1
1	6
2	3
3	2
4	5
5	7
6	2
7	10
8	8
9	0
Total	50

5   An octahedral (eight-sided) die is rolled 96 times with these results. Draw a frequency polygon of the data and then draw in the smooth curve for a true (not loaded) die.

Outcome	f
1	10
2	8
3	20
4	14
5	13
6	5
7	18
8	8
	96

Do you think this is a fair die? Why or why not?

6   A breakfast cereal company features a special offer by including one of four differently colored ball-point pens in a box. If a person buys 20 boxes of cereal, how many of each of the four pens would he or she expect to get? What is the largest number of any one color he or she could get? What is the smallest number? (To answer this question, we assume the company is fair. That is, each time we purchase a box of cereal, our chances of getting any of the four pens are equal.) Sample results are shown in Figure 6.2 for a person who bought 12 boxes of cereal.

Color	f
Blue	2
Yellow	4
Red	1
White	5
	12

FIGURE 6.2
NUMBER OF COLORED PENS OBTAINED WHEN 12 BOXES OF CEREAL ARE BOUGHT

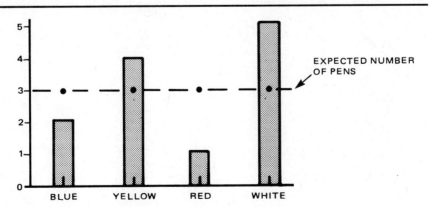

6.2   WHAT YOU EXPECT AND WHAT YOU GET

How well did the die we rolled in Section 6.1 meet our expectations? The graph in Figure 6.1 helps us answer the question. The dotted line shows the proportion of times each side would be expected to occur, if the die were fair. We now can make a table of the differences between how we *expected* the die to behave and the values we *obtained* when we actually rolled the die 60 times (Table 6.4).

If we find the total amount by which the expected and obtained outcomes were

different, we have

Total difference = 1 + 4 + 0 + 2 + 3 + 2 = 12

Let's now repeat our experiment with a die that we *know* to be loaded. This is accomplished by using a six-sided pencil and sandpapering one of the edges, as shown in Figure 6.3. Table 6.5 summarizes the outcomes of rolling this "loaded die" 60 times. The graph in Figure 6.4 illustrates the results. Table 6.6 shows the calculations for finding the total difference between obtained and expected results for the die under the assumption that the die is fair.

TABLE 6.4

OUTCOME	EXPECTED NUMBER	OBTAINED NUMBER	DIFFERENCE = EXPECTED – OBTAINED	ABSOLUTE VALUE OF DIFFERENCE[a]
1	10	9	10 – 9 = 1	1
2	10	14	10 – 14 = –4	4
3	10	10	10 – 10 = 0	0
4	10	8	10 – 8 = 2	2
5	10	7	10 – 7 = 3	3
6	10	12	10 – 12 = –2	2
				Total 12

[a]The absolute value of 1 is 1. The absolute value of –4 is 4, the absolute value of 0 is 0, and so on.

FIGURE 6.3

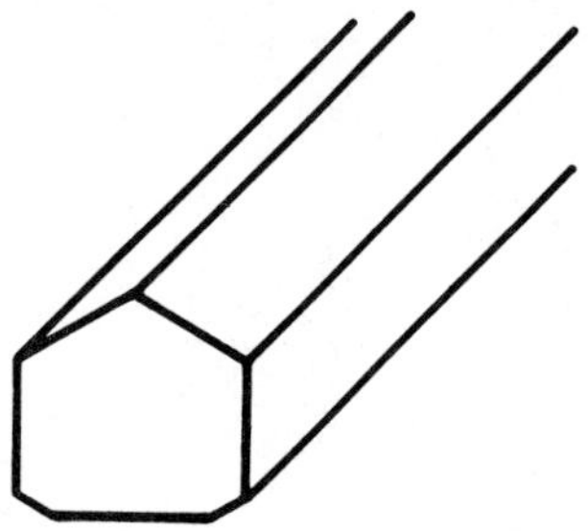

TABLE 6.5

NUMBER OF DOTS	f	P
1	3	.05
2	18	.30
3	9	.15
4	10	.17
5	11	.18
6	9	.15
Total	60	1.00

FIGURE 6.4

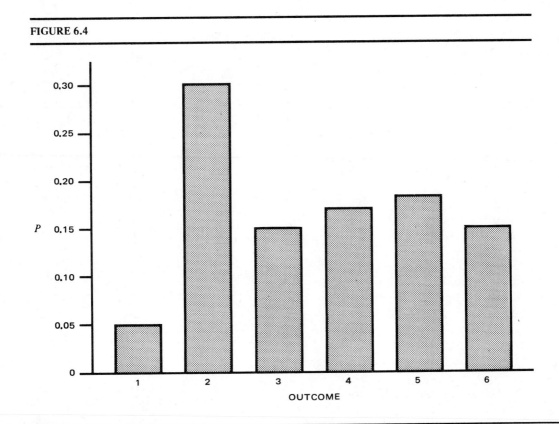

TABLE 6.6
OUTCOMES OF ROLLING A LOADED DIE 60 TIMES

OUTCOME	EXPECTED NUMBER	OBTAINED NUMBER	DIFFERENCE = EXPECTED − OBTAINED	ABSOLUTE VALUE OF DIFFERENCE
1	10	3	10 − 3 = 7	7
2	10	18	10 − 18 = −8	8
3	10	9	10 − 9 = 1	1
4	10	10	10 − 10 = 0	0
5	10	11	10 − 11 = −1	1
6	10	0	10 − 9 = 1	1
Totals	60	60	0	18

EXERCISES

1   A six-sided die was rolled 90 times; Table 6.7 shows the results. Complete the table by finding:

(a)   the expected number of outcomes for a fair die;

(b)   the difference between each expected and obtained outcome;

(c)   the absolute value of each difference, and

(d)   the total difference between obtained and expected frequencies (the sum of the right-hand column).

2   Roll a die 90 times and complete a table like the one in Exercise 1 of this section. By what total amount did the expected and obtained number of outcomes differ?

3   Do you think the die used in Exercise 1 of this section was fair? Explain your answer.

4   Do you think that the die you used in Exercise 2 of this section was fair?

5   Refer to Exercise 6 of Section 6.1, about the ball-point pen. Decide on a method of obtaining each of *four* outcomes with equal likelihood. (*Hint:* A six-sided die *can* be used, with two faces ignored. Or, use tetrahedral dice, which your teacher may be able to provide. Or, use playing cards. Shuffle the deck, deal a card, and record the suit—hearts, diamonds, spades, or clubs. Replace the card in the deck.) Do the experiment and record your outcomes in a table. Then draw a graph of your results. Draw a dotted line on your graph to show the expected number of each of the four outcomes.

6   By what total *amount* did the expected and obtained number of outcomes differ in Exercise 5 of this section?

7   Roll a four-sided die 12 times instead of 20. Now answer Exercises 5 and 6 of this section.

8   Table 6.8 tells how many of 40 persons prefer each of four kinds of orange juice. Complete the table, which is to show the

TABLE 6.7
OUTCOMES OF ROLLING A SIX-SIDED DIE 90 TIMES

OUTCOME	EXPECTED NUMBER	OBTAINED NUMBER	DIFFERENCE	ABSOLUTE VALUE OF DIFFERENCE
1		20		
2		16		
3		11		
4		12		
5		16		
6	___	15	___	___
Totals	90	90		

TABLE 6.8
DIFFERENCES BETWEEN OBTAINED AND EXPECTED NUMBER
OF PEOPLE PREFERRING EACH OF FOUR KINDS OF ORANGE JUICE

KIND OF ORANGE JUICE	EXPECTED NUMBER	OBTAINED NUMBER	ABSOLUTE DIFFERENCE
Fresh		13	
Freeze-dried		11	
Frozen		8	
Canned		8	
		Total 40	

expected and obtained numbers of people preferring each of the four kinds of orange juice, assuming that the outcomes were obtained by chance (and that the chances of choosing each of the four kinds of juice are the same).

9   Refer to the draft lottery outcomes in Table 6.2. What is the expected number of outcomes for each interval if the lottery is fair? What is the total number of outcomes that are different from expected?

10   Refer to the factory accident problem (the key problem). If the chances (likelihood) of accidents in each of the 8 hours of the shift are the same, how many accidents would be expected in each shift? What is the total number of accidents occurring that differed from expected (as was found, for example, in the right hand column of Table 6.4)?

PROJECT CORNER

Ask 30 or more people each to name a number between 1 and 10 inclusive. The people might be members of a class, but not the class which is using this book. Try to arrange things so that the people choose their numbers without knowledge of other peoples' choices; for example, you might ask everyone to write the number down quickly and quietly on a slip of paper. It is also a good idea to avoid discussion of the purpose of the project, since that might affect choices.

Make a frequency table of the numbers and see how much they depart from one-tenth of the number of respondents. That is, see how much the numbers depart from their expected frequency, if each number were chosen with equal probability.

It is usually found that numbers in this way appear to have different probabilities. People have likes and dislikes for numbers. A way of examining these apparent differences more closely is the chi-square test, discussed in this chapter.[1]

SPECIAL INTEREST
FEATURE

*Statistics and
the Laws
of Science*[2]

Broadly speaking, the laws of science can be divided into two types: *deterministic laws* and *probabilistic laws.* As an example of the first type, we have those of the German astronomer and mathematician Johannes Kepler (1571–1630). His laws of planetary motion enable one to predict, or determine, the motions of the planets around the sun. If we predict the position of Mars for 9:00 p.m. tomorrow, we expect it to be there then. If Mars is not there, we would not want to include Kepler's laws among the laws of science.

The laws of inheritance developed by the Austrian botanist and monk, Gregor Mendel (1822–1884), are an example of probabilistic laws. They cannot predict exactly the kind of offspring two parents will have, but they can be used to predict the *probability* of a certain kind of offspring.

The difference between these types is especially important when we try to test a law by checking it against a set of observations. As we said, if Mars is not where it is predicted to be, then we can reject the laws of planetary motion. (Unless, of course, there is an error in our calculation of its position or some catastrophe has destroyed the planet.) But Mendel's laws are not so easily rejected. For example, in one experiment Mendel predicted that a certain pea plant would have as offspring yellow peas with a probability of three-fourths. Out of 36 peas, 25 were yellow and the law was accepted, even though the number of peas predicted was 27 (since $27 = \frac{3}{4} \times 36$).

For probabilistic laws, it must be recognized that deviations from the law can occur by chance. The chi-square statistic provides a way of helping to decide the agreement of laws with real world data.

6.3   THE *D* STATISTIC

We are on the way to deciding whether a die is fair. In the last section, we rolled a die 60 times and found that outcomes differed from what is expected for a fair die by a total (ignoring plus and minus signs) of 12 (Table 6.4). We can use such a total to help us decide whether a die is fair. For example, a small total (close to zero) would indicate the die is giving results like those expected for a fair die. A large total would indicate outcomes rather different from those expected for a fair die.

There is a problem with using such a total number to decide fairness of a die. A large number of rolls can result in a large total number of differences between obtained and expected outcome, even when the *proportion* of expected outcomes is rather close to the proportion of obtained outcome. See Table 6.9. A fair die is rolled 600 times and the total difference between obtained and expected outcomes is 12. However, the proportion of obtained outcomes in each case is close to one-sixth, or .167.

So, we see dividing by the expected number (or frequency) for each outcome has the effect of compensating for the total number of rolls of the die. The larger the number of rolls, the larger the number of outcomes expected (for a fair die).

In this example, the total absolute difference score is

$$1 + 3 + 1 + 5 + 2 + 0 = 12$$

Dividing each difference by 100 (the expected number of outcomes for a fair die) gives

$$\frac{1}{100} + \frac{3}{100} + \frac{1}{100} + \frac{5}{100} + \frac{2}{100} + \frac{0}{100} = \frac{12}{100} = .12$$

We call this number *D,* for the *standard difference* between expected and obtained outcomes.

To make sure that you understand how to find *D,* let's find *D* for the outcomes of 150 rolls of a die. If the die is fair, the expected number of outcomes for each face is

$$\frac{150}{6} = 25$$

The number of outcomes obtained for 150 rolls of a fair die is as shown in Table 6.10.

TABLE 6.9

OUTCOME	EXPECTED NUMBER	OBTAINED NUMBER	P(OUTCOME)	ABSOLUTE DIFFERENCE
1	100	101	.168	1
2	100	97	.162	3
3	100	99	.165	1
4	100	105	.175	5
5	100	98	.163	2
6	100	100	.167	0
			Total	12

TABLE 6.10

OUTCOME	EXPECTED NUMBER	OBTAINED NUMBER	DIFFERENCE	ABSOLUTE DIFFERENCE
1	25	30	−5	5
2	25	24	1	1
3	25	20	5	5
4	25	29	−4	4
5	25	20	5	5
6	25	27	−2	2
Totals	150	150		22

For this experiment we have

Total difference = 5 + 1 + 5 + 4 + 5 + 2 = 22

And, dividing by the expected number for each trial, we have

$$D = \frac{5}{25} + \frac{1}{25} + \frac{5}{25} + \frac{4}{25} + \frac{5}{25} + \frac{2}{25} = \frac{22}{25} = .88$$

The number D turns out to be useful in describing outcomes of experiments, for it tells us an "averaged out" difference between expected and obtained outcomes.

How can we interpret D, though? For example, is our value of $D = .88$ large or small? We know that the larger D is, the more our dice rolls were different from what we would expect for a fair die. But how large a value of D is needed before we can say that, for a fair die, such a value will be obtained very seldom by chance?

In order to get some idea about how large these chance values for D can get and how often they occur for a fair die, we will do some more die rolling. We will carry out a total of 30 trials, where each trial is made up of rolling the die 150 times, and find the value of D, just as we did before. This is a lot of work, for we will have 29 more tables similar to Table 6.10, in which each trial consists of 150 rolls of a die. The results in Table 6.11 include the D value obtained above plus the D values from each of 29 other experiments! This is too much work to do on our own, so we could do it as a class project.

TABLE 6.11
STEM AND LEAF TABLE FOR VALUES OF D

Obtained from 30 trials, where a trial consists of rolling a fair six-sided die 150 times. (Values in table are to be multiplied by .01. For example, the first value is given as 32, so it is read as $D = .32$. The largest value in the table is $D = 1.54$.)

STEM	LEAF	f
0		0
1		0
2		0
3	2	1
4	0, 0, 8, 8, 8	5
5	6	1
6	4, 4, 4	3
7	2, 2, 2, 2, 2	5
8	0, 0, 0, 0, 0, 0	6
9	6, 6	2
10	4, 4	2
11	2	1
12	0	1
13	6	1
14	4	1
15	4	1
		30

The values of D that we get from this experiment tell us important information. We notice, first, that D varies quite a bit. Some values of D are very small, such as .32 or .40 or .48. (What if D were zero? This would mean that we obtained exactly what we expected for every outcome.) But D does get large, as well—notice the values 1.36, 1.44, and 1.54 for example.

What we really need to know from the table of values is how *often* we get large values of D, since large values indicate that the outcomes differ greatly from what is expected for a fair die. For example, if $D = 1.54$, how often can we expect to get a value as large as this or larger from rolling a fair die? We can tell from the table that this estimated probability is

$$P(D \geqslant 1.54) = \frac{1}{30} \doteq .03$$

So we estimate that only about 3 times in 100 will D be 1.54 or greater, due to chance, if the die is fair.

EXERCISES

Calculate D for the results given in each of the tables indicated. Show your work in each case.

1   Find D for the rolls of the die obtained in Exercise 1 of Section 6.2 (Table 6.7).

2   Find D for the outcomes for the orange juice data in Exercise 8 of Section 6.2 (Table 6.8).

3   Students in the statistics class at Forest View rolled a die and obtained the following results. Calculate D, assuming the die to be fair.

Outcome	Expected Number	Obtained Number	Difference
1		15	
2		32	
3		31	
4		9	
5		30	
6		33	
	Total	150	

4   Using Table 6.11, find these estimated probabilities:

(a) $P(D \geqslant 1.20)$
(b) $P(D \geqslant .80)$

5   Using Table 6.11, find these estimated probabilities:

(a) $P(D \geqslant 1.44)$
(b) $P(D \leqslant 2.0)$

6   It was noted in Section 6.3 that when $D = 0$, this means that the outcomes of the rolls of the die were exactly what would be expected for a fair die. From Table 6.11, what is $P(D = 0)$?

7   Refer to the data obtained at Forest View by rolling a die (Exercise 3 of this section). Use Table 6.11 to help you decide whether the die they rolled really was fair. Explain your decision.

8   Table 6.12 was obtained from 50 trials of rolling a fair six-sided die. Each trial consisted of 90 rolls. Using this data, find

(a) $P(D \geqslant 1.60)$

(b) $P(D \geqslant 1.87)$

TABLE 6.12

Multiply values in table by .01.

STEM	LEAF	f
1		0
2	7	1
3		0
4	0, 0	2
5		0
6	7, 7, 7, 7	4
7		0
8	0, 0, 0, 0, 0	5
9	3, 3, 3, 3, 3, 3, 3	7
10	7, 7, 7, 7, 7, 7, 7, 7, 7, 7	10
11		0
12	0, 0, 0, 0	4
13	3, 3, 3, 3, 3	5
14	7, 7, 7, 7	4
15		0
16	0, 0, 0, 0	4
17	3, 3	2
18	7	1
19		0
20		0
21		0
22	7	1
		50

9   Using the data of Table 6.12, find:

(a) $P(D \geqslant 1.40)$

(b) $P(D \leqslant 2.0)$

10   Six different colors of the same style of ball point pen are for sale in a store. For 90 sales, here is the number of each kind sold. Is there a distinct preference for one color over another, or do you think people are buying the colored pens at random, with each pen equally likely to be picked? Use the D statistic and Table 6.12 to help answer this question.

Color of Pen	Number Bought
Sky blue	17
Passionate pink	31
Deep purple	7
Burnt orange	10
Boring brown	9
Anemic ash	16
	90

11   Consider the problem of purchasing 20 boxes of cereal, each containing one of four different colored ballpoint pens with equal likelihood. Calculate D for this experiment. This gives the results for one trial. Now repeat for a total of 40 trials, computing the value of D for each trial; draw a graph of your results. What is the largest D value you obtained in the 40 trials? (This is a good class project.)

12   How often would a D value as large (or larger) as the one you obtained in Exercise 11 of this section, occur by chance (assuming all along that on each purchase the chances of getting any of the four pens remain equal)?

13   Take a die from a game that you play. Roll it 90 times and record how often you get each of the six sides. Calculate D for this set of dierolls. Do you think this die is loaded? Why or why not? Use Table 6.12 to help you decide.

6.4   HERE'S CHI-SQUARE

Calculating D for a set of outcomes of an experiment is a convenient way of telling how far the results differed from what was expected. However, a better (and usual) way of describing the results is to calculate a value called chi-square, written χ^2.

Let's see how chi-square is computed and compare that with the way D is computed. Table 6.13 shows both statistics for the same data, 60 rolls of a six-sided die.

Recall that to find D,

$$D = \text{sum of } \frac{|\text{expected number} - \text{obtained number}|}{\text{expected number}}$$

$$= \frac{1}{10} + \frac{4}{10} + \frac{1}{10} + \frac{3}{10} + \frac{3}{10} + \frac{2}{10} = \frac{14}{10} = 1.4$$

To find χ^2, we do something only slightly different:

$$\chi^2 = \frac{1^2}{10} + \frac{4^2}{10} + \frac{1^2}{10} + \frac{3^2}{10} + \frac{3^2}{10} + \frac{2^2}{10}$$

$$= \frac{1}{10} + \frac{16}{10} + \frac{1}{10} + \frac{9}{10} + \frac{9}{10} + \frac{4}{10} = \frac{40}{10} = 4.0$$

That is,

$$\chi^2 = \text{sum of } \frac{(\text{expected number} - \text{obtained number})^2}{\text{expected number}}$$

Now let's ask the same sort of question we asked involving the D statistic: How often will the expected and observed outcomes differ enough to produce χ^2 values as large as or larger than 4.0, by chance, for a fair six-sided die? To answer this question, we repeat the experiment of rolling a die many times, computing χ^2 for each set of 60 rolls. We then prepare a table of χ^2 values, as shown in Table 6.14.

TABLE 6.13
OUTCOMES OF 60 ROLLS OF A SIX-SIDED DIE

OUTCOME	EXPECTED NUMBER	OBTAINED NUMBER	DIFFERENCE = EXPECTED - OBTAINED	ABSOLUTE VALUE OF DIFFERENCE	DIFFERENCE SQUARED
1	10	9	$10 - 9 = 1$	1	$(1)^2 = 1$
2	10	14	$10 - 14 = -4$	4	$(-4)^2 = 16$
3	10	11	$10 - 11 = -1$	1	$(-1)^2 = 1$
4	10	7	$10 - 7 = 3$	3	$(3)^2 = 9$
5	10	7	$10 - 7 = 3$	3	$(3)^2 = 9$
6	10	12	$10 - 12 = -2$	2	$(-2)^2 = 4$
	60	60			

TABLE 6.14
STEM AND LEAF TABLE FOR χ^2

Obtained from 30 trials of rolling a fair six-sided die. Each trial consisted of 60 rolls. (Multiply values in table by .01.)

STEM	LEAF	f
0	6, 8	2
1	0, 4, 6	3
2	2, 2	2
3	6, 8	2
4	0, 0, 2, 4	4
5	0, 0, 0, 6, 8, 8	6
6	0, 4, 4, 6	4
7	0, 0, 0, 8	4
8		0
9	2	1
10	6	1
11		0
12	4	1
		$\overline{30}$

$\chi^2 \geqslant 4.0$ (bracketing stems 4 through 12)

From Table 6.14 we see that a χ^2 of 4.0 or larger was obtained in 21 of the 30 trials. Therefore

$$P(\chi^2 \geqslant 4.0) = \frac{21}{30} = .70$$

This estimated probability is represented by the shaded region of the graph in Figure 6.5.

FIGURE 6.5
GRAPH OF THE χ^2 VALUES REPORTED IN TABLE 6.14

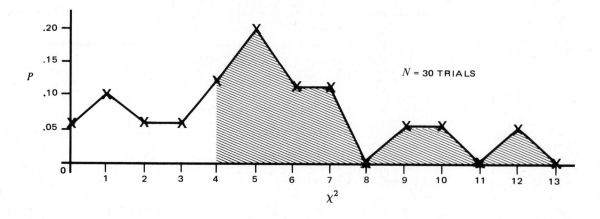

$N = 30$ TRIALS

EXERCISES

1   Use Table 6.14 to estimate the indicated probabilities:

(a) $P(\chi^2 \geqslant 2.2)$
(b) $P(\chi^2 \leqslant 1.6)$

2   Use Table 6.14 to estimate the indicated probabilities:

(a) $P(\chi^2 \geqslant 5.0)$
(b) $P(\chi^2 \leqslant 5.0)$

3   Use Table 6.14 to estimate the indicated probabilities:

(a) $P(\chi^2 \geqslant 7.8)$
(b) $P(\chi^2 \leqslant 10.0)$

4   A six-sided die was rolled 60 times with these results:

Outcome	Expected Number	Obtained Number
1		8
2		7
3		13
4		11
5		15
6		6
		60

Calculate D and chi-square for these data.

5   Using Table 6.14, estimate the probability of getting a χ^2 as large or larger than the value you obtained in Exercise 4. Do you think the die which was in Exercise 4 is fair? Why or why not?

6   In the colored pen problem (Exercise 10 of Section 6.3) these numbers of pens were bought:

Color of Pen	Number Bought
Sky blue	17
Passionate pink	31
Deep purple	7
Burnt orange	10
Boring brown	9
Anemic ash	16
Total	90

Use chi-square (Table 6.15) to help you decide if people tended to prefer one color of pen over others, or whether they were picking colors at random.

TABLE 6.15
STEM AND LEAF TABLE FOR χ^2

Obtained from 50 trials of rolling a fair six-sided die. Each trial consists of 90 rolls. (Multiply values by 0.1.)

STEM	LEAF	f
0	5, 7, 8	3
1	1, 3, 5, 6, 7	5
2	3, 5, 7, 7, 8	5
3	1, 1, 1, 2, 2, 3, 5, 6, 6, 6, 7, 7, 9	13
4	1, 3, 4, 4, 5, 5, 6, 9	8
5	1, 2, 3, 6, 6, 6, 7	7
6	0, 3, 3, 9	4
7	2, 3, 5, 9	4
8		0
9		0
10	3	1
		50

7 During a busy day in a large city, 90 traffic tickets were issued at six locations:

Location	Number of Tickets Given
A	12
B	7
C	21
D	15
E	11
F	24
	Total 90

Use chi-square (Table 6.15) to help you decide if the tickets are being given with equal likelihood at each location. (Assume all locations have the same amount of traffic.)

8 Roll a six-sided die (which you believe to be fair) 30 times and record the outcomes in a table like Table 6.13. Calculate D and chi-square for your results.

9 Table 6.16 gives the outcomes of rolling a fair four-sided die 40 times. The chi-square for these data is calculated below the table. Calculate D for these same data.

10 Now refer to Table 6.17 which gives another set of outcomes when rolling a fair four-sided die 40 times. Calculate χ^2 for the data of Table 6.17.

TABLE 6.17
OUTCOMES OF ANOTHER 40 ROLLS OF A 4-DIE

OUTCOME	EXPECTED NUMBER	OBTAINED NUMBER
1	10	12
2	10	7
3	10	14
4	10	7
	Total 40	40

TABLE 6.16
OUTCOMES OF ROLLING A 4-DIE 40 TIMES

OUTCOME	EXPECTED NUMBER	OBTAINED NUMBER	DIFFERENCE	(DIFFERENCE)^2
1	10	12	2	4
2	10	6	4	16
3	10	10	0	0
4	10	12	2	4
	Total 40	40		

$$\chi^2 = \frac{4}{10} + \frac{16}{10} + \frac{0}{10} + \frac{4}{10} = 2.4$$

11   A fair four-sided die was rolled for a total of 50 sets of 40 rolls each. Chi-square was calculated for each set of 40 rolls and the results are shown in Table 6.18. The graph of these data is shown in Figure 6.6. From Figure 6.6, find $P(\chi^2 \geqslant 3.0)$. (Estimate using areas).

12   Use Table 6.18 to find these estimated probabilities:

(a) $P(\chi^2 \geqslant 2.0)$
(b) $P(\chi^2 \leqslant 3.0)$

13   Use Table 6.18 to find these estimated probabilities:

(a) $P(\chi^2 \geqslant 3.0)$
(b) $P(\chi^2 \leqslant 6.2)$

14   Use Table 6.18 to find these estimated probabilities:

(a) $P(\chi^2 \geqslant 5.0)$
(b) $P(\chi^2 \leqslant 0.6)$

15   Use Table 6.18 to find these estimated probabilities:

(a) $P(\chi^2 \geqslant 8.0)$
(b) $P(\chi^2 \leqslant 3.6)$

TABLE 6.18
STEM AND LEAF TABLE FOR χ^2

Obtained from 50 trials of rolling a four sided die. Each trial consists of 40 rolls. (Multiply values in table by 0.1.)

STEM	LEAF	f
0	0, 4, 4, 6, 6, 6, 6	7
1	0, 0, 0, 0, 2, 4, 4, 4, 4, 6, 8, 8	12
2	0, 0, 2, 2, 4, 5, 5, 5, 5, 5	10
3	0, 2, 4, 4, 4, 6, 6, 8, 8	9
4	0, 0, 0, 2, 6, 6	6
5	0, 2, 4	3
6	2	1
7	2	1
8		0
9		0
10	0	1
	Total	50

FIGURE 6.6
GRAPH OF THE χ^2 VALUES REPORTED IN TABLE 6.18

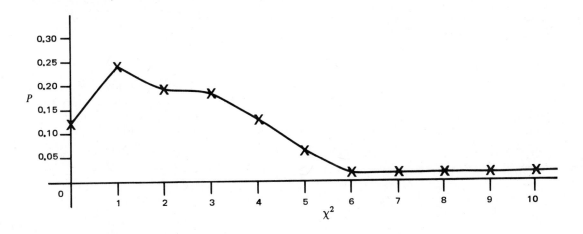

16   The table below (from Exercise 8 of Section 6.2) tells how many of 40 persons prefer each of four kinds of orange juice. Find D and chi-square for these data. Then use Table 6.18 to help you decide if there is convincing evidence that some kinds of orange juice are preferred over others.

Kind of Orange Juice	Number of Persons Preferring
Fresh	13
Freeze-dried	11
Frozen	8
Canned	8
Total	40

PROJECT CORNER

Choose a white page of telephone numbers from your home telephone directory. Starting at the top, write down the last digit of 100 consecutive telephone numbers. That is, take one number after another, and don't leave out any. Prepare a frequency table of the 10 digits and calculate D and chi-square for the results.

Suppose you want to see whether your phone book appears to be a good source of random digits. You need to estimate how likely the chi-square you got will occur by chance. One way to do this is for your classmates to each get 100 digits from a table of random digits, calculate chi-square, and record it in a table like 6.18. Or, if you can get the help of a computer, the BASIC program in Appendix G will calculate the D and chi-square values for you, for as many trials as you like.

6.5   SIX STEPS TO CHI-SQUARE

We learned six steps to decision making in Chapter Five. Now let's see how they can be used with chi-square.

■ **Example:**   We will demonstrate the six steps using the key problem (Accidents at Irongate) for this chapter. In that problem we were to decide whether there was a statistical evidence that accidents are more likely to happen during some hours of the shift rather than others. We use the data from Table 6.1.

Hour of Shift	Number of Accidents
1	19
2	17
3	15
4	24
5	20
6	26
7	22
8	25
Total	168

Solution:

In order to determine whether there is statistical evidence that accidents are more likely to happen during some hours of the shift rather than others, we *assume* accidents are equally likely to happen during each hour. So our model is a fair die.

Step 1—*Model:*   We will use an eight-sided fair die (an octahedral die) where a side corresponds to each of the eight hours of the shift. With such a model, we would *expect* one-eighth of the accidents to happen during each hour. That is, the expected number of accidents during each hour is:

$$\frac{1}{8} \times 168 = 21$$

Number of Accidents

Hour of Shift	Expected	Obtained	(Expected − Obtained)^2
1	21	19	$(2)^2 = 4$
2	21	17	$(4)^2 = 16$
3	21	15	$(6)^2 = 36$
4	21	24	$(-3)^2 = 9$
5	21	20	$(1)^2 = 1$
6	21	26	$(-5)^2 = 25$
7	21	22	$(-1)^2 = 1$
8	21	25	$(-4)^2 = 16$
Totals	168	168	

$$\chi^2 = \frac{4}{21} + \frac{16}{21} + \frac{36}{21} + \frac{9}{21} + \frac{1}{21} + \frac{25}{21} + \frac{1}{21} + \frac{16}{21} + \frac{108}{21} \doteq 5.1$$

Now we must determine whether this obtained value of chi-square ($\chi^2 = 5.1$), or greater, occurs rarely, or often. To do so, we go to the next steps.

Step 2—*Trial*:   A trial consists of rolling the fair eight-sided die 168 times, one for each accident which was reported in Table 6.1.

Step 3—*Outcomes of trial*:   The outcomes are recorded in the usual way. Here are the results for the first trial. The outcomes correspond to accidents occurring in each shift.

Number of Accidents

Hour of Shift	Expected	Obtained	(Expected − Obtained)^2
1	21	20	1
2	21	24	9
3	21	23	4
4	21	18	9
5	21	15	36
6	21	22	1
7	21	21	0
8	21	25	16
Totals	168	168	

$$\chi^2 = \frac{1}{21} + \frac{9}{21} + \frac{4}{21} + \frac{9}{21} + \frac{36}{21} + \frac{1}{21} + \frac{0}{21} + \frac{16}{21} = \frac{76}{21} \doteq 3.6$$

Step 4—*Repeat trials:* We obtain a total of 50 trials of rolling an octahedral die and record the chi-squares in Table 6.19.

TABLE 6.19

Each trial consists of 168 rolls. (Multiply values in table by .01.)

STEM	LEAF	f	
0		0	
1	5, 7, 8	3	
2	7	1	
3	0, 4, 6, 8	4	
4	1, 2, 3, 3, 3, 4, 6, 6, 8	9	
5	0, 0, 1, 2, 3, 8	6	
6	1, 4, 5, 5, 6, 6, 9	7	
7	0, 2, 6, 9	4	
8		0	
9	0, 0, 4, 6, 9, 9	6	
10	0, 9	2	
11	2, 6	2	$\chi^2 \geqslant 5.1$
12	7	1	
13	5, 5, 7	3	
14		0	
15		0	
16	2	1	
17		0	
18	0	1	

Total 50

Step 5—*Find the probability of the obtained statistic (or greater):* From Table 6.1 we obtained a chi-square of 5.1. From Table 6.19 we see that a chi-square of 5.1 or greater was obtained 31 times in 50 trials. So,

$$P(\chi^2 \geqslant 5.1) = \frac{31}{50} = .62$$

Step 6—*Make a decision:* We found in Step 5 that a chi-square of 5.1 or more occurred rather often (over one-half of the time) by chance. So we conclude that there is no statistical evidence of some hours of the shift being more accident-prone than others. The model of the fair die used in Step 1 for representing accidents in each hour of the shift is acceptable. The differences in occurrence of accidents are those we would expect due to chance variation.

Note: You should remind yourself from Chapter 5 what decision we would make if the estimated probability we found in Step 5 were small (less than 0.10). In that case, we would conclude that the obtained statistic happened only rarely by chance. So, the model we used in Step 1 would not be acceptable. The results we obtained from our data would not likely have been produced by the model. (In terms of the key problem, a small probability for the obtained chi-square—*not* what we found, to be sure—would be evidence of some hours of the shift being more likely to have accidents than others.) ∎

EXERCISES

1 Use Table 6.19 to find these probabilities:

(a) $P(\chi^2 \geqslant 13.5)$
(b) $P(\chi^2 \geqslant 18.0)$

2 Use Table 6.19 to find these probabilities:

(a) $P(\chi^2 \leqslant 5.1)$
(b) $P(\chi^2 \leqslant 3.0)$

3   Table 6.20 is a stem-and-leaf table for chi-square based on 50 trials of spinning a ten-sided fair spinner.

TABLE 6.20

Each trial consists of 100 spins. (Multiply each value in the table by 0.1.)

STEM	LEAF	f
0		0
1	2	1
2	8	1
3	2, 4, 8	3
4	0, 4, 8	3
5	0, 4, 6, 8	4
6	0, 4, 4	3
7	0, 0, 4, 6, 6	5
8	2, 2, 4, 4	4
9	0, 0, 0, 2, 4, 6	6
10	0, 4, 8, 8, 8	5
11	0, 0, 2, 2, 6	5
12	0, 4, 4, 4, 6, 8, 8	7
13		0
14		0
15	2, 4, 8	3
		50

Use Table 6.20 to find these probabilities:

(a) $P(\chi^2 \geqslant 6.0)$
(b) $P(\chi^2 \leqslant 4.8)$

4   Using Table 6.20 from Exercise 3 of this section, find these probabilities:

(a) $P(\chi^2 \geqslant 15.2)$
(b) $P(\chi^2 \leqslant 4.4)$

Use the six steps to solve Exercises 5 through 8 which follow.

5   The statistics class at Forest View High rolled an octahedral die 168 times and got these results. Use Table 6.19 to help decide if their die was loaded.

Outcome	f
1	29
2	22
3	18
4	19
5	20
6	23
7	12
8	25
	168

6   The statistics class at Buffalo Grove College interviewed 100 persons at random and asked them to give their favorite number between 0 and 9. Here are the results. Use Table 6.20 to help you decide if the students prefer some numbers over others.

Outcome	f
0	5
1	3
2	11
3	10
4	19
5	9
6	11
7	15
8	13
9	4
Total	100

7   A clothing store stocks men's ties which are identical except that they are in four different colors. After 40 sales, inventory shows the following purchases. Use Table 6.18 to help decide whether some colors are preferred over others.

Color of Tie	Number Sold
Amber	7
Blue	9
Orange	14
Maroon	10
Total	40

8   The last digit of each of 100 telephone numbers taken (in order) from one page of a telephone book, were as follows:

Digit	f
0	3
1	8
2	15
3	14
4	10
5	7
6	8
7	9
8	11
9	15
Total	100

Calculate chi-square for these data. Do you think the telephone book is a good source of random data? Why or why not? Use Table 6.20 to help you in your answer.

6.6 SMOOTH CHI-SQUARE CURVES

We have already seen that a way in which to improve the accuracy of our probability estimates is to increase the number of trials. Table 6.21 shows three estimates of $P(\chi^2 \geqslant 9.0)$ based on 30, 60, and 100 trials. (Each trial consists of 90 rolls of a fair six-sided die.)

Recall that we have used a relative frequency polygon to represent data like the chi-square data here (see, for example, Figure 1.10). Figure 6.7 gives the relative frequency polygon for the results of 100 trials where each trial consists of rolling a fair six-sided die 90 times (data from Table 6.21).

TABLE 6.21

χ^2	N TRIALS = 30 f	N TRIALS = 60 f	N TRIALS = 100 f
0 - .9	0	1	0
1.0 - 1.9	3	4	9
2.0 - 2.9	2	8	19
3.0 - 3.9	6	12	14
4.0 - 4.9	5	8	13
5.0 - 5.9	5	8	11
6.0 - 6.9	5	8	7
7.0 - 7.9	0	0	10
8.0 - 8.9	1	5	8
9.0 - 9.9	2	3	4
10.0 - 10.9	0	0	0
11.0 - 11.9	1	3	2
12.0 - 12.9			0
13.0 - 13.9			2
14.0 - 14.9			0
15.0 - 15.9			0
16.0 - 16.9			1

$\left. \begin{array}{} \\ \\ \\ \\ \\ \\ \end{array} \right\} \chi^2 \geqslant 9.0$

| Mean χ^2 | 5.24 | 4.67 | 4.79 |

$$P(\chi^2 \geqslant 9.0) = \frac{3}{30} \qquad P(\chi^2 \geqslant 9.0) = \frac{6}{60} \qquad P(\chi^2 \geqslant 9.0) = \frac{9}{100}$$

$$= .10 \qquad\qquad\qquad = .10 \qquad\qquad\qquad = .09$$

FIGURE 6.7
GRAPH OF THE χ^2 VALUES REPORTED IN TABLE 6.21 (N TRIALS = 100)

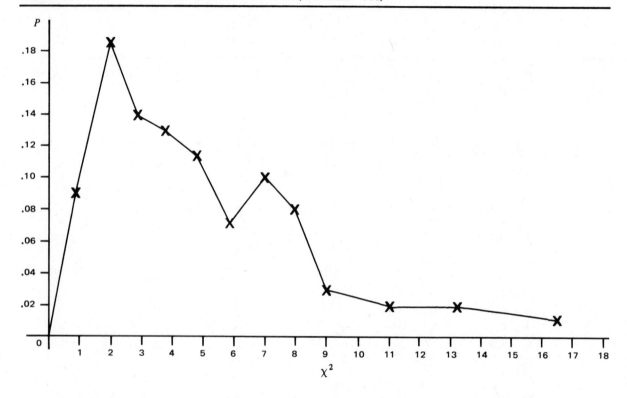

From this figure, you can read the probability of finding, for example, the probability of getting a chi-square of 9.0 or more $[P(\chi^2 \geq 9.0)]$.

As we conduct more and more trials, this polygon will get closer and closer to a smooth curve. When we draw in a smooth curve, as in Figure 6.8, we can get an estimate of the

FIGURE 6.8
SMOOTH CURVE FOR DATA OF TABLE 6.21
(X MARKS DATA POINTS FOR 100 TRIALS. * MARKS DATA POINTS FOR 500 TRIALS.)

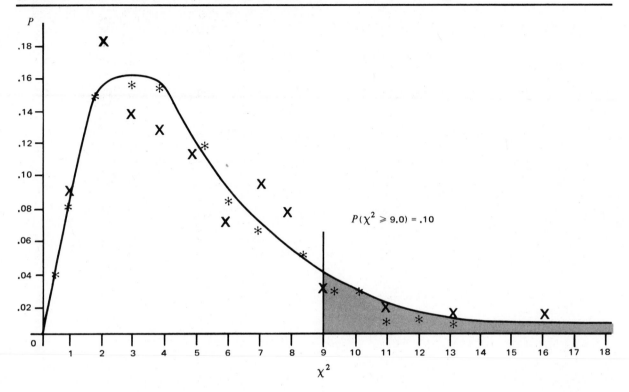

theoretical probability of obtaining a particular chi-square value. Figure 6.8 shows the estimated probability for $P(\chi^2 \geqslant 9.0)$. The 'X's show the data points for 100 trials and the '*'s show the data points for 500 trials.

You will see in the next section that tables can be used to give the theoretical probabilities for a given value of chi-square corresponding to rolls of a die with a given number of sides.

Table 6.22 and Figure 6.9 show χ^2 values for 100 trials of rolling a six-sided die, where the number of rolls per trial varies.

The three graphs in Figure 6.9 are similar. The ranges of chi-square values are comparable, as are their means. These results suggest that the value of chi-square is not affected much by the total number of observations involved in calculating the statistic. (Can you suggest reasons for this?) In more advanced books,

this property is treated more thoroughly. The D statistic does not have this property, nor does it have other important properties of χ^2. For now, we assume that the χ^2 can be used as long as a reasonable amount of data is available (By "reasonable," we mean that there are five or more expected observations in each cell of the chi-square table. This value is used by professional statisticians in their work.)

TABLE 6.22

χ^2	100 TRIALS 30 ROLLS PER TRIAL f	100 TRIALS 60 ROLLS PER TRIAL f	100 TRIALS 90 ROLLS PER TRIAL f
0 – .9	2	5	3
1.0 – 1.9	13	8	16
2.0 – 2.9	20	15	13
3.0 – 3.9	12	17	18
4.0 – 4.9	11	9	9
5.0 – 5.9	8	10	11
6.0 – 6.9	17	14	6
7.0 – 7.9	3	5	6
8.0 – 8.9	5	7	7
9.0 – 9.0	3	5	2
10.0 – 10.9	3	2	2
11.0 – 11.9	2	1	3
12.0 – 12.9	0	1	1
13.0 – 13.9	1	0	1
14.0 – 14.9	0	0	0
15.0 – 15.9	0	0	1
16.0 – 16.9	0	1	1
Mean χ^2	4.70	4.94	5.01

FIGURE 6.9

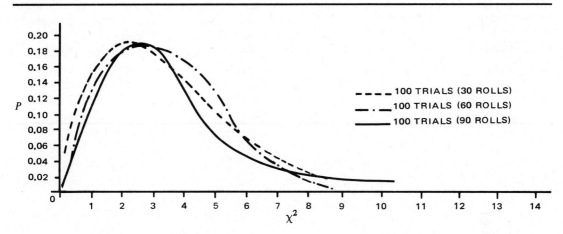

DEGREES OF FREEDOM

We now explore the effect on the values of chi-square of the number of sides of the die being used. We carry out 50 trials of rolling a fair die. Each trial consists of 60 rolls of a die.

But for the first experiment, we use a four-sided die (Table 6.23); for the second experiment, we use a six-sided die (Table 6.24); and for the third experiment, we use an isocahedral die which serves as a ten-sided die (Table 6.25).

TABLE 6.23
TABLE OF χ^2 FOR A FOUR-SIDED DIE

Fifty trials. Each trial consists of rolling the die 60 times. (Multiply each value in the table by 0.1).

STEM	LEAF
0	1, 3, 4, 4, 7, 7, 9, 9, 9, 9
1	1, 2, 2, 2, 3, 3, 5, 7, 7, 7, 7
2	0, 0, 0, 3, 3, 3, 4, 5, 8, 8, 8, 9
3	1, 2, 3, 3, 5, 5, 6, 6
4	1, 4, 4, 7, 8
5	7
6	5
7	
8	
9	9
10	3

Mean chi-square = 2.7
Standard deviation = 2.1

TABLE 6.24
TABLE OF χ^2 FOR A SIX-SIDED DIE

Fifty trials. Each trial consists of rolling the die 60 times. (Multiply each value in the table by 0.1.)

STEM	LEAF
0	4, 8
1	0, 4, 6, 8
2	0, 4, 6, 8, 8
3	0, 2, 2, 2, 2, 4, 4, 6, 6, 8
4	0, 0, 2, 2, 4, 4, 6, 6, 6, 6
5	2, 4, 4, 6, 6, 8, 8
6	2, 4, 4
7	0, 4, 8
8	2, 4, 8
9	6
10	
11	2
12	
13	6

Mean chi-square = 4.7
Standard deviation = 2.8

TABLE 6.25
TABLE OF χ^2 FOR A TEN-SIDED DIE

Fifty trials. Each trial consists of rolling the die 60 times. (Multiply each value in the table by 0.1.)

STEM	LEAF
0	
1	7
2	7, 7
3	3, 3
4	0, 0, 3, 7, 7
5	0, 0, 3, 3
6	3, 3, 7
7	0, 0, 0, 0, 0, 3, 7, 7, 7
8	0, 0, 0, 7, 7
9	0, 7
10	0, 3, 3, 3, 7, 7
11	0
12	0, 3
13	0, 3
14	0, 3
15	0
16	3
17	0
18	0

Mean chi·square = 8.4
Standard deviation = 4.0

Notice first that as the number of sides on the die gets larger, so does the mean chi-square value. Notice that the standard deviation of the chi-squares gets larger as well. That is, as the number of sides of the die gets larger, the chi-square values also tend to become more spread out.

Tables 6.23 through 6.25 suggest that the number of sides of a die (that is, the number of possible outcomes when rolling a die) has an effect on the chi-square produced. This *number of possible outcomes* has to do with the idea of *degrees of freedom* (sometimes abbreviated *df*). We can look at the idea of degrees of freedom in terms of tables of outcomes for rolling a die.

Tables 6.26 through 6.28 each give results for one trial of rolling a fair die. Each trial consists of 60 rolls.

TABLE 6.26
OUTCOMES OF ROLLING A FOUR-SIDED DIE

OUTCOME	EXPECTED	OBTAINED	
1	15	12 ⎫	
2	15	17 ⎬ 3 degrees of freedom	
3	15	14 ⎭	
4	15	◄—17 more needed to make 60 observations	
	60	60	

TABLE 6.27
OUTCOMES OF ROLLING A SIX-SIDED DIE

OUTCOME	EXPECTED	OBTAINED	
1	10	9	
2	10	14	
3	10	11	} 5 degrees of freedom
4	10	7	
5	10	7	
6	10	←—— 12 needed to make 60 observations	
	60	60	

TABLE 6.28
OUTCOMES OF ROLLING A TEN-SIDED DIE

OUTCOME	EXPECTED	OBTAINED	
1	6	8	
2	6	2	
3	6	10	
4	6	9	
5	6	12	} 9 degrees of freedom
6	6	5	
7	6	5	
8	6	1	
9	6	4	
0	6	←——4 needed to make 60 observations	
	60	60	

In Table 6.26, we see that there were only four possible outcomes for the die. But once we knew how many times a one, a two, and a three were obtained, we knew exactly how many fours were obtained, since the total number of rolls was 60. We say that Table 6.26 has three degrees of freedom (one less than the number of sides of the die).

Similarly, in Table 6.27, once we know how many ones, twos, threes, fours, and fives were obtained, we know how many sixes were obtained. We say that Table 6.27 has five degrees of freedom.

Since the number of degrees of freedom affects the size of the chi-squares produced by tables of outcomes, it is important to know how many degrees of freedom are associated with a given chi-square. Therefore, we indicate that a chi-square has, for example, three degrees of freedom by writing

$$\chi^2_3$$

When we wish to find a probability of a certain chi-square, we must first know its degrees of freedom.

■ **Example 1:**   Find $P(\chi^2_3 \geqslant 5.7)$.

Solution:

We use Table 6.23, because the chi-square has three degrees of freedom.

$$P(\chi^2_3 \geqslant 5.7) = \frac{4}{50} = .08 \ ■$$

■ **Example 2:**   Find $P(\chi^2_5 \geqslant 7.8)$.

Solution:

We use Table 6.24, because the chi-square has five degrees of freedom.

$$P(\chi^2_5 \geqslant 7.8) = \frac{7}{50} = .14 \ ■$$

■ **Example 3:**   Find $P(\chi^2_9 \geqslant 17.0)$.

Solution:

We use Table 6.25, because the chi-square has nine degrees of freedom.

$$P(\chi^2_9 \geqslant 17.0) = \frac{2}{50} = .04 \ ■$$

An important property of the chi-square statistic is that the mean of the chi-squares produced by a set of trials, such as given in Table 6.23, is an estimate of the number of degrees of freedom in the table outcomes from which the chi-squares were produced. Notice that in Table 6.23 the mean of the 50 chi-squares is 2.7. These outcomes were produced by a four-sided die. Table 6.29 summarizes the mean chi-squares for Tables 6.23–6.25.

TABLE 6.29
MEAN VALUES OF CHI-SQUARE FOR TABLES 6.23–6.25

TABLE	MEAN χ^2	DEGREES OF FREEDOM IN TABLE
6.23	2.7	3
6.24	4.7	5
6.25	8.4	9

As the number of trials increases, the mean of the chi-squares will get closer and closer to the degrees of freedom in the table.

EXERCISES

1   Find the estimated probabilities. You must first decide which table of values to use, Table 6.23, Table 6.24, or Table 6.25.

(a) $P(\chi^2_3 \geqslant 5.7)$

(b) $P(\chi^2_3 \geqslant 9.9)$

2   Find the estimated probabilities. You must first decide which table of values to use, Table 6.23, Table 6.24, or Table 6.25.

(a) $P(\chi^2_5 \geqslant 6.0)$

(b) $P(\chi^2_5 \geqslant 8.0)$

3   Find the estimated probabilities. You must first decide which table of values to use, Table 6.23, Table 6.24, or Table 6.25.

(a) $P(\chi^2_5 \geqslant 11.2)$

(b) $P(\chi^2_5 \geqslant 9.6)$

4   Find the estimated probabilities. You must first decide which table of values to use, Table 6.23, Table 6.24, or Table 6.25.

(a) $P(\chi_9^2 \geqslant 10.0)$

(b) $P(\chi_9^2 \geqslant 15.0)$

5   A six-sided die was rolled 90 times and the following outcomes were obtained.

Outcome	f
1	17
2	13
3	17
4	10
5	16
6	17

(a)   Calculate χ^2 for these data.
(b)   How many degrees of freedom are associated with this chi-square? Explain.
(c)   What is the probability of obtaining a chi-square as large or larger than the value you obtained in part (a), assuming that a fair die was used? [That is, for whatever the value v you obtained in part (a), what is $P(\chi^2 \geqslant v)$?]
(d)   Do you believe the die used in this exercise was a fair die? Why or why not?

6   You can use chi-square to solve the key problem for this chapter. How many degrees of freedom does that chi-square have? Explain.

7   Sue uses her telephone book as a source of random digits. She takes the first 300 telephone numbers in the book, in order, and writes down the last digit of each number. If she uses a chi-square to help her decide if the digits are random, how many degrees of freedom will the chi-square have? Explain.

8   A record store stocks albums for the top 20 artists. The manager of the store wants to determine whether the artists are equally popular (in terms of sales of albums) or if some artists are definitely favored over others. If a chi-square is used, how many degrees of freedom will it have? Explain.

6.7   THE CHI-SQUARE TABLE OF AREAS

We have now reached the point in our study of statistics where we can be more accurate in finding probabilities of a chi-square as large or larger than a given chi-square. Tables have been produced that provide theoretical probabilities, based on theoretical chi-square curves, or distributions.

We begin by finding an experimental or estimated probability.

■ **Example 1**:   Find $P(\chi_5^2 \geqslant 1.6)$.

Solution:

We are to find the estimated probability of a chi-square greater than or equal to 1.6. This chi-square has 5 degrees of freedom. So, we use Table 6.24:

$$P(\chi_5^2 \geqslant 1.6) = \frac{46}{50} = .92 \ ■$$

Now let us find the corresponding theoretical probability for this chi-square.

■ **Example 2:** Find $p(\chi_5^2 \geqslant 1.6)$.

Solution:

We use a table of areas for chi-square. The theoretical curve for a chi-square with 5 degrees of freedom is shown in Figure 6.10. The shaded area can be thought of as a probability.

Table 6.30 gives areas under the chi-square curve. There are rows for degrees of freedom from 1 through 5. Look at row 5, which gives areas for various regions under the curve in Figure 6.10. (An expanded version of this table is in Appendix A.)

Look along row five until you find 1.6. (Actually 1.61 is in the table, but this is 1.6 when rounded to the nearest tenth.) Now look at the top of the column in which 1.6 is located. You find 0.90 which is the shaded area in Figure 6.10. (The total area under the curve is 1.0.) So, we have:

$$p(\chi_5^2 \geqslant 1.6) = .90 \ ■$$

Compare this theoretical probability with the estimated probability of 0.92 which we found from Table 6.24 (Example 1). As we have noted before, larger numbers of trials in Table 6.24 would give us better estimates of the desired probabilities.

FIGURE 6.10

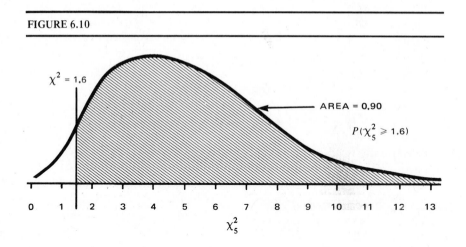

TABLE 6.30

DEGREES OF FREEDOM	PROBABILITY (AREA)								
	0.99	0.95	0.90	0.50	0.20	0.10	0.05	0.01	0.001
1	0.00	0.00	0.02	0.45	1.64	2.71	3.84	6.73	10.83
2	0.02	0.10	0.21	1.39	3.22	4.61	5.99	9.21	13.82
3	0.11	0.35	0.58	2.37	4.64	6.25	7.81	11.34	16.27
4	0.30	0.71	1.06	3.36	5.99	7.78	9.49	13.28	18.47
5	0.55	1.15	1.61	4.35	7.29	9.24	11.07	15.09	20.52

■ **Example 3:**   Find $p(\chi^2_5 \geq 8.0)$.

Solution:

Again we use Table 6.30, referring to the row for 5 degrees of freedom. The value of 8.0 does not appear in this row. We have, instead, 7.29 and 9.24. The probability (area) given for 7.29 (see the top of the column) is 0.20. The probability given for 9.24 is 0.10. So the probability for 8.0 is between 0.20 and 0.10.

Figure 6.11 shows the area which we are trying to find. Here is what we can do to get a good estimation of this area.

  i.   Find at what relative distance from 7.29 to 9.24 the value of 8.0 is located.

Distance from 7.29 to 9.24 = 9.24 − 7.29 = 1.95
(almost 2.0)

Distance from 7.29 to 8.00 = 8.00 − 7.29 = 0.71
(almost 0.75)

  ii.   Before we do any calculating, we can estimate that 8.0 is roughly one-half (or 0.5) of the distance between 7.29 and 9.24. That

is, 8.0 is roughly half-way between 7.29 and 9.24. So, we estimate:

$$p(\chi^2_5 \geq 8.0) \doteq .15$$

For many purposes, this estimated probability is adequate. However, for a better approximation, use this rule called *linear interpolation.* (See Appendix D for further discussion and formulas.)

  iii.   Calculate the value of the fraction:

$$\frac{\text{distance between 7.29 and 8.00}}{\text{distance between 7.29 and 9.24}} =$$

$$\frac{8.00 - 7.29}{9.24 - 7.29} = \frac{0.71}{1.95} = .36$$

Then find the difference between the probability (area) for 7.29 and for 9.24 = .20 − .10 = .10. So,

desired probability = .20 − (.36)(.10) =

.20 − .036 = .164 ≐ .16

and $p(\chi^2_5 \geq 8.0) \doteq .16$

As we use the table of chi-square areas in solving problems, we can use the six steps, as well. ■

FIGURE 6.11

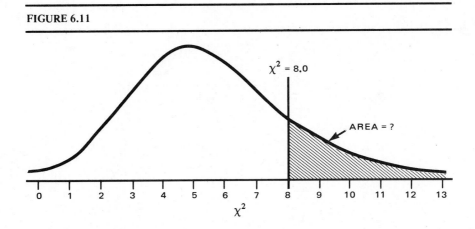

■ **Example 4**:   In 1967, the weekly numbers of animal bites reported to the Chicago Board of Health were:

Week Ending	Number of Animal Bites[3]
October 26	268
November 2	189
November 9	199

Do you think that the weekly variation in the number of bites reported is due to chance alone? Use the chi-square table of areas to help you decide.

Solution:

We assume that each of the three weeks is equally likely to have animal bites reported.

Step 1—*Model*:   We could use a three-sided spinner.

1: Week 1
2: Week 2
3: Week 3

With such a model, we would expect one-third of the bites to be reported each week. So, the *expected* number of bites is

$$\frac{1}{3} \times 656 \doteq 218.7$$

Since the table has three rows, the x^2 has two degrees of freedom. We need to find out if the chances of getting a $x_2^2 \geqslant 16.9$ are large or small.

With a table of chi-square areas, we do not need to actually do steps 2–4. But by thinking through what we would do (if we had to) we can get a better understanding of how to use chi-square.

Step 2—*Trial*:   A trial would consist of doing 656 spins, one for each bite reported.

Step 3—*Outcomes of trial*:   We would record how many times each of the three sides of the spinner was obtained. That is, we would find out how many of the 656 bites were reported each week, by chance, in one trial.

Number of Bites

Week Ending	Expected	Reported	(Expected − Reported)^2
October 26	218.7	268	$(218.7 - 268)^2 = (-49.3)^2$
November 2	218.7	189	$(218.7 - 189)^2 = (29.7)^2$
November 9	218.7	199	$(218.7 - 199)^2 = (19.7)^2$
	656.1	656	

$$\chi^2 = \frac{(218.7 - 268)^2}{218.7} + \frac{(218.7 - 189)^2}{218.7} + \frac{(218.7 - 199)^2}{218.7} = 16.9$$

(*Note:* Expected values do not add to 656 due to rounding.)

Step 4—*Repeat trials*:   We would do many trials. The more trials, the better the estimate of the probability we are trying to find.

But since we have a table of chi-square areas, we don't have to do the trials. We now can give theoretical instead of estimated probabilities.

Step 5—*Find probability of chi-square greater than or equal to obtained value*:   We need to find $p(\chi^2_2 \geqslant 16.9)$.

We use Table 6.30. We go to row 2 since this chi-square has two degrees of freedom (as we noted in Step 1).

A chi-square of 16.9 does not appear in row 2. The largest value in row 2 is 13.82. Because 16.9 is greater than 13.82, we know that $P(x^2_2 \geqslant 16.9)$ is smaller than .001. We could estimate what the probability actually is, but we do not need to know this for Step 6.

Step 6—*Make decision*:   We found that the probability of getting a chi-square greater than or equal to 16.9 is less than .001. This is a very small probability.

That is, the chances are less than 1 out of 1000 (.001) that a chi-square of 16.9 would be obtained by chance. So, we conclude that the differences in numbers of bites reported to the Chicago Board of Health are *not* due to chance alone. It may be the task of someone to try to find out *why* the numbers reported are different from week to week. Can you think of any reasons? ∎

EXERCISES

For each of the following, find the estimated probabilities (use Table 6.24) and then find the theoretical probability from Table 6.30.

1    $P(\chi^2_5 \geqslant 4.4)$

2    $P(\chi^2_5 \geqslant 7.4)$

3    $P(\chi^2_5 \geqslant 9.2)$

4    $P(\chi^2_5 \geqslant 11.2)$

For Exercises 5 and 6, find the estimated probability (use Table 6.23) and then find the theoretical probability from Table 6.30.

5    $P(\chi^2_3 \geqslant 4.6)$

6    $P(\chi^2_3 \geqslant 6.5)$

7   Find these probabilities. Use the expanded table of chi-square areas in Appendix A.

(a) $p(\chi^2_4 \geqslant 9.5)$

(b) $p(\chi^2_4 \geqslant 13.3)$

8   Find these probabilities. Use the expanded table of chi-square areas in Appendix A.

(a) $p(\chi^2_7 \geqslant 12.0)$

(b) $p(\chi^2_7 \geqslant 14.1)$

9   Find these probabilities. Use the expanded table of chi-square areas in Appendix A.

(a) $p(\chi^2_{10} \geqslant 12.0)$

(b) $p(\chi^2_{10} \geqslant 20.0)$

10   Find these probabilities. Use the expanded table of chi-square areas in Appendix A.

(a) $p(\chi^2_{20} \geqslant 18.0)$

(b) $p(\chi^2_{20} \geqslant 40.0)$

For Exercises 11 through 14, use the chi-square table of areas in Appendix A.

11   The statistics class at Forest View High rolled a die and obtained the following results.

Outcome	f
1	26
2	7
3	17
4	19
5	14
6	13
	96

Calculate χ^2 for this experiment. How likely are you to get a χ^2 value as large or larger than the one obtained here, by chance, if the die they used was fair? Do you believe the die used in this question was fair? Explain.

12 The statistics class at New Trier High School interviewed 380 students (at random, we assume) and asked them for their favorite number between zero and nine. Here are the results:

Outcome	f
0	0
1	12
2	27
3	56
4	37
5	40
6	58
7	103
8	31
9	16
	380

Calculate χ^2 for these results. Do you think that the New Trier students prefer some numbers over others? Why?

13 The number of eights in the fraction on a closing stock price for low-priced stocks from the American Stock Exchange are shown below.[4] Why are some fractions more common than others? Explain.

Fraction	Frequency
0	60
1/8	30
2/8 (1/4)	29
3/8	27
4/8 (1/2)	47
5/8	49
6/8 (3/4)	37
7/8	38
	317

14 In a study of rounding errors made by shoppers when they estimate the cost of articles, 174 cash register slips were examined. The last digit of the price registered on the slip was recorded and gave these results:[5]

Last Digit	Frequency
0	11
1	10
2	10
3	34
4	7
5	25
6	4
7	18
8	6
9	49
	174

Would you conclude that the costs of articles are more likely to have some last digits rather than others?

CHAPTER 7

Statistics In Two Variables

KEY PROBLEM

Crickets, Anyone? (Or, "Did you hear what the temperature is today?")

Crickets make their chirping sounds by rapidly sliding one wing over the other. The faster they move their wings, the higher the chirping sound that is produced. Scientists have noticed that crickets move their wings faster in warm temperatures than in cold temperatures. Therefore, by listening to the pitch of the chirp of crickets, it is possible to tell (reasonably accurately) the temperature of the air. Table 7.1 gives the recorded pitch (in number of vibrations per second) of a cricket chirping recorded at 15 different temperatures.[1]

In this chapter we will learn ways of finding the relationship between two measures (like pitch and temperature) so that for a certain pitch of a cricket chirp we can calculate the corresponding temperature to be expected from our data. Hence, instead of a thermometer, crickets, anyone?

TABLE 7.1

CHIRPS PER SECOND	20	16	20	18	17	16	15	17	15	16	15	17	16	17	14
TEMPERATURE (°F)	89	72	93	84	81	75	70	82	69	83	80	83	81	84	76

7.1 RELATING TWO MEASURES

So far, we have been dealing with statistics like *average* and *range* in only one variable at a time, such as rainfall or temperature or weight. But in real life, we are often concerned with more than one variable at a time. For example, we may wish to know relationships between two variables.

Table 7.2 shows the number of traffic violations for various age groups.

TABLE 7.2

AGE GROUP	TRAFFIC VIOLATIONS*
16–25	4.8
26–35	4.0
36–45	2.9
46–55	2.1
56–65	1.0
66–	.4

*Data are mean number of violations per 1000 persons in age group.)

The data here suggest a relationship between the *variables,* age and number of traffic violations. That is, the older the drivers are, the fewer traffic violations they commit. (Or, younger drivers commit more violations.) There are, of course, other questions to ask before we jump to the conclusion that young drivers are "worse" than older drivers. For example, are they watched more carefully than older drivers? Do they do more driving? But for now, we will focus only on the data as we see them. There does appear to be a relationship between two *variables,* age of driver and number of traffic violations.

Other examples of questions about two measures are:

1. *Jogging:* Is there a relationship between amount of jogging (say, minutes per day) and length of life (in years)?

2. *Smoking:* Is there a relationship between amount of smoking (for instance, packs per day) and illness (number of days per year)?

3. *Cloud-seeding:* Is there a relationship between amount of cloud-seeding (for example, number of pounds or kilograms of chemicals applied) and rainfall (in inches or centimeters)?

4. *Fertilizing fields:* Does increasing amount of fertilizer increase crop yield (in, for instance, bushels per acre)?

5. *Studying for examinations:* Do people who study more do better on examinations?

EXERCISES

1 What variables are of interest when each of these questions is asked?

(a) Do children who view a lot of television do poorly in school work?
(b) Does a copper bar expand when heated?
(c) Do persons who are vaccinated for polio get the disease less often?
(d) Do tires with higher inflation (more air in them) give better mileage than those with lower inflation?

2   Professor Eron of the University of Illinois did a ten-year study of the television viewing of 875 third-grade children. The following excerpt from *Time* reports his conclusions.[2]

> [Eron] now believes that a "direct, positive relation" exists between television viewing by small boys and aggressive behavior.

What two variables are of interest in this study? (It is also noted in the article that little girls who watch television do not show such an increase in aggressive behavior.)

3   The following article reports on relationships between variables.[3] What are the variables and what is the relationship? (*Hint:* Pick what pair of variables you like.)

HARDENED ARTERIES

As people get older there is a marked decline in sports involvement—except for viewing sports on television. The average time devoted to active sports per week is 5.1 hours for those aged 18 to 24 compared to 1.7 hours for those 55 and over. But the time spent watching TV sports increases somewhat for all those over 35.

---

4   Identify three sets of variables whose relationships are discussed in the following article.[4]

CITE CORRELATIONS ON TRAFFIC DEATHS

During the months immediately following the 1974 gasoline shortage it was noted that deaths from vehicular accidents were on a significant decline for the first time in decades. The phenomenon was hailed at the time as one of the few benefits from the fuel crunch.

More recently notice has been paid to other statistics from that period which indicate a drop in deaths from causes other than highway accidents, but which some researchers believe could have a tie-in to reduced driving habits. San Francisco county, for one, noted a 13 per cent decline in deaths from all causes during a three-month period at the height of the gasoline shortage.

A decrease of almost 33 per cent in deaths from chronic lung disease was noted, along with a 16 per cent decline in deaths from cardiovascular diseases. It is possible, of course, that a combination of factors brought about these unexpected results.

But a number of driving-related causes also have been suggested. Less stress from driving less, fewer pollutants in the air, more walking and simply more opportunity to relax at home are among suggested possibilities. No definite conclusions have been reached, but the correlations are intriguing.

---

5   The graph in Figure 7.1 shows the number of stolen motor vehicles (in thousands) in the United States during 1957–1982. For example, in 1957 about 250,000 motor vehicles were stolen.

(a)   According to the graph, about how many motor vehicles were stolen in 1967? In 1982?

(b)   What two variables are shown in this graph?

(c)   What relationship between these two variables is suggested by the graph?

6   The graph in Figure 7.2 reports the percentage of people favoring cuts in Federal aid for college students, based on those peoples' incomes.[6] For example, for those whose income is less than $15,000 a year, 31% favored cuts in Federal aid.

(a)   What two variables are shown in this graph?

(b)   What relationship between these variables is suggested?

7   Find an article in a newspaper or a magazine that deals with a relationship between two variables. What are those variables? What relationship between those two variables is presented or discussed?

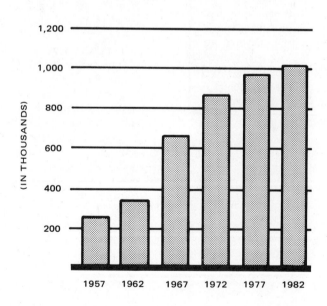

FIGURE 7.1[5]
STOLEN MOTOR VEHICLES IN THE UNITED STATES

FIGURE 7.2[6]
THOSE WHO SUPPORT CUTS IN FEDERAL AID FOR COLLEGE STUDENTS, BY INCOME

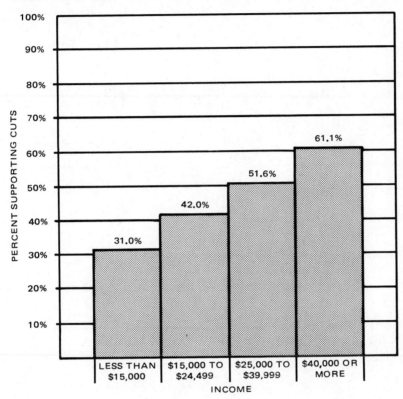

(NOTE: GRAPH ALSO SUPPLIED BY SOURCE.)

7.2 WHAT'S MY RULE?

One way of thinking of how two variables can be related is to play the game, "What's my rule?" The game goes as follows. I am thinking of pairs of numbers, N and M. Try to find the rule that will give me the second number, M, if I know the first number, N (see Table 7.3).

TABLE 7.3

FIRST NUMBER N	SECOND NUMBER M
2	4
3	6
4	8
6	12
7	?

If you get 14 for M when N is 7, you know my rule. It is,

$$M = 2 \times N$$

or

$$M = 2N$$

That is, the second number, M, is twice the first number, N.

Table 7.4 gives one that is a little more difficult.

TABLE 7.4

FIRST NUMBER Q	SECOND NUMBER R
1	1
2	3
3	5
5	9
8	?

My rule for this game gives 15 as the second number when the first number is 8. So, my rule is

$$R = 2 \times Q - 1$$

In other words, my rule is: The second number is one less than twice the first number. If you are not sure how the rule works for this example, work it out for each value of Q shown in Table 7.5.

TABLE 7.5

Q	RULE $2 \times Q - 1$		R
0	$2 \times 0 - 1$	$= 0 - 1$	-1
1	$2 \times 1 - 1$	$= 2 - 1$	1
2	$2 \times 2 - 1$	$= 4 - 1$	3
3	$2 \times 3 - 1$	$= 6 - 1$	5
5	$2 \times 5 - 1$	$= 10 - 1$	9
8	$2 \times 8 - 1$	$= 16 - 1$	15

This game shows how two quantities can be related using a little bit of mathematics. The rule is a way of stating this relationship. For example, suppose I have in mind the rule

$$Y = \frac{1}{2}X + 1.5$$

Use the rule to find what second number, Y, will correspond to the first number, X, in Table 7.6.

TABLE 7.6

X	Y
1	1.0
2	2.5
3	3.0
6	4.5
9	?

We use the rule to obtain the results, shown in Table 7.7.

TABLE 7.7

X	RULE $\frac{1}{2}X + 1.5$		Y
2	$\frac{1}{2}(2) + 1.5$	$= 1 \quad + 1.5$	2.5
3	$\frac{1}{2}(3) + 1.5$	$= 1.5 + 1.5$	3.0
6	$\frac{1}{2}(6) + 1.5$	$= 3.0 + 1.5$	4.5
9	$\frac{1}{2}(9) + 1.5$	$= 4.5 + 1.5$	6.0

The following rule shows the relationship between the Fahrenheit and Celsius temperature scales:

$$F = 1.8C + 32$$

Table 7.8 shows corresponding Fahrenheit and Celsius temperatures for a few values.

TABLE 7.8

°C	RULE $1.8C + 32$	°F
10	$F = 1.8(10) + 32$	50
15	$F = 1.8(15) + 32$	59
100	$F = 1.8(100) + 32$	212
0	$F = 1.8(0) + 32$	32
−10	$F = 1.8(−10) + 32$	14

As a final example, let's look at the pairs of numbers in Table 7.9.

TABLE 7.9

FIRST NUMBER S	SECOND NUMBER T
6	0
5	1
4	2
3	3
2	4

Here we see that as the first number gets smaller, the second gets larger. The rule is

$$T = 6 - S$$

The quantities S and T are said to be *negatively* related. In the other examples given in this section, the quantities have been *positively* related, because both increase together.

EXERCISES

1 What's my rule for each of these tables?

R	?	S	X	?	Y	W	?	X
−1		−3	−2		−1	−1		0
0		0	0		0	0		2
1		3	2		1	1		4
2		6	4		2	2		6
4		12	5		2.5	3		8
5		15	10		5	5		12

2 Use the given rules to complete the accompanying table.

Rule: $M = N + 1$		Rule: $Y = X - 3$	
N	M	X	Y
0		10	
1		6	
3		5	
7		3	
10		1	

Rule: $W = 2X + \dfrac{1}{2}$

X	W
0	
1	
1.25	
2	
3	

3 The following equation can be used (in some region of the country) to calculate the expected yield of corn (Y) in bushels per acre, given the July rainfall in inches (X):

$$Y = 3.3X + 21.9$$

What yield can be expected for these July rainfalls:

(a) 5 inches
(b) 2.1 inches
(c) 10 inches

4   A rule, known as Clark's Rule, has been developed for determining a child's dosage of medication based on the recommended adult dose. The rule is to be used for children over 2 years of age.

$$\text{Child's dose} = \frac{\text{weight of child in pounds}}{150} \times \text{adult dose}$$

[*Note:* 150, the divisor, is taken as the mean (average) weight of an adult.]

Use this rule to determine the child's dose for each of the adult doses if a child weighs 75 pounds.

Adult Dose (teaspoons)	Clark's Rule	Child's Dose
1		
2		
3		

5   Recalculate the child's dose for a child of:

(a)   50 pounds;
(b)   100 pounds.

Adult Dose (teaspoons)	Clark's Rule	Child's Dose
1		
2		
3		

6   The cooking time t (in hours) for a turkey that weighs w pounds is given by

$$t = \frac{1}{4}w + 2.5$$

Find the cooking time for each turkey.

w	Rule	t
4		
8		
16		
18		
20		

7   The speed v (in feet per second) of an object after t seconds is given by

$$v = 10t + 50$$

Complete the table by finding the speed for each time given.

t	Rule	v
0		
1		
5		
10		
25		

8   The distance traveled by a freely falling object at time t may be given by

$$s = 16t^2$$

(a)   Suppose that we know that an object has fallen 1600 feet. How long has it been falling?

(b)   Joe wanted to use a barometer to measure the height of a building. So, he went to the roof of the building, dropped the barometer, and measured how long it took to reach the ground. It took 5 seconds. Assuming the barometer fell freely (no wind resistance), how high is the building?

9   The number of pushups N that can be done by a certain group of persons is given by the relationship:

$$N = 100 - 2A$$

where A is person's age in years. Using this relationship, estimate the number of pushups that can be done by a person whose age is

(a)   20 years   (b)   45 years

Is this a realistic rule?

SPECIAL INTEREST FEATURE

Statistics and Human Growth[7]

The relationship between a person's height and weight depends upon many factors, including race, nationality, sex, nutrition, age and so on. However, an equation has been found that relates height and weight for children and youth in several countries. The equation is:

$$\log w = .02h + .76$$

where: w is in pounds
  h is in inches and
  $\log w$ is the logarithm of w.

(*Note:* Logarithms are expressed in terms of powers of a base. So, for example, $\log 100 = 2$, since $10^2 = 10 \times 10 = 100$. Assume here that we are using base 10.)

The conditions under which this relationship holds well are listed below.

SUMMARY OF CONDITIONS UNDER WHICH THE HEIGHT/WEIGHT RELATIONSHIP ($\log w = .02h + .76$) HAS SO FAR BEEN FOUND TO HOLD

Race:	White, Black, Chinese (in the West Indies)
Countries:	United Kingdom, Ghana, Katanga, West Indies, France, Canada
Time:	1880–1970 approximately
Age:	2–18 years
Sex:	Male (2–18 years) and Female (2–13 years)
Socio-Economic Class:	Various in United Kingdom, France, and Canada

The investigation of conditions is still continuing. However, the conditions are so varied that some have already called the equation

$$\log w = .02h + .76$$

a "lawlike relationship" for human growth patterns.

Notice that the United States is not referred to in this table. Would you expect the relationship to hold for the United States as well?

7.3   GRAPHS OF RELATIONSHIPS

We can draw pictures of relationships by using the idea of a map of a city. Suppose we want to draw a picture (a graph) of the relationship between two variables, X and Y.

We can think of the values of X as streets— First Street, Second Street, and so on—and the values of Y as avenues—First Avenue, Second Avenue and so forth (see Figure 7.3). Notice that Zero Street and Zero Avenue are marked in heavy lines. The intersection of Zero Street and Zero Avenue, written as $(0,0)$ is called the *origin*.

In the map in Figure 7.3, Bill is standing at the corner of First Street and Third Avenue. We write this point as $(1,3)$ and show it on the map using the square dot. Mary is at Second Street and First Avenue. Her location is written as $(2,1)$ and is shown on the map as a round dot.

FIGURE 7.3

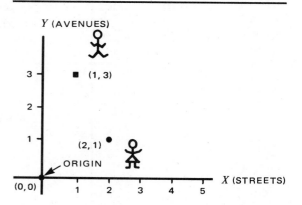

Now we show each of the values in the table for X and Y in Figure 7.4.

Notice that the values for X and Y lie in a straight line. Such a relationship between two variables X and Y is called a *linear relationship* because the graph of the relationship is a straight line.

FIGURE 7.4

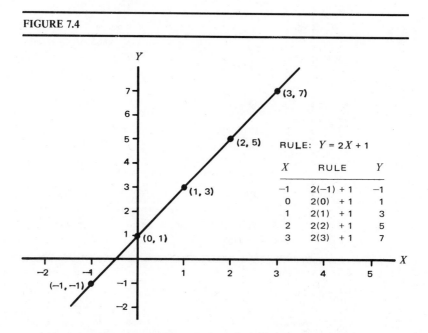

RULE: $Y = 2X + 1$

X	RULE	Y
−1	2(−1) + 1	−1
0	2(0) + 1	1
1	2(1) + 1	3
2	2(2) + 1	5
3	2(3) + 1	7

EXERCISES

1   Suppose the rule relating two quantities is: $Y = X + 2$. Complete this table, and draw a graph of the values of X and Y.

X	Rule $Y = X + 2$		Y
−2	$Y = X + 2$	$= -2 + 2$	0
−1	$Y = X + 2$	$= -1 + 2$	1
1			
2			
5			

2   The relationship between Fahrenheit and Celsius temperature is

$$C = \frac{5}{9}(F - 32)$$

Complete this table of temperatures and draw the graph of your results.

F	Rule $C = \frac{5}{9}(F - 32)$	C
212		
32		
100		
0		
−18		

3   A graph between two relationships can be very useful. For example, the graph in Figure 7.5 can be used to estimate the number of gallons needed to fill an automobile's gas tank for various readings on the fuel gauge. Use the graph to find out about how many gallons will fill the tank for readings of

(a) $\dfrac{3}{4}$ (b) $\dfrac{1}{2}$

(c) $\dfrac{1}{4}$ (d) $\dfrac{1}{8}$

FIGURE 7.5

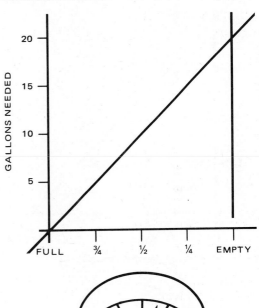

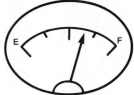

4   The graph in Figure 7.6 can be used for a certain automobile to estimate how many miles can be driven before it runs out of gas, for a given reading on the gas gauge. According to this rule, how far will the auto run when the gauge reads

(a) $\dfrac{1}{2}$

(b) $\dfrac{3}{8}$

(c) $\dfrac{1}{4}$

(d) $\dfrac{1}{8}$

FIGURE 7.6

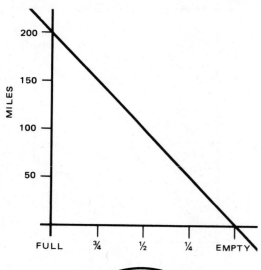

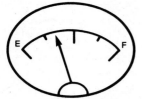

5   The following rule has been found to be useful in predicting the weight of 5- to 13-year-old Ghanaian boys, given their height, in inches.

$$w = 3h - 90$$

where h = height in inches and w = weight in pounds.

Find the predicted weight of boys having heights as shown in the table.

		Rule		
h		$w = 3h - 90$		w
43				
50				
55				
60				

6   Draw a graph of the prediction rule given in Exercise 5 of this section. Label the axes and title the graph.

7   The following rule in physics gives the velocity, v, of a free-falling body, given its time of fall, t, and an initial velocity of 30 meters per second:

$$v = 10t + 30$$

where t is expressed in seconds and v in meters.

Find the displacement of the falling body after the times indicated in the table, and draw a graph of the rule. Label the axes and provide a table for the graph.

t		$v = 10t + 30$		v
0				
5				
10				
15				
20				

8   The following table is used by paper-hangers to estimate the number of rolls, R, of wallpaper required to do a room, given the distance, P (in feet), around the room. (This table assumes a 9-foot ceiling height.) What rule gives R when P is known?

Distance Around Room P	Number of Rolls of Wallpaper R
30	9
40	12
50	15
60	18
70	21
80	24
90	27

7.4   THE EQUATION OF A LINE

The relationship between two variables can be described by a rule. Another way of thinking about the relationship between two variables is in terms of an equation. For example,

$$Y = 2X + 3$$

is the equation of the straight line in Figure 7.7. Several data points on this line are given in Table 7.10.

TABLE 7.10

X	EQUATION $Y = 2X + 3$	Y
−2	$Y = 2(-2) + 3$	−1
0	$Y = 2(0) \ \ + 3$	3
1	$Y = 2(1) \ \ + 3$	5
2	$Y = 2(2) \ \ + 3$	7
3	$Y = 2(3) \ \ + 3$	9

FIGURE 7.7

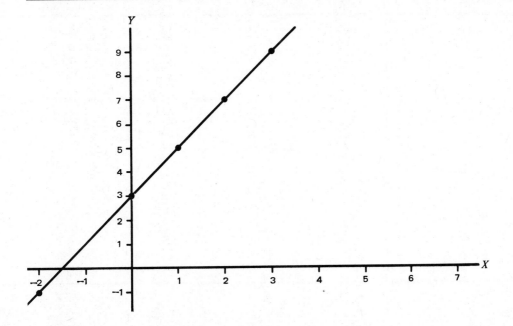

Let's compare the straight line we have just drawn with that of Table 7.11, which has equation

$$Y_2 = \frac{1}{4}X_2 + 5$$

and the graph shown in Figure 7.8.

If we draw the two lines on the same graph paper, we can compare them more easily (Figure 7.9).

(The small 2s on the X and Y in the second equation are called *subscripts*. They remind us that there are two different equations being considered here.)

TABLE 7.11

	EQUATION	
X_2	$Y_2 = \frac{1}{4}X_2 + 5$	Y_2
−4	$Y = \frac{1}{4}(-4) + 5$	4
0	$Y = \frac{1}{4}(0) + 5$	5
1	$Y = \frac{1}{4}(1) + 5$	5.25
2	$Y = \frac{1}{4}(2) + 5$	5.5
3	$Y = \frac{1}{4}(3) + 5$	5.75
4	$Y = \frac{1}{4}(4) + 5$	6.0
8	$Y = \frac{1}{4}(8) + 5$	7.0

FIGURE 7.8

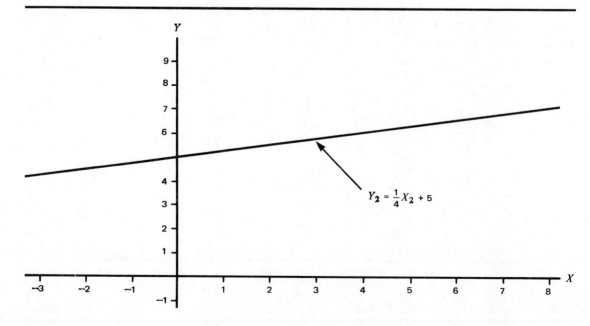

$$Y_2 = \frac{1}{4}X_2 + 5$$

FIGURE 7.9

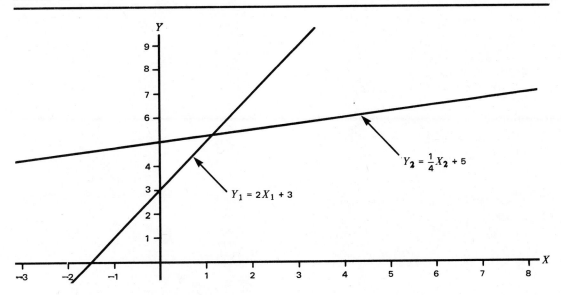

Notice that in Figure 7.9 the first line

$$Y_1 = 2X_1 + 3$$

is steeper than the second. We say that the first line has greater *slope* than the second. For the first line, as X increases by 1, Y increases by 2. This is the *rate of increase* of Y with respect to X. If we take two points on the first line, we can see, using the points $(3, 9)$ and $(0, 3)$, that

$$\text{Slope}_1 = \frac{\text{amount of change in } Y_1}{\text{amount of change in } X_1} = \frac{9-3}{3-0} = \frac{6}{3} = 2.0$$

The second line, on the other hand, shows little change in Y as X increases. Consider the points $(4, 6)$ and $(0, 5)$.

$$\text{Slope}_2 = \frac{\text{amount of change in } Y_2}{\text{amount of change in } X_2} = \frac{6-5}{4-0} = \frac{1}{4}$$

In each case, you can see that the slope is included in the equation of the line when it is written in the form

$$Y = mX + C$$

The slope is simply the coefficient of X. We can tell another thing about the two lines. The first line crosses the Y-axis at 3. The second line crosses higher up, at 5. The point at which a line crosses the Y-axis is called its *Y-intercept*. Both of these values can be read from the equation of a line when it is written in the form

$$Y = mX + C$$

Notice that line 1 has the equation

$$Y_1 = 2X_1 + 3$$

and line 2 has the equation

$$Y_2 = \frac{1}{4}X_2 + 5$$

The Y-intercept for a line is just the value of C when the equation is written in the form

$$Y = mX + C$$

Since this form of the equation of a line is so important, it is given a special name: *the slope-intercept form of the equation of a line.* Sometimes it is necessary to change the equation of a line so that it appears in this form, which most easily tells us the slope and intercept of the line.

- **Example 1:** A line has the equation

$$2Y = 3(X + \frac{2}{3})$$

What is its slope? What is its Y-intercept?

Solution:

Change the equation to the slope-intercept form.

$$2Y = 3(X + \frac{2}{3})$$

$2Y = 3X + 2$ (multiply through by 3 on right)

$Y = 1.5X + 1$ (divide both sides by 2)

So, $m = 1.5$ and $C = 1$. The line has a slope of 1.5 and crosses the Y-axis at 1 (see Figure 7.10). ∎

TABLE 7.12

X	Y
−1	−.5
0	1
1	2.5
3	5.5

FIGURE 7.10

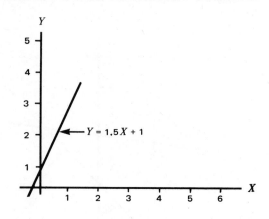

NEGATIVE SLOPE

In most of our examples so far, the lines have been uphill. That is, if a person were to move from left to right, it would be necessary to climb up the line. But lines may be downhill, as in Figure 7.11.

TABLE 7.13

X	Y
0	6
1	5.5
2	5
4	4
3	4.5
6	3

This line has the equation

$$Y = -\frac{1}{2}X + 6$$

FIGURE 7.11

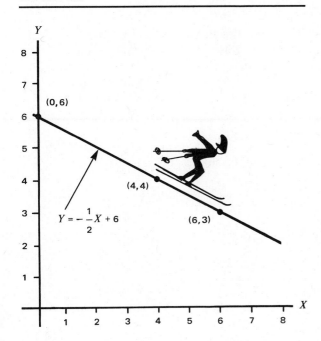

As X gets larger, Y gets smaller. To use the formula for finding the slope of the line, take two points on the line, $(6,3)$ and $(0,6)$.

$$\text{Slope} = \frac{\text{change in } Y}{\text{change in } X} = \frac{3-6}{6-0} = \frac{-3}{6} = -\frac{1}{2} = -.5$$

The Y-intercept is 6. Notice the graph crosses the Y-axis where $Y = 6$.

■ **Example 2:**   A line has a slope of $-.75$ and Y-intercept of 2. Find its equation.

Solution:

The form is $Y = mX + C$. So, the equation is

$$Y = -.75X + 2 \;\blacksquare$$

EXERCISES

1   What is the slope of each line whose equation is given below?

(a) $Y = 3X + 2$ (b) $Y = 2X + 3$

(c) $Y = 5X - 3$ (d) $Y = \frac{1}{2}X + 6$

(e) $Y = \frac{1}{4}X - 1$ (f) $Y = X + 1$

(g) $Y = X$ (h) $Y = 2X$

(i) $Y = -3X + 2$ (j) $Y = -4X - 3$

(k) $Y = -\frac{1}{2}X$

2   For what value of Y does each of the lines in Exercise 1 of this section cross the Y-axis? (That is, what is the Y-intercept for each line?)

3   Write each of these equations in the slope-intercept form. Then tell the slope and Y-intercept of each line.

(a) $Y = 2(X + 3)$ (b) $Y = 4 + 3X$

(c) $Y = -3(X + 3)$ (d) $Y = -\frac{1}{2}(X - 4)$

(e) $Y = 6X + 9$ (f) $3X = 6Y$

(g) $Y - X = 5$ (h) $Y - 2X = 7$

(i) $5Y + 2X = 10$ (j) $\frac{1}{2}Y + \frac{1}{4}X = 6$

(k) $X = 2Y - 4$

4   Complete the table of values for each of these equations.

(a) $Y = 3X + 2$ (b) $Y = 5X - 3$

(c) $Y = X$ (d) $Y = -\dfrac{1}{2}X$

X	Equation	Y
−1		
0		
1		
2		
3		
4		
5		

5   Draw each line whose equation is given in Exercise 4 of this section.

6   Find the slope and Y-intercept of each line in Exercise 4 of this section, using the graph drawn in Exercise 5, above. Compare your answers with those found by reading the values from the slope-intercept form of the equation.

7   Complete the table of values for each of these equations.

(a) $Y = 2(X + 3)$ (b) $Y = -\dfrac{1}{2}(X - 4)$

(c) $Y - X = 5$ (d) $X = 2Y$

X	Equation	Y
−2		
−1		
0		
1		
2		
3		
4		
5		

8   Draw each line whose equation is given in Exercise 7 of this section.

9   Find the slope and Y-intercept of each line in Exercise 7 of this section, using the graph drawn in Exercise 8, above. Compare your answers with those found by first changing the equation to the slope-intercept form.

10   Write the equation of each line, given the following slopes and Y-intercepts.

	Slope	Y-intercept
(a)	2	1
(b)	−2	3
(c)	$\dfrac{1}{4}$	3
(d)	3	0
(e)	1	1
(f)	1	0
(g)	0	2

11   Figure 7.12 shows several lines. Find their slopes and intercepts from the drawings. Then write an equation for each line.

FIGURE 7.12

a)   LINE 1

FIGURE 7.12 (continued)

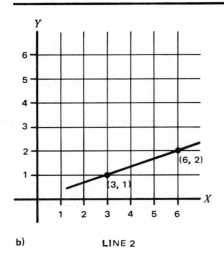

b)   LINE 2

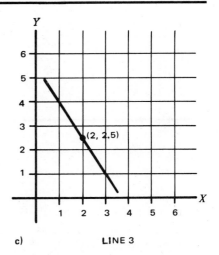

c)   LINE 3

7.5   FITTING RULES TO DATA

Many of our examples so far have used data that follow a rule exactly. However, in the real world, data are not so well behaved. Errors of measurement, for example, can help produce data which do not exactly obey a given rule. In this section, we will see a way for finding rules for relating real world data.

The following data were obtained from an experiment that you could easily do yourself. A statistics class collected cylindrical objects like coffee cans, juice cans, and a rolled oats container. They then measured the circumference (distance around) and the diameter (distance across) of each object and prepared a table like Table 7.14.

The class next prepared a graph to show how the data are related, as shown in Figure 7.13.

TABLE 7.14

OBJECT	DIAMETER (CENTIMETERS)	CIRCUMFERENCE (CENTIMETERS)
Orange juice can	3	10
Coffee can (small)	5	16
Tomato juice can	10.8	32.5
Coffee can (large)	13	40
Rolled oats container	10	32.3
Soup can	6.8	21
Candle	4.5	18

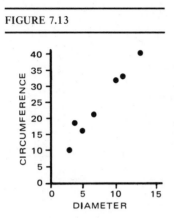

FIGURE 7.13

They looked for a rule which would fit the data they found. Since they knew that a straight line has an equation which looks like

$$Y = mX + C$$

they tried drawing a line through the data in such a way that it would pass through as many of the points as possible. Once they found such a line, they could write its equation, since they could find its slope and intercept (approximately) from the graph.

Many of the students had ideas how to do this. We will discuss a few different approaches in various parts of this book. But we report here only one of the simple suggestions. Barb took a piece of thread and held it taut over the data points in such a way that it was possible to see where the "best" line should be drawn. When the thread appeared to pass over, or closest to, the greatest number of points, she held the thread so that Ron could use it as a guide to draw in that best straight line. Figure 7.14 shows how they did it.

Once the line was drawn in, they estimated the slope of the line and the intercept.

Here are the values they found, using the two points (3, 10) and (13, 40):

Change in $Y = 40 - 10 = 30$ units

Change in $X = 13 - 3 = 10$ units

Slope of best line $= \dfrac{30}{10} = 3.0$

Intercept $= 0$

The intercept is 0 because the line goes through the origin.

From this experiment, they were able to produce the following rule, which relates the circumference of a circular object to its diameter.

$$Y = 3.0X + 0$$

or

Circumference $= 3.0 \times$ diameter

You may know from other mathematics courses that the circumference of a circle is pi times its diameter, and the value of pi is approximately 3.1416. Therefore, this experiment has provided a way of estimating the value of pi.

However, you also may have noticed that the experiment of the class did not produce perfect results. Some of the points were rather far from the line that Barb found. A major reason for this is the difficulty of measuring accurately around and across cylindrical objects. You may think of other reasons why some of the points are away from the line. However, most of the points are rather close to the line, and in this case we were able to find a line that fits the data fairly well.

In order to show that the rule found is intended as a fit to the data and that some (or many, or all) points may not fall exactly on the line drawn, the equation of a line that

FIGURE 7.14

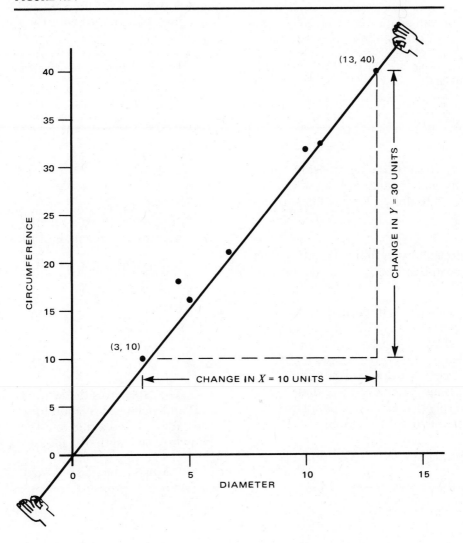

is fit to data is written using Y' as:

$$Y' = mX + C$$

The prime on the Y shows that the Y-value is estimated by the rule that was fit to the data. But the value of Y' may not be the same as the value of Y. Notice that for $X = 5$ centimeters, we have:

$$Y' = 3.0 \times 1.0$$
$$= 3.0(5) = 15.0 \text{ centimeters}$$

In the class experiment, however, a can with a diameter of 5 centimeters was found to have a circumference of 16.0 centimeters. The error of estimation in this case is

$$Y - Y' = 16.0 - 15.0 = 1.0 \text{ centimeter}$$

EXERCISES

1   The following data were collected at a slot-racing track on a slot car going full speed.

Racing Time (Minutes)	Distance Traveled (Meters)
2	520
3	770
4	1050
6	1560

Draw a graph to show the relationship between time and distance. Then use the thread method to draw a line that best fits the data.

2   What rule did you find for relating time and distance in Exercise 1 of this section?

3   A psychologist is interested in the relationship between athletic ability and popularity. This table gives physical education scores (X) and popularity ratings (Y) of 10 high school boys. (A rating of 10 means "very popular.") Draw a graph of these data and then find a rule that estimates popularity (Y') when athletic ability (X) is given.

Athletic Ability (X)	Popularity (Y)
8	7
4	5
7	8
7	6
3	4
5	8
9	8
7	10
6	6
8	7

4   This table records the time required to cook a turkey. Plot the data and find a rule that gives the cooking time for a turkey for a given weight.

Weight (Pounds)	Cooking Time (Hours)
5	3.5
7	4.0
10	4.5
14	5.5
18	6.75
22	8.5

5   Taxi fares recorded for 5 trips are given. Plot the data and find a rule that gives the fare in terms of the length of the trip in miles.

Length of Trip (Miles)	Total Fare
2	1.20
3	1.30
4.5	1.45
7	1.80
10	2.00

6   The table gives the length (from nose to tail) and weight of 8 laboratory mice. Plot the data and find a rule to estimate weight knowing the length of a mouse.

Length (Centimeters) X	Weight (Grams) Y
16	32
15	26
20	40
13	27
15	30
17	38
16	34
21	43

7   These data are final exam scores obtained by 12 students in their senior year of high school mathematics and their first year of college mathematics. Plot the data. Then use the thread method to find a rule for predicting college mathematics scores from high school mathematics scores.

High School Final Exam Score	First Year of College Final Exam Score
13	23
27	28
18	29
17	27
21	29
26	26
28	31
19	20
23	19
7	18
21	26
19	30

7.6   HOW GOOD IS THE RULE?

In the last section you saw that different rules for relating data can be obtained from the same set of data. We now will give a few steps for finding rules and show how the "goodness" of rules may be compared.

Table 7.15 is a height-weight chart based on data collected on boys from 5 to 13 years of age in Birmingham, England.

TABLE 7.15

HEIGHTS (INCHES)	WEIGHT (POUNDS)
43	42
46	46
48	51
50	57
52	62
54	68
55	73
58	82
59	88
Totals 465	569

$$\text{Mean height} = \frac{465}{9} \doteq 51.7 \text{ inches}$$

$$\text{Mean weight} = \frac{569}{9} = 63.2 \text{ pounds}$$

The data are plotted on the graph in Figure 7.15. Notice that we show only that part of the graph that we are using.

FIGURE 7.15

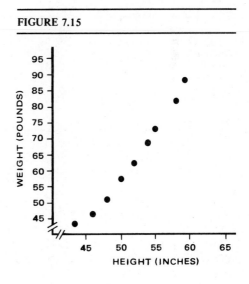

To find a rule, use these steps.

Step 1:   We assume that whatever rule we use will give an average (mean) weight for an average (mean) height. That is, the line we draw will pass through the point represented by the means of the X- and Y-values. We call this the *mean data point*. The graph in Figure 7.16 shows this point.

FIGURE 7.16

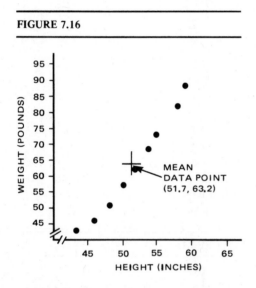

Draw in the mean data point, (51.7, 63.2), as shown on the graph.

Step 2:   Using a thread, a ruler, or just eye estimating, draw in the line through the mean data point so that it passes through or close to as many of the data points as possible.

We used a straightedge to draw in the line shown. Notice that since the origin is not given on this graph, we cannot read the value of the Y-intercept from the graph. This value will have to be calculated (see Step 4).

FIGURE 7.17

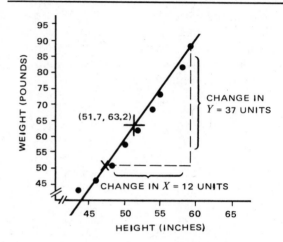

Step 3:   Find the slope of this line from the graph. We need to know the change in Y and corresponding change in X. Choose parts of the graph that are easiest to count. We chose the part of the graph as follows: Upper right point is (59, 88) and, by estimation, lower left point is (47, 51).

$$\text{Slope} = \frac{88 - 51}{59 - 47} = \frac{37}{12} \div 3.08$$

If other parts of the graph had been chosen, slightly different values of the slope may have been found.

Step 4:   Use the slope-intercept form of the equation of a line to find the Y-intercept. The equation has the form

$$Y' = mX + C$$

But since we require a line to go through the mean data point, we have

$$\bar{Y} = m\bar{X} + C$$

And, we know

$$\bar{X} = 51.7$$
$$\bar{Y} = 63.2$$
$$m = 3.08$$

So,

$$63.2 = 3.08(51.7) + C$$

or

$$63.2 = 159.3 + C$$

Therefore

$$C = -95.9$$

and the equation is

$$Y' = 3.08X - 95.9$$

Step 5:   Find how closely the line fits the data by adding up the differences between Y and Y' (the deviation scores for Y) for the rule, ignoring signs. Then find the mean (average) of these estimation errors.

We use the rule in Step 4

$$Y' = 3.08X - 95.9$$

in Table 7.16. From Table 7.16, the mean estimation error is 1.75.

Our five steps give us a way of comparing rules. We find how close, on the average, each rule is to the data points themselves.

The mean error is mean (average) distance of points from the line.

TABLE 7.16
DATA TABLE: HEIGHTS AND WEIGHTS OF BIRMINGHAM BOYS

Estimation rule for this table: $Y' = 3.08X - 95.9$

HEIGHT (X)	WEIGHT (Y)	ESTIMATED (Y')	$Y - Y'$ (ERROR)	$|Y - Y'|$
43	42	36.53	5.47	5.47
46	46	45.77	.23	.23
48	51	51.93	− .93	.93
50	57	58.09	−1.09	1.09
52	62	64.25	−2.25	2.25
54	68	70.41	−2.41	2.41
55	73	73.49	− .49	.49
58	82	82.73	− .73	.73
59	88	85.81	2.91	2.91
465	569			15.79

$$\text{Mean } X = \frac{465}{9} \qquad \text{Mean } Y = \frac{569}{9} \qquad \text{Mean error} = \frac{15.79}{9}$$

$$\doteq 51.7 \qquad\qquad \doteq 63.2 \qquad\qquad \doteq 1.75$$

To illustrate, we use a different rule

$$Y' = 4X - 143.4$$

on the same data and find the mean (average) of the deviations of the estimated scores (Y') from the actual Y-scores (see Table 7.17). We call this quantity the *mean error deviation*. You should turn back to Chapter Two and compare this quantity to the mean deviation.

The ***mean deviation*** is the mean (average) distance of all the scores from the mean for the set of scores. The ***mean error deviation*** is the mean (average) distance of all of the data points, in terms of Y-scores, from the line representing the prediction rule. At times, we will use "mean error" instead of "mean error deviation."

Table 7.17 illustrates mean error deviation for our prediction rule

$$Y' = 4X - 143.4$$

With this new rule, the mean error deviation is 4.84. That is, on the average, the data points are 4.84 units away from the line representing this rule. Recall that for the previous rule ($Y' = 3.08X - 95.9$), the mean error deviation was 1.75. Since the previous rule gave a smaller deviation, we can say that it fits the data better than does the rule $Y' = 4X - 143.4$.

TABLE 7.17
DATA TABLE: SAME HEIGHT AND WEIGHT DATA AS TABLE 7.16

New estimation rule: $Y' = 4X - 143.4$

HEIGHT (X)	WEIGHT (Y)	ESTIMATED (Y')	$Y - Y'$	$|Y - Y|$
43	42	28.6	13.4	13.4
46	46	40.6	5.4	5.4
48	51	48.6	2.4	2.4
50	57	56.6	.4	.4
52	62	64.6	−2.6	2.6
54	68	72.6	−4.6	4.6
55	73	76.6	−3.6	3.6
58	82	88.6	−6.6	6.6
59	88	92.6	−4.6	4.6
465	569			43.6

$$\text{Mean } X = \frac{465}{9} \qquad \text{Mean } Y = \frac{569}{9} \qquad \text{Mean error} = \frac{43.6}{9}$$

$$= 51.7 \qquad\qquad = 63.2 \qquad\qquad \doteq 4.84$$

EXERCISES

Use the five steps of this section to find the equation of a line that fits the data referred to in Exercises 1 through 3 which follow.

1   The data in Exercise 1 of Section 7.5 (the slot car data).

2   The data in Exercise 3 of Section 7.5 (the athlete's popularity data).

3   The data in Exercise 4 of Section 7.5 (the turkey data).

4   Find the mean error deviation for the line you found in Exercise 1 of this section.

5   Find the mean error deviation for the line you found in Exercise 2 of this section.

6   Find the mean error deviation for the line you found in Exercise 3 of this section.

7   Refer to the height-weight data for the Birmingham boys in Table 7.16. Use the rule

$$Y' = 2X - 40.2$$

to relate weight (Y) and height (X). Find the mean error deviation for this rule. In terms of mean error deviation, which rule is better; this one or the rules considered in Section 7.6? That is,

$$Y' = 3.08X - 95.9$$

or

$$Y' = 4X - 143.4$$

8   Find the equation of the line best fitting the cricket data for the key problem using the method of this section.

9   Use the method of this section to find the best-fitting line for the laboratory mice data of Exercise 6 of Section 7.5.

10   Use the method of this section to find the best-fitting line for the mathematics exam data of Exercise 7 of Section 7.5.

7.7   THE MEDIAN FIT

This section gives a more precise way to find a rule that fits a set of data. It requires you to know how to find the median of a set of data. You may want to review medians in Chapter Two.

■ **Example—Crickets, Anyone?**   This is the key problem for this chapter. We are searching for a rule that relates cricket chirps per second to air temperature.

Solution:

We first draw a graph of the data (Figure 7.18). You should *always* draw a graph of data if you are looking for patterns or relationships. Professional statisticians usually draw graphs of their data.

FIGURE 7.18

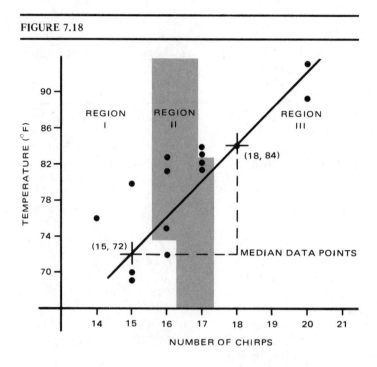

Notice that drawing the graph has helped order the data. In particular, you see the chirps per second in increasing order from left to right along the horizontal axis. Also, the corresponding temperatures are ordered along the vertical axis. Table 7.18 shows these data ordered in this way.

The data on the graph are then divided into three groups, which are roughly the same size. We have divided our set of 15 data points into three groups of five. How convenient! If the number of data points is not divisible by three, you should make the groups as equal as possible and have the same *odd* number of data points in the top group and the bottom group. (The reason for an odd number is that you are going to find medians. And you may remember that with an odd number, the median of the data is easier to find.)

TABLE 7.18

CHIRPS PER SECOND	TEMPERATURE (°F)
14	76
15	80
15	69
15	70
16	72
16	75
16	81
16	83
17	81
17	82
17	83
17	84
18	84
20	89
20	93

Figure 7.18 shows the data divided into the three groups, labeled I, II, and III.

Now look at part I of the graph. We find the median of the data values for the horizontal axis (in this case, the cricket chirps) for these five data points. The five values are:

14, 15, 15, 15, 16

and their median is 15. This value is marked on the graph by a vertical line. We then find the median (midway point) of the corresponding values on the vertical axis (in this case, the air temperatures). The five values are:

76, 80, 69, 70, 72

Reordered, these give

69, 70, 72, 76, 80

The median is 72. This value is marked by a horizontal line as shown, to intersect the vertical line drawn previously.

So, for part I of this graph, the *median data point* (15,72) is determined by the intersecting lines in part I of the graph. Notice that the median data point is not a data point in the real data of our example. This is usually the case with real data.

In part III, we repeat the process and find the median data point for these five data points.

For the horizontal axis, the data values are:

17, 17, 18, 20, 20

with a median of 18. Draw a vertical line at 18 in part III. For the vertical axis, the data values are:

83, 84, 84, 89, 93

and the median is 84. Draw the horizontal line at 84. The intersecting lines mark the median data point (18,84).

Now draw a straight line through the two median data points (15,72) and (18,84). This line determines the rule that relates cricket chirps and air temperature by the Method of *Median Fit*.

The formula for this rule is found by finding the equation of a line which contains two given points.

First, find the slope of the line through the two points.

$$\text{Slope} = \frac{\text{change in vertical scale values } (Y)}{\text{change in horizontal scale values } (X)}$$

The horizontal scale, air temperature, changes in value a total of $84 - 72 = 12$ degrees between the two median data points.

The vertical scale, cricket chirps, changes $18 - 15 = 3$ chirps per second. So,

$$\text{Slope} = \frac{84 - 72}{18 - 15} = \frac{12}{3} = 4$$

We know the rule has the form

$$Y' = mX + C$$

To find C, we provide the coordinates of a point known to satisfy the rule. One such point is the median data point in part I of the graph, $(15, 72)$. So, in the equation

$$Y = mX + C$$

we let $X = 15$ and
$Y = 72$.

$$72 = m15 + C$$

But we already know that $m = 4$. So,

$$72 = 4(15) + C$$

or

$$72 = 60 + C$$

Solving for C, we have

$$C = 72 - 60 = 12$$

Thus, the Method of Median Fit gives the following rule for the cricket data:

$$Y' = 4X + 12$$

Let's try this rule for a couple of the given data points to see how well it works (Table 7.19). The median fit has an important advantage in that it is quick and easy to find.

TABLE 7.19

X CHIRPS PER SECOND	Y ACTUAL TEMPERATURE	RULE	Y' ESTIMATED TEMPERATURE	Y − Y' ERROR
14	76	4(14) + 12 = 68	68	76 − 68 = 4
15	80	4(15) + 12 = 72	72	80 − 72 = 8

EXERCISES

1   Complete Table 7.19 to find the estimated temperature for all of the cricket data points (given in the key problem), using the rule

$$Y' = 4X + 12$$

2   Refer to the wallpaper data in Exercise 8 of Section 7.3. Use the Method of Median Fit to find a rule that fits those data.

3   Refer to the data in Table 7.14. Use the Method of Median Fit to find a rule to fit those data.

4   Refer to Table 7.12. Use the Method of Median Fit to find a rule to fit those data. Prepare a table like the one below to estimate weights of boys, given their heights. Is the rule determined by the Method of Median Fit better than the one found following the steps in Section 7.6? (Find the mean error for this rule.)

Height	Weight	Estimated Weight	Error
___	___	_____	___
___	___	_____	___
___	___	_____	___

5   Refer to Exercise 7 of Section 7.5. Use the Method of Median Fit to estimate college mathematics scores given high school mathematics scores.

Regression and Correlation

KEY PROBLEM

Radiation and the Environment

There is a lot of concern these days about the possible harmful effects on our health from pollution due to nuclear reactors. In one such study, data were gathered on the amount of exposure of nine Oregon counties to radioactive waste from the Atomic Energy Commission's plant in Hanford, Washington, and the rate of deaths due to cancer in those counties. The data are given in Table 8.1.[1]

TABLE 8.1

COUNTY	INDEX OF EXPOSURE	CANCER MORTALITY (PER 1,000,000 PERSONS)
Unatilla	2.49	147.1
Morrow	2.57	130.1
Gilliam	3.41	129.9
Wasco	1.25	113.5
Hood River	3.83	162.3
Portland	11.64	207.5
Columbia	6.41	177.9
Clatsop	8.34	210.3

In this chapter, we'll learn a method of finding the relationship between the degree of exposure to radioactive substances in these counties and the cancer mortality rate. The method will also enable us to estimate the mortality rate for any other county with a given degree of radioactive exposure.

8.1   LINEAR REGRESSION

We have already used the *idea* of linear regression in Chapter Seven. Now we will use the name that was given to this idea by statisticians almost 100 years ago. *Linear* means a straight line. *Regression* means going back or returning.

The advertisement in Figure 8.1 provides data designed to show a relationship between time spent in the slenderizing program and the weight loss of nine persons. The information is summarized in Table 8.2. (Weight loss is rounded to the nearest whole number.)

TABLE 8.2

WEEKS ATTENDED PROGRAM (X)	TOTAL WEIGHT LOSS (Y)
12	19
4	18
8	8
4	16
12	26
7	12
5	12
9	29
14	31
Total 75	Total 171
Mean (X) = 8.33	Mean (Y) = 19.0

FIGURE 8.1 [2]

The data are plotted in Figure 8.2. To find a relationship between time spent in the slenderizing program and amount of weight loss, we try to fit a line to these data by following the five steps outlined in Section 7.6. Finding such a straight line is called the Method of Linear Regression. The equation of the line which we are looking for is called the *regression equation.*

As a review of Chapter Seven, but using our new terms, let's find a line that fits the weight-loss data.

Step 1:   Since the regression line will pass through the mean data point, find $\overline{X}$ and $\overline{Y}$ and draw in $(\overline{X}, \overline{Y})$, as in Figure 8.3.

FIGURE 8.2

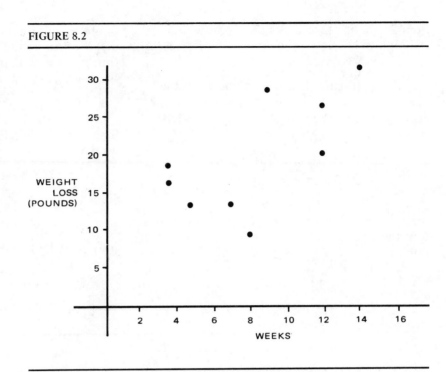

FIGURE 8.3

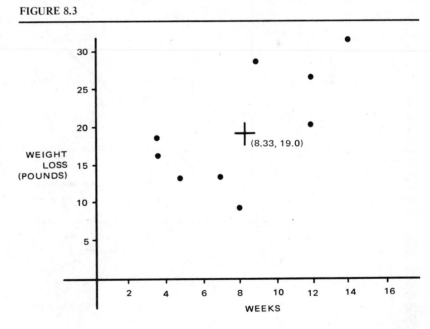

Step 2: Using thread as a guide, draw in line that appears to best fit the data (Figure 8.4).

Step 3: Estimate the slope of line from the graph. (If graph paper is used, this will help.) See Figure 8.4. The points (14, 31) and (4, 8) are chosen.

$$m = \text{slope} = \frac{\text{change in } Y}{\text{change in } X} = \frac{31-8}{14-4} = \frac{23}{10} = 2.3$$

Step 4: Find the Y-intercept by solving for C in the equation

$$\bar{Y} = m\bar{X} + C$$

Suppose that the line of best fit has a slope of 2.3. We require the regression line to pass through the mean data point, so we can find the Y-intercept as follows.

$$\bar{Y} = m\bar{X} + C$$

Since $\bar{X} = 8.33$ and $\bar{Y} = 19.0$, and $m = 2.3$, we have

$$19 = 2.3(8.33) + C$$

So,

$$C = 19 - 19.16 = -.17$$

and the regression equation is

$$Y' = 2.3X - .17$$

FIGURE 8.4

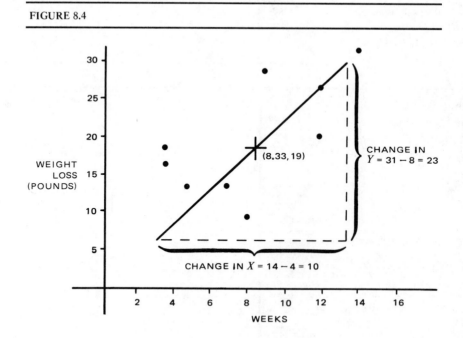

Step 5:   Calculate the mean error, as shown below.

We now use the equation

$$Y' = 2.3X - .17$$

to estimate the weight loss for each of the persons, based on how long each spent in the slenderizing program. See Figure 8.5. For example, M. Andolsek spent 4 weeks in the program. So, Andolsek's estimated weight loss is

$$Y' = 2.3(4) - .17 = 9.2 - .17 = 9.03$$

From the data table, we see that Andolsek's actual weight loss was 18 pounds. So, the prediction error for Andolsek was

$$Y - Y' = 18 - 9.03 = 8.97 \text{ pounds}$$

The prediction error is a positive number. In Figure 8-5 you can see that the point representing Andolsek is about 9 points *above* the regression line. Andolsek lost 8.97 pounds *more* than was predicted.

Now we find the prediction error for Price, who was in the program for 8 weeks.

Price's predicted weight loss is

$$Y' = 2.3(8) - .17 = 18.4 - .17 = 18.23 \text{ pounds}$$

The prediction error is

$$Y - Y' = 8 - 18.23 = -10.23 \text{ pounds}$$

Price was predicted to lose 18.23 pounds, and actually lost only 8 pounds—about 10 pounds *less* than predicted (see Figure 8.5).

FIGURE 8.5

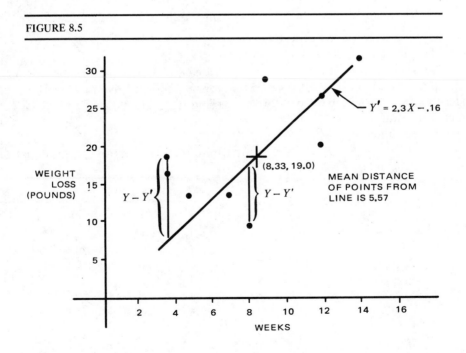

You should check Table 8.3 and see how Y' was calculated and interpret $Y - Y'$ in each case.

TABLE 8.3

Regression Equation: $Y' = 2.3X - .17$

X	Y	Y'	$Y - Y'$	$|Y - Y'|$
12	19	27.43	−8.43	8.43
4	18	9.03	8.97	8.97
8	8	18.23	−10.23	10.23
4	16	9.03	6.97	6.97
12	26	27.43	−1.43	1.43
7	12	15.93	−3.93	3.93
5	12	11.33	.67	.67
9	29	20.53	8.47	8.47
14	31	32.03	−1.03	1.03
				50.13

$$\text{Mean error} = \frac{50.13}{9} = 5.57$$

As we saw in Chapter Seven, the fit of a line to data may be described in terms of the *mean error deviation*, or, *mean error*. The mean error is the mean (average) of the absolute values of the estimation errors. [What would be the mean error (within rounding error) if we did not ignore the minus signs?]

In this example, the mean error for the equation

$$Y' = 2.3X - .17$$

is found from Table 8.3 to be 5.57.

EXERCISES

1   Table 8.4 gives the length (from nose to tail) and weights of eight laboratory mice. Plot the data and the mean data point. Draw in your estimated regression line.

TABLE 8.4

LENGTH (CENTIMETERS) X	WEIGHT (GRAMS) Y
16	32
15	26
20	40
13	27
15	30
17	38
16	34
21	43

2   Suppose the regression equation for the mice data in Exercise 1 is

$$Y' = 2X + .5$$

Table 8.5 shows the estimated weight in grams for each mouse, given its length in centimeters, using the regression equation

$$Y' = 2X + .5$$

Find the mean error for this equation.

TABLE 8.5

X	Y	Y'	$Y - Y'$
16	32	32.5	− .5
15	26	30.5	−4.5
20	40	40.5	− .5
13	27	26.5	.5
15	30	30.5	− .5
17	38	34.5	3.5
16	34	32.5	1.5
21	43	42.5	.5

3   Suppose the regression line for the weight-loss data (Table 8.2) had a slope of 1.5. Find the equation of this line, assuming that it goes through the mean data point. You should first construct a table of values like the one in Exercise 2 of this section.

4   The estimated weight loss Y' (in pounds) for each person given the amount of time in the program K (in weeks), is given in Table 8.6. If the regression equation

$$Y' = 1.5X + 6.5$$

is used, a table of values is obtained as shown in Table 8.6.

TABLE 8.6

X	Y	Y'	Y − Y'	|Y − Y'|
12	19	24.5	− 5.5	5.5
4	18	12.5	5.5	5.5
8	8	18.5	−10.5	10.5
4	16	12.5	3.5	3.5
12	26	24.5	1.5	1.5
7	12	17.0	− 5.0	5.0
5	12	14.0	− 2.0	2.0
9	29	20.0	9.0	9.0
14	31	27.5	3.5	3.5

What is the mean error for this equation?

5   The data in Table 8.7 are the final examination scores obtained by 12 students in their senior year of high school mathematics and their first year of college mathematics. Plot the data and the mean data point. Draw in your estimated regression line. What is the equation of the line you found?

TABLE 8.7

HIGH SCHOOL FINAL EXAM SCORE X	FIRST-YEAR COLLEGE FINAL EXAM SCORE Y
13	23
27	28
18	29
17	27
21	29
16	26
28	31
19	20
23	19
7	18
21	26
19	30

6   Suppose the regression equation for the final exam data (Exercise 5 of this section) is

$$Y' = .4X + 17.7$$

Use this equation to estimate a student's score on the final examination in first-year college mathematics, given that student's score on the final exam in the senior year of high school mathematics. Prepare a table like the one in Exercise 4 of this section. What is the mean error for this equation?

7   Suppose the regression equation for the cricket data (key problem, Chapter Seven) is

$$Y' = 4X + 12$$

Find the mean error for this equation.

8.2 LEAST MEAN ERROR

The mean error for a regression line tells us how far, on the average, the data points are from that line. For example, consider the weight-loss data from Section 8.1. We showed there that the regression line

$$Y' = 2.3X - .16$$

produces a mean error of 5.57. That is, on the average, the data points are 5.57 units from this line.

We will now see how well two other regression lines fit these same data, one with a slope of 1.5 and another with a slope of 2.5 (see Figure 8.6).

Tables 8.8 and 8.9 show how we calculated the mean estimation error for each of the two lines. (Check the first row with a calculator.)

TABLE 8.8

Regression Equation: $Y' = 1.5X + 6.5$

X	Y	Y'	Y'	$\|Y - Y'\|$
12	19	24.5	− 5.5	5.5
4	18	12.5	5.5	5.5
8	8	18.5	−10.5	10.5
4	16	12.5	3.5	3.5
12	26	24.5	1.5	1.5
7	12	17.0	− 5.0	5.0
5	12	14.0	− 2.0	2.0
9	29	20.0	9.0	9.0
14	31	27.5	3.5	3.5
				46.0

$$\text{Mean error} = \frac{46.0}{9} = 5.11$$

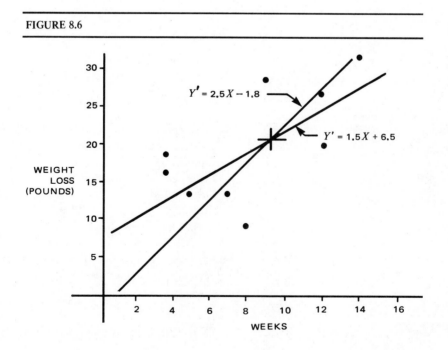

FIGURE 8.6

$Y' = 2.5X - 1.8$

$Y' = 1.5X + 6.5$

WEIGHT LOSS (POUNDS)

WEEKS

TABLE 8.9

Regression Equation: $Y' = 2.5\,Y - 1.83$

X	Y	Y'	Y − Y'	|Y − Y'|
12	19	28.17	−9.17	9.17
4	18	8.17	9.83	9.83
8	8	18.17	−10.17	10.17
4	16	8.17	7.83	7.83
12	26	28.17	− 2.17	2.17
7	12	15.67	− 3.67	3.67
5	12	10.67	1.33	1.33
9	29	20.67	8.33	8.33
14	31	33.17	− 2.17	2.17
				54.67

$$\text{Mean error} = \frac{54.67}{9} = 6.07$$

If we draw a graph that shows the mean error produced by each of the three lines (Table 8.10) it looks like Figure 8.7.

TABLE 8.10
ESTIMATION ERRORS

SLOPE OF REGRESSION LINE	MEAN ERROR
1.5	5.11
2.3	5.57
2.5	6.07

FIGURE 8.7

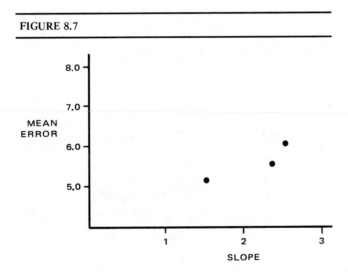

With the help of a computer, we calculate the mean error for lines with several slopes (Table 8.11) and fill in the graph some more (Figure 8.8).

(*Note:* See the program LINREG in Appendix G.)

TABLE 8.11
ESTIMATION ERRORS

SLOPE OF REGRESSION LINE	MEAN ERROR	
1.0	5.26	
1.1	5.23	
1.2	5.20	
1.3	5.17	
1.4	5.14	
1.5	5.11	
1.6	5.08	
1.7	5.05	
1.8	5.02	
1.9	4.99	← Least error
2.0	5.04	
2.3	5.57	
2.5	6.07	
3.0	7.33	

FIGURE 8.8

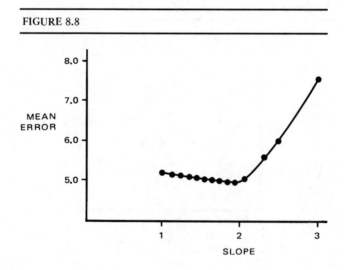

From the graph, it appears as if the mean error reaches its smallest value for a regression line with a slope of about 1.9. With the help of our computer, we can use a "magnifying glass" and calculate the mean error for lines with slopes close to 1.9 (Table 8.12). With these data, the graph can be filled in even more (Figure 8.9).

From Table 8.12, we can state that the slope of the regression line that gives the least mean error is, to the nearest hundredth, 1.91.

Without the help of a computer (or at least a calculator), finding the slope of the best-fitting line can be very time-consuming by this method. If you do not have a calculator or computer, you should concentrate on the meaning of best-fitting line without doing the calculations.

TABLE 8.12
MAGNIFIED ESTIMATION ERRORS

SLOPE OF REGRESSION LINE	MEAN ERROR
1.80	5.02222
1.81	5.01926
1.82	5.01630
1.83	5.01333
1.84	5.01037
1.85	5.00741
1.86	5.00444
1.87	5.00148
1.88	4.99852
1.89	4.99556
1.90	4.99259
1.91	**4.99037** ← Least error
1.92	4.99556
1.93	5.00074
1.94	5.00593
1.95	5.01111
1.96	5.01630
1.97	5.02148
1.98	5.02667
1.99	5.03185
2.00	5.03704

FIGURE 8.9

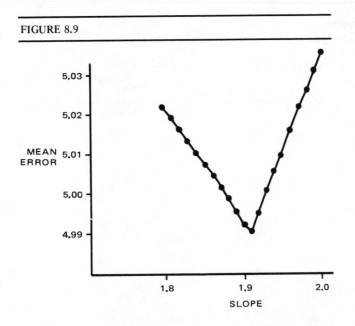

EXERCISES

1   Table 8.13 gives the prediction errors for the mice data of Exercise 2 of Section 8.1, but this time for the equation

$$Y' = X + 17.13$$

TABLE 8.13

		Regression Equation: $Y' = X + 17.13$		
X	Y	Y'	Y − Y'	|Y − Y'|
16	32	33.13	−1.13	1.13
15	26	32.13	−6.13	6.13
20	40	37.13	2.87	2.87
13	27	30.13	−3.13	3.13
15	30	32.13	−2.13	2.13
17	38	34.13	3.87	3.87
16	34	33.13	0.87	0.87
21	43	38.13	4.87	4.87
			Total	25.00

What is the mean error for this equation?

2   Which of the two regression equations for the mice data (Exercise 2 of Section 8.1 and Exercise 1 of this section) is a better fit of the data, in terms of mean error,

$$Y' = 2X + .5 \quad \text{or} \quad Y' = X + 17.13?$$

Why?

3   Refer to the cricket data in the key problem of Chapter Seven. For the regression line

$$Y' = 3X + 30.33$$

Table 8.14 of estimation errors is given. What is the mean error for this equation?

TABLE 8.14

		Regression Equation: $Y' = 3X + 30.33$		
X	Y	Y'	Y − Y'	|Y − Y'|
20	89	90.33	−1.33	1.33
16	72	78.33	−6.33	6.33
20	93	90.33	2.67	2.67
18	84	84.33	− .33	.33
17	81	81.33	− .33	.33
16	75	78.33	−3.33	3.33
15	70	75.33	−5.33	5.33
17	82	81.33	.67	.67
15	69	75.33	−6.33	6.33
16	83	78.33	4.67	4.67
15	80	75.33	4.67	4.67
17	83	81.33	1.67	1.67
16	81	78.33	2.67	2.67
17	84	81.33	2.67	2.67
14	76	72.33	3.67	3.67

4 Table 8.15 gives the estimation error for a line having a slope of 1.9. What is the mean error for this line? Compare this result with the value you get by reading Figure 8.7.

TABLE 8.15

Regression Equation: $Y' = 1.9X + 3.16$

X	Y	Y'	Y − Y'	|Y − Y'|
12	19	25.97	−6.97	6.97
4	18	10.77	7.23	7.23
8	8	18.37	−10.37	10.37
4	16	10.77	5.23	5.23
12	26	25.97	.03	.03
7	12	16.47	−4.47	4.47
5	12	12.67	− .67	.67
9	29	20.27	8.73	8.73
14	31	29.77	1.23	1.23

---

5 Ten students took a spelling test (X) and a reading test (Y) and got the scores in Table 8.16. Use the regression equation

$$Y' = 2X - 3.8$$

to estimate each student's reading score knowing his or her spelling score.

TABLE 8.16

SPELLING X	READING Y
12	20
10	12
10	18
8	10
7	12
6	14
6	6
5	7
4	3
2	1

---

6 Table 8.17 estimates reading scores (Y'), given spelling scores (X), for the ten students in Exercise 5 of this section using the equation

$$Y' = 2.3X - 5.8.$$

What is the mean error for this equation?

TABLE 8.17

X	Y	Y'	Y − Y'	|Y − Y'|
12	20	21.8	−1.8	1.8
10	12	17.2	−5.2	5.2
10	18	17.2	.8	.8
8	10	12.6	−2.6	2.6
7	12	10.3	1.7	1.7
6	14	8.0	6.0	6.0
6	6	8.0	−2.0	2.0
5	7	5.7	1.3	1.3
4	3	3.4	− .4	.4
2	1	−1.2	2.2	2.2

---

7 Table 8.18 gives the mean error for the mice data (Table 8.4) where regression lines having slopes from 1.0 to 3.0 are used. What is the slope of the regression line that best fits the data, according to this table? What is the equation of this best-fitting line?

TABLE 8.18

SLOPE	MEAN ERROR
1.0	3.12
1.2	2.75
1.4	2.37
1.6	2.00
1.8	1.62
2.0	1.50
2.2	1.56
2.4	1.80
2.6	2.07
2.8	2.35
3.0	2.65

8   Using the best-fitting line that you found in Exercise 7 of this section, estimate the weight of mice having these lengths:

(a)   15 cm
(b)   18 cm
(c)   20 cm

9   Table 8.19 summarizes the mean error for the cricket data when regression lines with slopes from 2.5 to 3.5 are used. What is the slope of the regression line that best fits the data, according to this table? Find the equation of this line.

TABLE 8.19

SLOPE	MEAN ERROR
2.5	3.155
2.6	3.102
2.7	3.090
2.8	3.089
2.9	3.100
3.0	3.111
3.1	3.121
3.2	3.132
3.3	3.143
3.4	3.153
3.5	3.164

10   Using the best-fitting regression line found in Exercise 9 of this section, estimate the air temperature when the following chirps are heard:

(a)   16 per second
(b)   20 per second
(c)   24 per second

11   Eleven students were given an attitude inventory to see how interested they were in mathematics. They also took their final mathematics test. Table 8.20 shows their interest in mathematics (X) and their scores on the test (Y). Use the median fit to find a regression line for estimating test scores from interest scores. Find the mean error for this line.

TABLE 8.20

X	Y
27	23
33	40
32	22
30	21
27	16
21	15
37	18
23	24
30	26
28	22
24	33

8.3 MEAN SQUARE ERROR

Another way of describing how well a line fits a set of data is to *square* the errors $Y - Y'$ and find the mean (average) of all the squared errors. This should remind you of how you found the variance of a set of scores in Chapter Two. You found the deviation scores (distance for each score is from the mean), squared them, and averaged them. Here, we are finding the mean (average) of the squared distances of the data points from the regression line.

In Chapter Seven we found the mean error for the spelling and reading data in Table 8.21. Now we find the mean square error for the same regression line by first adding another column to the table we used in Section 8.2.

TABLE 8.21

Regression Equation: $Y' = 1.5X - .2$

X	Y	Y'	$|Y - Y'|$	$|Y - Y'|^2$
12	20	17.8	2.2	4.84
10	12	14.8	2.8	7.84
10	18	14.8	3.2	10.24
8	10	11.8	1.8	3.24
7	12	10.3	1.7	2.89
6	14	8.8	5.2	27.04
6	6	8.8	2.8	7.84
5	7	7.3	.3	.09
4	3	5.8	2.8	7.84
2	1	2.8	1.8	3.24
		Totals	24.6	75.10

$$\text{Mean error} = \frac{24.6}{10} = 2.46$$

$$\text{mean square error} = \frac{75.10}{10} = 7.510$$

We now find the mean square error for the regression line (see Table 8.22)

$$Y' = 2.0X - 3.7$$

TABLE 8.22

Regression Equation: $Y' = 2.0X - 3.7$

X	Y	Y'	$|Y - Y'|$	$|Y - Y'|^2$
12	20	20.3	.3	.09
10	12	16.3	4.3	18.49
10	18	16.3	1.7	2.89
8	10	12.3	2.3	5.29
7	12	10.3	1.7	2.89
6	14	8.3	5.7	32.49
6	6	8.3	2.3	5.29
5	7	6.3	.7	.49
4	3	4.3	1.3	1.69
2	1	.3	.7	.49
			Total	70.10

Mean square error = 7.01

This regression line fits the data a little better than did the previous one, since the mean (average) square error has been reduced from 7.51 to 7.01. In the next section, we will discuss a method for finding the best-fitting regression line in terms of the mean square errors. In the following section, we will give a formula for finding the slope of the best-fitting line (in terms of mean square errors).

We are looking for the line that best fits the data when we are judging fit in terms of the mean (average) square error. We found the mean square errors for two regression lines, one with a slope of 1.5 and one with a slope of 2.0. They are shown in Table 8.23.

TABLE 8.23

SLOPE OF REGRESSION LINE	MEAN (AVERAGE) SQUARE ERROR
1.5	7.510
2.0	7.010

With the help of a computer, we found the mean square errors for many more regression lines, with slopes of from 1.5 to 2.5. Table 8.24 and the graph of the values in the table (Figure 8.10) help to give us an idea of what is taking place.

TABLE 8.24

SLOPE	MEAN SQUARE ERROR	
1.5	7.510	
1.6	7.074	
1.7	6.806	
1.8	6.706	◄──── Least error
1.9	6.774	
2.0	7.010	
2.1	7.414	
2.2	7.986	
2.3	8.726	
2.4	9.634	
2.5	10.710	

As the regression lines become steeper than a slope of 1.7, the mean square error becomes smaller and smaller—for a while. Then this error gets larger again. A line with slope somewhere between 1.6 and 1.9 is the best-fitting line. It is *best* because the mean square error for this line is the *least* possible. Such a line is called the *least squares best-fitting line* for a set of data.

 The least squares best-fitting line for a set of data in two variables is the regression line that has the smallest possible mean square error for estimating one variable from the other.

This smallest possible mean square error is called the *error variance* when Y is estimated from X.

With the help of a computer, we can put a magnifying glass on the interval between 1.70 and 1.90 to attempt to find the slope that

FIGURE 8.10

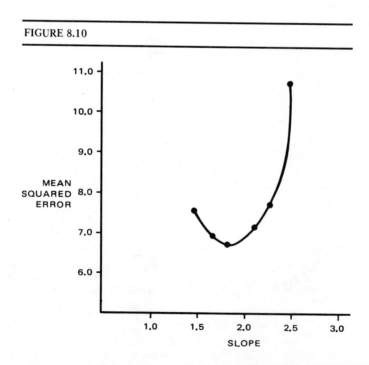

gives the smallest mean square error (that is, the error variance). See Table 8.25.

TABLE 8.25

SLOPE	MEAN SQUARE ERROR	
1.70	6.80600	
1.71	6.78844	
1.72	6.77256	
1.73	6.75836	
1.74	6.74584	
1.75	6.73500	
1.76	6.72584	
1.77	6.71836	
1.78	6.71256	
1.79	6.70844	
1.80	6.70600	
1.81	**6.70524**	← Least error (error variance)
1.82	6.70616	
1.83	6.70876	
1.84	6.71304	
1.85	6.71900	
1.86	6.72664	
1.87	6.73596	
1.88	6.74696	
1.89	6.75964	
1.90	6.77400	

We see from this table that a line with a slope of 1.81 gives the smallest mean square error, 6.705. The smallest mean square error is the *error variance* when estimating reading scores from spelling scores. That is, the error variance of the reading scores is 6.705.

It is not necessary to have a computer to find the best-fitting line in this way. But a calculator would be very helpful! A good way to do this would be to have a group of students, perhaps six or eight, work together and have each student calculate the mean square error for one regression line. However, with a large number of data, this is a big job, too. In the next section, we will give a formula for the slope of the least squares best-fitting regression line.

EXERCISES

1 Table 8.26 gives the length (from nose to tail) and weights of eight laboratory mice. The regression equation

$$Y' = 1.5X + 8.81$$

was used to estimate the weight of a mouse given it length. What is the mean square error for estimating weight using this equation? (X = Length in centimeters; Y = Weight in grams.)

TABLE 8.26

X	Y	Y'	$|Y - Y'|$	$|Y - Y'|^2$
16	32	32.81	.81	.6561
15	26	31.31	5.31	28.1961
10	40	38.81	1.19	1.4161
13	27	28.31	1.31	1.7161
15	30	31.31	1.31	1.7161
17	38	34.31	3.69	13.6161
16	34	32.81	1.19	1.4161
21	43	40.31	2.69	7.2361

2 The regression equation

$$Y' = 1.9X + 2.16$$

is used in Table 8.27 to estimate the weight of a mouse given its length. What is the mean square error for this equation?

TABLE 8.27

X	Y	Y'	$|Y - Y'|$	$|Y - Y'|^2$
16	32	32.56	.56	.3136
15	26	30.66	4.66	21.7156
20	40	40.16	.16	.0256
13	27	26.86	.14	.0196
15	30	30.66	.66	.4356
17	38	34.46	3.54	12.5316
16	34	32.56	1.44	2.0736
21	43	42.06	.94	.8836

3   Which of the two equations

(a) $Y' = 1.5X + 8.8$ (Exercise 1 of
          or   this section)
(b) $Y' = 1.9X + 2.16$ (Exercise 2 of
                              this section)

fits the mice data better, in terms of mean square error?

4   Refer to the cricket data in the key problem, Chapter Seven. For the regression equation

$$Y' = 4X + 13.7$$

what is the mean square error? See Table 8.28.

TABLE 8.28

X	Y	Y'	$Y - Y'$	$|Y - Y'|$	$|Y - Y'|^2$
20	89	93.73	−4.73	4.73	22.37
16	72	77.73	−5.73	5.73	32.83
20	93	93.73	− .73	.73	.53
18	84	85.73	−1.73	1.73	2.99
17	81	81.73	− .73	.73	.53
16	75	77.73	−2.73	2.73	7.45
15	70	73.73	−3.73	3.73	13.91
17	82	81.73	.27	.27	.07
15	69	73.73	−4.73	4.73	22.37
16	83	77.73	5.27	5.27	27.77
15	80	73.73	6.27	6.27	39.31
17	83	81.73	1.27	1.27	1.61
16	81	77.73	3.27	3.27	10.69
17	84	81.73	2.27	2.27	5.15
14	76	69.73	6.27	6.27	39.31

5   Ten students took a spelling test (X) and a reading test (Y) and got the scores in Table 8.29. The regression equation

$$Y' = X + 3.3$$

was used to estimate a student's reading score given his or her spelling score. What is the mean square error for this equation?

TABLE 8.29

X	Y	Y'	$Y - Y'$	$|Y - Y'|$	$|Y - Y'|^2$
12	20	15.3	4.7	4.7	22.09
10	12	13.3	−1.3	1.3	1.69
10	18	13.3	4.7	4.7	22.09
8	10	11.3	−1.3	1.3	1.69
7	12	10.3	1.7	1.7	2.89
6	14	9.3	4.7	4.7	22.09
6	6	9.3	−3.3	3.3	10.89
5	7	8.3	−1.3	1.3	1.69
4	3	7.3	−4.3	4.3	18.49
2	1	5.3	−4.3	4.3	18.49

6   Suppose that the regression equation

$$Y' = 1.5X - .2$$

was used in the data of Exercise 5 of this section to estimate reading scores from spelling scores. Complete Table 8.30 and find the mean square error for this equation

TABLE 8.30

X	Y	Y'	Y − Y'	|Y − Y'|	|Y − Y'|²
12	20	17.8	2.2	2.2	_____
10	12	14.8	−2.8	2.8	_____
10	18	14.8	3.2	3.2	_____
8	10	11.8	−1.8	1.8	_____
7	12	10.3	1.7	1.7	_____
6	14	8.8	5.2	5.2	_____
6	6	8.8	−2.8	2.8	_____
5	7	7.3	− .3	.3	_____
4	3	5.8	−2.8	2.8	_____
2	1	2.8	−1.8	1.8	_____

7   Of the two regression equations

(a) $Y' = 1.5X - .2$ (Exercise 6 of
                or   this section)
(b) $Y' = X + 3.3$ (Exercise 5 of
                                this section)

which fits the spelling test and reading data best? Why?

8   Table 8.31 gives the mean square errors for the spelling and reading data for regression equations having different slopes from 1.0 to 2.3. According to this table, what is the slope of the line that best fits these data? What is the equation of this line?

TABLE 8.31

SLOPE	MEAN SQUARE ERROR
1.0	12.21
1.1	10.93
1.2	9.82
1.3	8.88
1.4	8.11
1.5	7.51
1.6	7.07
1.7	6.80
1.8	6.70
1.9	6.77
2.0	7.01
2.1	7.41
2.2	7.98
2.3	8.72

9   Refer to the weight-loss data (Section 8.2). Table 8.32 gives the estimation errors for the regression line

$$Y = 1.5X + 6.5$$

Find the mean square error for this equation.

TABLE 8.32

X	Y	Y'	Y − Y'	|Y − Y'|	|Y − Y'|²
12	19	24.5	− 5.5	5.5	30.25
4	18	12.5	5.5	5.5	30.25
8	8	18.5	−10.5	10.5	110.25
4	16	12.5	3.5	3.5	12.25
12	26	24.5	1.5	1.5	2.25
7	12	17.0	− 5.0	5.0	25.00
5	12	14.0	− 2.0	2.0	4.00
9	29	20.0	9.0	9.0	81.00
14	31	27.5	3.5	3.5	12.25

10   Table 8.33 gives the mean square errors for the weight loss data for regression equations having different slopes. According to this table, what is the slope of the best-fitting line for these data? Find the equation of this line.

TABLE 8.33

SLOPE	MEAN SQUARE ERROR
1.0	36.00
1.1	35.14
1.2	34.53
1.3	34.16
1.4	34.04
1.5	34.16
1.6	34.53
1.7	35.14
1.8	36.00
1.9	37.10
2.0	38.44

11   Table 8.34 gives the mean square errors for regression equations having several different slopes. What is the slope of the least squares best-fitting line for the cricket data? Compare this with the equation for the best-fitting line found in Exercise 7 of Section 8.2, where we found the least mean error. What is the error variance of the temperatures?

TABLE 8.34

SLOPE	MEAN SQUARE ERROR
2.5	14.8489
2.6	14.4793
2.7	14.1652
2.8	13.9065
2.9	13.7033
3.0	13.5556
3.1	13.4633
3.2	13.4265
3.3	13.4452
3.4	13.5193
3.5	13.6489

12   Table 8.35 gives the mean square error (for the mice data) for estimation equations having several different slopes from 1.5 to 2.5. What slope gives the least mean square square error? What is the equation of the regression line for the mice data?

TABLE 8.35

SLOPE	MEAN SQUARE ERROR
1.5	6.99
1.6	6.24
1.7	5.62
1.8	5.10
1.9	4.71
2.0	4.49
2.1	4.37
2.2	4.38
2.3	4.50
2.4	4.74
2.5	5.11

What is the error variance for the weights of the mice?

13   What is the estimated weight, using the equation found in Exercise 10 of this section, for mice having these lengths?

(a)   18 centimeters
(b)   22 centimeters
(c)   26 centimeters

8.4   NEGATIVE SLOPE

So far in this chapter, most of our examples of lines fitted to data have had *positive*, or *uphill*, slopes. That is, as one variable increases, the other tends to increase as well. But two variables may be related in another way, as the following example suggests. A class is given a questionnaire about mathematics anxiety, which has questions like those in Figure 8.11.

FIGURE 8.11

IT MAKES ME NERVOUS TO TAKE MATHEMATICS TESTS.

□ STRONGLY AGREE
□ STRONGLY DISAGREE

WHENEVER I TAKE A MATHEMATICS TEST, I CAN'T THINK CLEARLY.

□ STRONGLY AGREE
□ STRONGLY DISAGREE

TABLE 8.36

	ANXIETY SCORE Y	EXAM SCORE Y	
Robert	15	37	
Juan	20	33	
Margo	23	5	
Debi	30	31	I
Phil	34	40	
Jo	34	35	
Harry	35	32	
Mark	36	30	
Polly	38	14	
Jim	40	9	II
Pete	42	20	
Mary	44	24	
Sue	46	18	
Roger	49	21	
Tim	50	13	
Jose	53	14	
Sue	59	7	III
John	60	9	
Jill	62	7	
Jon	63	12	

Persons who get high scores on this questionnaire are those who worry a lot about mathematics tests. Table 8.36 shows the scores obtained by the 20 students on the anxiety questionnaires and on their mathematics exam.

It appears as if the best-fitting line for these data slopes downward as one goes from left to right (see Figure 8.12).

Relationships between two variables having a regression line with a negative slope can be interpreted as relationships in which one variable increases as the other decreases. In our example, it appears that people with more anxiety about mathematics do less well on tests in mathematics.

We will use the median-fit method to find a regression line for these data. Then we will use the slope of this line as an estimate for the slope of the line that gives the smallest mean square error.

We divide the graph and table into three regions, as shown. Then, we find the median of the horizontal scores (anxiety) in Region I.

FIGURE 8.12

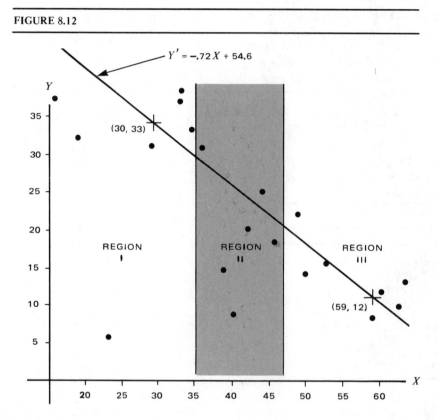

$Y' = -.72X + 54.6$

The median of 15, 20, 23, 30, 34, 34, and 35 is 30. For the corresponding vertical scores, the median of 5, 31, 32, 33, 35, 37, and 40 is 33. So, the median data point in Region I is (30, 33).

For Region III, the median of the anxiety scores 49, 50, 53, 59, 60, 62, and 63 is 59. The median of the exam scores 7, 7, 9, 12, 13, 14, and 21 is 12. The median data point for Region III is (59, 12).

The median-fit line passes through these two points and has a slope of

$$\frac{33 - 12}{30 - 59} = \frac{21}{-29} = -.72$$

The Y-intercept of this line is found as follows. We know

$$Y' = mX + C$$

Since one point on the line is (30, 33) and the slope is $-.72$, we have

$$33 = (-.72)(30) + C$$
$$33 = -21.6 + C$$

which gives

$$C = 54.6$$

So, the equation of the median-fit line is

$$Y' = -.72X + 54.6$$

To find Mark's estimated test score, we observe that his anxiety score was 36. So, his estimated test score, Y', is

$$Y' = -.72(36) + 54.6 = -25.92 + 54.6$$

$$= 28.7$$

His actual test score was $Y = 30$, so the estimation error is

$$Y - Y' = 30 - 28.7$$

$$= 1.3$$

We can calculate the mean error and the mean square error for the regression line

$$Y' = -.72X + 50.53$$

as shown in Table 8.37.

TABLE 8.37

X	Y	Y'	$Y - Y'$	$|Y - Y'|$	$|Y - Y'|^2$
42	20	20.29	− .3	.29	.08
15	37	39.73	− 2.74	2.73	7.49
34	40	26.05	13.94	13.94	194.37
53	14	12.37	1.62	1.62	2.63
36	30	24.61	5.38	5.38	28.96
23	5	33.97	−28.98	28.97	839.72
40	9	21.73	−12.74	12.73	162.25
34	35	26.05	8.94	8.94	79.95
50	13	14.53	− 1.54	1.53	2.36
62	7	5.89	1.10	1.10	1.21
−49	21	15.25	5.74	5.74	32.97
46	18	17.41	.58	.58	.33
20	33	36.13	− 3.14	3.13	9.84
60	9	7.33	1.66	1.66	2.76
35	32	25.33	6.66	6.66	44.38
−30	31	28.93	2.06	2.06	4.25
44	24	18.85	5.14	5.14	26.44
59	7	8.05	− 1.06	1.05	1.11
63	12	5.17	6.82	6.82	46.54
38	14	23.17	− 9.18	9.17	84.23

Mean error =   5.96
Mean square error = 78.59

Let's use the method of Section 8.3 and see if we can find a line that fits the data better. We try lines with slopes from −1.0 to −.2. See Table 8.38.

TABLE 8.38

SLOPE	MEAN ERROR	MEAN SQUARE ERROR	
−1.00	7.84	112.66	
− .95	7.42	104.47	
− .90	7.00	97.20	
− .85	6.68	90.84	
− .80	6.36	85.40	
− .75	6.10	80.87	
− .70	5.87	77.26	
− .65	5.67	74.56	
− .60	5.64	72.77	Least mean
− **.55**	5.84	71.90	square error
− .50	6.12	71.94	
− .45	6.40	72.90	
− .40	6.69	74.77	
− .35	7.08	77.55	
− .30	7.46	81.25	
− .25	7.85	85.87	
− .20	8.24	91.39	

Recall also how we find the equation of a line when we know its slope:

$$Y' = mX + C$$

But we know from Table 8.38 that the slope of the least-squares best-fitting line is:

$$m = -.55$$

Also, the line passes through the mean data point for the anxiety and test scores.

$$\text{Mean anxiety score} = \frac{\text{sum of scores}}{\text{number}} = \frac{833}{20} = 41.65$$

$$\text{Mean test score} = \frac{411}{20} = 20.55$$

So,

$$20.55 = (-.55)41.65 + C$$

$$20.55 = -22.91 + C$$

$$C = 43.46$$

The equation of the least squares best-fitting line is, therefore,

$$Y' = -.55X + 43.46$$

WHICH METHOD FOR FINDING THE REGRESSION LINE?

We have now used several methods for finding the regression line (and will learn a couple more later in this chapter). If you need just a general idea of what the regression line is, then the thread method of Section 8.1 is usually adequate. The median-fit method is just a little more effort and can provide a regression equation that performs rather well. The most usual method is to find the line that makes the mean square error as small as possible (method of least squares). But a table of slopes and errors is not typically used. Instead, formulas like the ones in Sections 8.5 and 8.6 are employed. We do not object to using formulas, but we recommend that you use a table of slopes and errors until you are sure you understand the way in which the least squares method works. The formulas are just a shortcut for finding the slope of the line you are seeking.

In certain cases, you should really find the smallest mean error instead of the smallest mean square error. But for now, we will not go into the reasons why. If you are not told which method to use, use your own judgment or preference.

EXERCISES

1   From Table 8.38, find the slope of the regression line that gives the smallest mean error, instead of the smallest mean square error. Find the equation of this line.

2   Find the mean error when the regression line used to estimate test scores from anxiety scores has a slope of −.7. You should construct a table like Table 8.37.

3   Interest in science is measured by an attitude scale (this measure is variable X in Table 8.39). The final score on a physics exam is called Y. This table gives science interest and physics exam scores for 20 students. Find the median-fit line for estimating a student's physics exam score, knowing that student's interest score. (*Note:* For this attitude scale, a high score means "don't like science.")

TABLE 8.39

DATA VALUES

X	Y
41	40
32	48
38	50
44	37
39	45
34	33
40	34
38	48
43	37
47	34
43	40
42	39
33	47
47	34
38	46
37	45
41	42
46	34
48	36
39	37

4   Using the equation

$$Y' = -1.0X + 81$$

find the mean error when estimating physics scores from science interest scores (data in Table 8.39).

5   Use the method of Section 8.1 to try to find a line that has a better fit of the data in Table 8.39 than the median-fit line (found in Exercise 3 of this section).

6   The data in Table 8.40 are the number of push-ups that could be done by a sample of 12 male teachers at Oak High. (*Note:* The 38-year-old teacher was the gym instructor.)

Draw a graph of these data. Find the median-fit line for these data.

TABLE 8.40

AGE	NUMBER OF PUSH-UPS
21	10
25	8
22	11
28	6
30	7
38	15
22	9
27	6
44	4
48	3
35	8
48	5

7   Suppose the equation of the regression line in Exercise 6 of this section is

$$Y' = -X + 40$$

Calculate the estimated number of push-ups that could be done by each of the 12 teachers and then calculate the mean error and the mean square error. Complete a table with these heads:

$$X \quad Y \quad Y' \quad Y - Y' \quad |Y - Y'|^2$$

8   Table 8.41 gives slopes and errors for the push-ups data (Exercise 6 of this section). From this table, obtain the necessary information to find the equation of the least squares best-fitting line for these data.

TABLE 8.41

SLOPE	MEAN SQUARE ERROR
−.5	18.72
−.4	13.27
−.3	9.68
−.2	7.94
−.1	8.07
0.0	10.05
.1	13.89
.2	19.59
.3	27.14
.4	36.56

9 Table 8.42 shows infant mortality rates (deaths under one year per 1000 live births) in the state of Massachusetts at the beginning of the thirteen decades 1850–1970.[3]

TABLE 8.42

DECADE X	INFANT MORTALITY RATE Y
1	131
2	143
3	170
4	161
5	163
6	141
7	116
8	78
9	54
10	34
11	23
12	22
13	17

Plot these data and draw in the straight line which appears to best fit the data.

10 Suppose that the equation of the regression line for the mortality data (Exercise 9 of this section) is

$$Y' = -14X + 194$$

Using this regression equation, estimate the infant mortality rate for each of the 13 decades and calculate the mean estimation error in the mortality rate. If it can be assumed that the trend in these data continues, estimate the infant mortality rate for the fourteenth and fifteenth decades (the 1980s and 1990s) in Massachusetts.

11 The winning time for a certain event at an annual track meet is recorded in Table 8.43 for the past 12 years. Use the median fit to find a regression line for these data. If the trend continues, estimate the winning time for this event for the next two annual track meets.

TABLE 8.43

YEAR	TIME (SECONDS)
1	300
2	299
3	290
4	290
5	285
6	283
7	282
8	280
9	276
10	277
11	270
12	270

8.5 COVARIANCE

You no doubt have concluded by now that it can be a pretty long and complicated business to find the least squares best-fitting line for a set of data. Well, there is a shortcut for you to use, now that you understand what the method is about. This shortcut is helped by a new measure, though, that of *covariance*. We learned about variance in Chapter Two. Variance is a measure of the amount that a set of data varies about its mean, or average. Covariance, another measure of variation, tells how much two variables change, or vary, with respect to one another. (Think of *co*operate or *co*pilot.)

We have seen examples of how data vary. If the slope of the regression line is positive, as one variable increases, so does the other. Such data have a positive covariance.

Figure 8.13 is a graph of the data in the key problem for this chapter: exposure to radioactive waste and death rate due to cancer. It appears from the graph that as the exposure to radiation increases, so does the cancer death rate.

If the slope of the regression line is negative, as one variable increases, the other decreases. Such data have a negative covariance. Figure 8.14 is a graph of the mortality data from Exercise 9 of Section 8.4. As time increases, the death rate of infants decreases. (Can you think of reasons for this?)

FIGURE 8.13

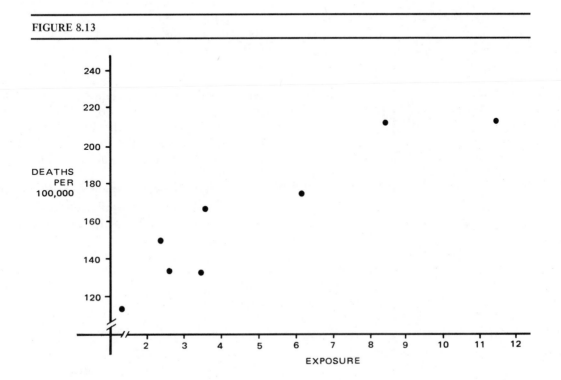

DEATHS PER 100,000

EXPOSURE

FIGURE 8.14

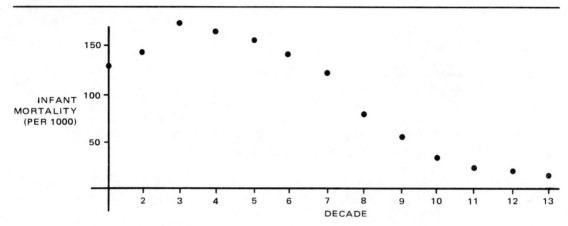

Now let's see how to calculate covariance. We use the spelling and reading data of Section 8.3 as an example.

Step 1:   Find the mean of the X (spelling) scores and the Y (reading) scores. As is seen from Table 8.44:

$$\bar{X} = \text{mean of } X = \frac{70}{10} = 7.0$$

$$\bar{Y} = \text{mean of } Y = \frac{103}{10} = 10.3$$

The covariance of X and Y (that is, the average of the sum of the products of the deviation scores) is

$$\frac{152.0}{10} = 15.2$$

The interpretation of covariance is not as straightforward as that of variance. But, we can say that, since the covariance of the spelling and reading data is positive, the slope of the least squares best-fitting regression line for these data is positive. That is, as spelling scores (X) increase, reading scores (Y) tend to increase.

We also calculate the variance of X, since we need that in our formula to find the slope of the best-fitting line for the data. Refer to Chapter Two, Section 5, for a reminder of how to calculate the variance of a set of data.

To find the variance of the X scores, we will add a column to our covariance table (see Table 8.45). This will give us the squares of the X deviation scores.

TABLE 8.44
FINDING THE COVARIANCE OF A SET OF DATA

X	Y	$X - \bar{X}$	$Y - \bar{Y}$	$(X - \bar{X})(Y - \bar{Y})$
12	20	$12 - 7.0 = 5$	$20 - 10.3 = 9.7$	$5 \times 9.7 = 48.5$
10	12	$10 - 7.0 = 3$	$12 - 10.3 = 1.7$	$3 \times 1.7 = 5.1$
10	18	$10 - 7.0 = 3$	$18 - 10.3 = 7.7$	$3 \times 7.7 = 23.1$
8	10	$8 - 7.0 = 1$	$10 - 10.3 = - .3$	$1 \times - .3 = - .3$
7	12	$7 - 7.0 = 0$	$12 - 10.3 = 1.7$	$0 \times 1.7 = 0$
6	14	$6 - 7.0 = -1$	$14 - 10.3 = 3.7$	$-1 \times 3.7 = - 3.7$
6	6	$6 - 7.0 = -1$	$6 - 10.3 = -4.3$	$-1 \times -4.3 = 4.3$
5	7	$5 - 7.0 = -2$	$7 - 10.3 = -3.3$	$-2 \times -3.3 = 6.6$
4	3	$4 - 7.0 = -3$	$3 - 10.3 = -7.3$	$-3 \times -7.3 = 21.9$
2	1	$2 - 7.0 = -5$	$1 - 10.3 = -9.3$	$-5 \times -9.3 = 46.5$
70	103			152.0

Sum of positive products   156.0
Sum of negative products -4.0

Total   152.0

TABLE 8.45

X	Y	$X-\bar{X}$	$Y-\bar{Y}$	$(X-\bar{X})(Y-\bar{Y})$	$(X-\bar{X})^2$
12	20	5.0	9.7	48.5	25.0
10	12	3.0	1.7	5.1	9.0
10	18	3.0	7.7	23.1	9.0
8	10	1.0	- .3	- .3	1.0
7	12	0.0	1.7	0.0	0.0
6	14	-1.0	3.7	- 3.7	1.0
6	6	-1.0	-4.3	4.3	1.0
5	7	-2.0	-3.3	6.6	4.0
4	3	-3.0	-7.3	21.9	9.0
2	1	-5.0	-9.3	46.5	25.0
70	103			152.0	84.0

As we recall from Chapter Two,

$$\text{Variance of } X \text{ scores} = \frac{\text{sum of squares of deviations of } X}{\text{number of scores}}$$

$$= \frac{84.0}{10}$$

$$= 8.4$$

THE SLOPE
OF THE LEAST SQUARES LINE

It is shown in more advanced statistics courses that the slope of the least squares best-fitting line is

$$\text{Slope} = \frac{\text{covariance of } X \text{ and } Y}{\text{variance of } X}$$

For our spelling and reading data, this formula gives

$$\text{Slope} = \frac{15.2}{8.4} = 1.81$$

Note that a slope of 1.81, as given by the formula, is the same as we found in Table 8.25 (page 207). Sometimes the table of least mean square errors will not have enough decimal places for the required slope. The formula, on the other hand, gives as many decimal places as required.

EXERCISES

1   Calculate the covariance of X and Y for the cricket data (key problem, Chapter Seven). Also find the variance of X, the cricket chirps per second. Use a table like the one developed in this section (Table 8.45).

2   A table of slopes and errors for the cricket data is given in Exercise 11 of Section 8.3 (Table 8.34). What slope, according to that table, gave the smallest mean square error? Now, use the formula of Section 8.5 to find the slope of the the least squares best-fitting line for the cricket data. Compare the value of the slope given by this formula with the one you found from the table of slopes and errors.

3   Find the covariance of the mice data (Exercise 1 of Section 8.3). Also find the variance of X, the lengths in centimeters of the mice.

4   Use the slope formula of Section 8.5 to find the slope of the best-fitting line for the mice data. Then compare this value with that given in the table of slopes and errors for the mice data (Exercise 12 of Section 8.3), Table 8.35.

5   Find the covariance of the mathematics anxiety data in Table 8.36. This covariance should be negative. Explain why. Also find the variance of the X scores.

6   Find the slope of the least squares best-fitting line for the anxiety data using the formula. Compare with the slope of the line from Table 8.38. They should be very close.

7   Find the covariance of the push-up data (Exercise 6 of Section 8.4, Table 8.40). Also find the variance of X, the age of the teachers doing the push-ups.

8   Find the slope of the least squares best-fitting line for the push-ups data (Table 8.40) using the formula of Section 8.5. Compare with the value given in Table 8.41.

9   Use the slope of the least squares best-fitting line for the push-ups data that you found in Exercise 8 of this section. Find the equation of the regression line with this slope. Then use the regression line to estimate the number of push-ups that could be done by teachers of each age.

(a)   26 years
(b)   39 years
(c)   50 years

What assumptions are you making when you make these estimates?

10   Find the covariance of the radiation data in Table 8.1 (key problem, Chapter Eight). Then find the variance of X, the index of exposure.

11   Use the formula of Section 8.5 to find the slope of the least squares best-fitting line for the radiation data (Table 8.1).

12   Find the equation of the regression line for the slope you found in Exercise 11 of this section. Then estimate cancer mortality rates for these index values.

(a)   1.0
(b)   5.0
(c)   10

8.6   CORRELATION

We have found ways of finding the line that best fits a set of data. But how good is this fit, for a given set of data? The goodness of fit is expressed in terms of estimation errors for Y scores, given X scores. Another way of describing the goodness of fit of data to a line is to ask how strongly the X and Y scores are related. If they are very strongly related, Y scores can be estimated from X scores very accurately (with little estimation error). If the X and Y scores are only weakly related, Y is estimated from X with considerable error.

THE PEARSON CORRELATION COEFFICIENT

Around the year 1900, a statistician by the name of Karl Pearson invented a statistic to describe the strength of the relationship between two variables. It is called the *Pearson Correlation Coefficient* and is given the label r.

The correlation coefficient is defined so that it has values between -1 and 1, inclusive.

Type of Relationship	Value of r
Strong, positive	Close to +1.0
Weak, positive	Positive, but close to 0
None	Rather close to 0
Weak, negative	Negative, but close to 0
Strong, negative	Close to −1.0

The sign of r (whether r is positive or negative) is the same as the sign of the slope of the least squares best-fitting line for the data.

Some examples will illustrate.

■ **Example 1—Strong Positive Relationship:** Figure 8.15 shows scores on mid-term (X) and final (Y) exams for a class of students. These scores are strongly related and the slope of the least squares best-fitting line for the data is positive. The correlation between X and Y is close to 1.0. ■

FIGURE 8.15

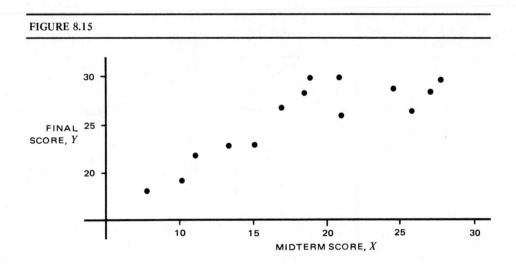

■ **Example 2—Weak Positive Relationship:**
Figure 8.16 shows interest in mathematics (X) and final mathematics test scores (Y) for a class of students. These scores are weakly related and the slope of the regression line is positive. This is called a weak, positive relationship between X and Y. The correlation between X and Y is positive, but close to zero. ■

■ **Example 3—No Relationship:**   Figure 8.17 shows student popularity (X) and mathematics test scores (Y) for a class of students. There is no relationship between these two variables. The slope of the best-fitting line is close to zero. ■

■ **Example 4—Weak Negative Relationship:**
Figure 8.18 shows anxiety about mathematics (X) and final mathematics test score (Y). These scores are weakly related and the slope of the regression line is negative. This is called a weak, negative relationship between X and Y. The correlation between X and Y is negative and close to zero. ■

■ **Example 5—Strong Negative Relationship:**
Figure 8.14 shows the year (X) and the infant mortality rate in Massachusetts (Y). (See Exercise 9 of Section 8.4.) These scores are strongly related (they fit rather closely to the regression line, so there is little estimation error), and the slope of the regression line is negative. The correlation between X and Y is close to -1.0. ■

HOW TO CALCULATE

There are several ways to calculate r. Here is one:

$$r = \frac{\text{covariance } X \text{ and } Y}{S_X \cdot S_Y}$$

That is, the correlation between X and Y is the covariance of X and Y divided by the product of the standard deviation of X and the standard deviation of Y.

FIGURE 8.16

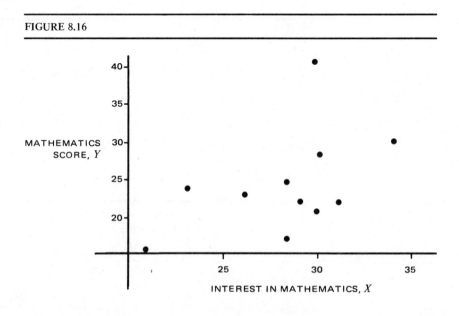

INTEREST IN MATHEMATICS, X

FIGURE 8.17

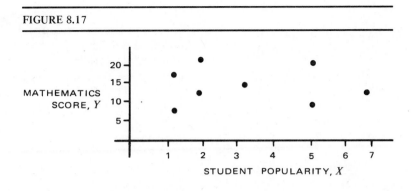

FIGURE 8.18

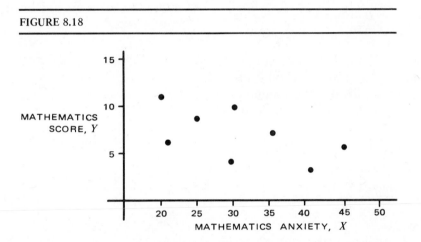

A table is helpful in our calculations. This table is not much different from the one used to find covariance.

■ **Example 6**:   Find the correlation between weeks in weight loss program (X) and weight lost by participants (Y) (refer to Table 8.2).

Solution:

We follow six steps, using data from Table 8.46. Compare this table with Table 8.45.

Step 1: Find the mean for X and for Y. We see from Table 8.46 that

$$\text{Mean of } X = \frac{75}{9} \doteq 8.33$$

$$\text{Mean of } Y = \frac{171}{9} = 19.0$$

Step 2: Find the deviation scores for each X and Y. (That is, $X-\bar{X}$ and $Y-\bar{Y}$.)

Step 3: Find the squares of the deviation scores for X and Y. Then find the total sums of squares for each and divide by the number of scores, N. This gives the variance of X and the variance of Y.

$$\text{Variance of } X = \frac{\text{sum of }(X-\bar{X})^2}{N \text{ scores}} = \frac{110.01}{9} = 12.22$$

$$\text{Variance of } Y = \frac{\text{sum of }(Y-\bar{Y})^2}{N \text{ scores}} = \frac{522}{9} = 58.0$$

Step 4: Find the square root of the variance for X and the variance for Y. This gives the standard deviation for X, called S_X, and the standard deviation of Y, called S_Y.

$$\text{Standard deviation of } X = S_X = \sqrt{\text{variance of } X}$$
$$= \sqrt{12.22} \doteq 3.50$$

$$\text{Standard deviation of } Y = S_Y = \sqrt{\text{variance of } Y}$$
$$= \sqrt{58.0} \doteq 7.62$$

Step 5: Find the product of the deviation scores, add them up, and divide by the number of scores N.

$$\text{Covariance of } X \text{ and } Y = \frac{\text{sum of }(X-\bar{X})(Y-\bar{Y})}{N \text{ scores}}$$

$$= \frac{154.0}{9} \doteq 17.11$$

Step 6: Find the correlation between X and Y.

$$r = \frac{\text{covariance } X \text{ and } Y}{S_X \cdot S_Y} = \frac{17.11}{(3.50)(7.62)} = .64 \ \blacksquare$$

TABLE 8.46

X	Y	$X-\bar{X}$	$Y-\bar{Y}$	$(X-\bar{X})(Y-\bar{Y})$	$(X-\bar{X})^2$	$(Y-\bar{Y})^2$
12	19	$12 - 8.33 = 3.67$	$19 - 19 = 0$	0	13.47	0
4	18	$4 - 8.33 = -4.33$	$18 - 19 = -1$	4.33	18.75	1
8	8	$8 - 8.33 = -.33$	$8 - 19 = -11$	3.63	.11	121
4	16	$4 - 8.33 = -4.33$	$16 - 19 = -3$	12.99	18.75	9
12	26	$12 - 8.33 = 3.67$	$26 - 19 = 7$	25.69	13.47	49
7	12	$7 - 8.33 = -1.33$	$12 - 19 = -7$	9.31	1.77	49
5	12	$5 - 8.33 = -3.33$	$12 - 19 = -7$	23.31	11.09	49
9	29	$9 - 8.33 = .67$	$29 - 19 = 10$	6.70	.45	100
14	31	$14 - 8.33 = 5.67$	$31 - 19 = 12$	68.04	32.15	144
75	171			154.00	110.01	522

FINDING CORRELATION FROM THE SLOPE

Another way to calculate the correlation r between two variables is to begin with the slope of the least squares best-fitting line for the data. We can find the slope from a table of slopes and errors (if we have a computer to help us) or we can calculate it using the formula above, that is

$$m = \frac{\text{covariance of } X \text{ and } Y}{\text{variance of } X}$$

Once we know the slope, m, of the line that predicts Y scores from X scores, we find r as follows:

$$r = \frac{(m)(S_X)}{S_Y}$$

■ **Example 7—Weight-loss Data (Table 8.2):** We need to find the slope of the least squares best-fitting line. We found it in Exercise 10 of Section 8.3 (Table 8.33) to be 1.4.

If we use the above formula, we get a slope of

$$m = \frac{\text{covariance of } X \text{ and } Y}{\text{variance of } X} = \frac{17.11}{12.22} = 1.40$$

(See Table 8.46.)

Since we know that

$$S_X = 3.50$$

and

$$S_Y = 7.62$$

we have

$$r = \frac{(1.40)(3.50)}{7.62} = .64 \ ■$$

CORRELATION: HOW GOOD IS THE FIT?

Remember that the least squares best-fitting line gives the smallest mean squared estimation error. This smallest error is called the *error variance* of Y and is sometimes written

$$S_e^2$$

We can use this information about the estimation error as another way to calculate r, the correlation between X and Y.

If the error variance is *small*, this means that there is little estimation error. In other words, the regression line fits the data well.

If the error variance is *large*, this means that there are large estimation errors. The regression line is not very good at estimating Y-values from X-values.

So, small error variance means a large correlation (close to +1 or −1). Large error variance means a small correlation (close to 0).

We then look at the ratio of the error variance of Y to the regular, or usual, variance of Y (the variance we studied in Chapter Two), or

$$\frac{S_e^2}{S_Y^2}$$

When this ratio is subtracted from 1, we have the square of the correlation between X and Y, that is,

$$r^2 = 1 - \frac{S_e^2}{S_Y^2}$$

■ **Example 8—Spelling and Reading Data**:
We found in Table 8.25 that the least mean square error was 6.705. This is the error variance for the reading scores. The variance of the reading scores is

$$S_Y^2 = 34.21$$

So,

$$r^2 = 1 - \frac{S_e^2}{S_Y^2}$$

$$= 1 - \frac{6.71}{34.21} = 1 - .196 = .804$$

We now take the square root of .804.

$$r \doteq \pm.90$$

How can we decide whether the correlation is positive or negative? For one thing, we can look at the slope of the regression line, which is 1.81 (see Table 8.25). Or, we can find the covariance of X and Y. The correlation has the same sign as the covariance and the slope. So,

$$r = +.90. ■$$

EXERCISES

1   Figure 8.19 on page 227 shows graphs of different collections of data in two variables. Tell, by inspection, whether the correlation between the two variables is

(a) strong positive
(b) weak positive
(c) weak negative
(d) strong negative.

2   Tell whether you would expect the correlation between the two variables in each

of these sets of data to be strong positive, weak positive, weak negative, or strong negative.

(a) Height and weight of children from ages 4 to 12 years.
(b) Time spent studying for an exam and score on that exam.
(c) Attitudes toward mathematics (whether you like it or not) and final grade in mathematics.
(d) Popularity in school or college and academic achievement.
(e) Price of a watch and how many minutes it gains or loses.
(f) Age of a car and number of trouble-free miles per month.
(g) Pressure applied to a container of a gas (like hydrogen) and the volume which the gas occupies.
(h) Air pressure kept in automobile tires and mileage obtained from the tires.

Use the covariance and standard deviation method of this section to find the correlation in Exercises 3 through 7 that follow.

3   Find the correlation between the spelling and reading data, Table 8.44, Section 8.5.

4   Find the correlation between age and number of push-ups, as given in Exercise 6 of Section 8.4.

5   Find the correlation between the length and body weight of mice, as given in Exercise 1 of Section 8.3.

6   Find the correlation between cricket chirps and temperature as given in the key problem, Chapter Seven.

7   Find the correlation between radiation in the environment and cancer rate, as given in the key problem, Chapter Eight.

FIGURE 8.19

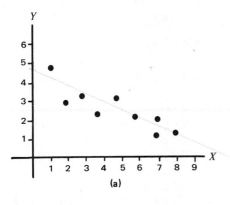

(a)

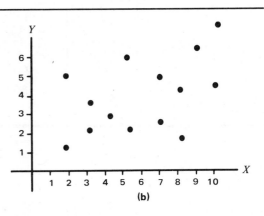

(b)

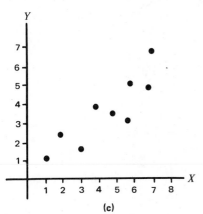

(c)

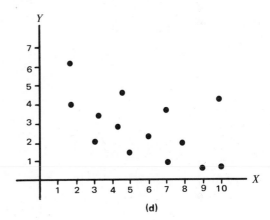

(d)

8 Figure 8.1 (page 192) gives not only weight loss for each person in the program, but total inches lost. Draw a graph to show these "slenderizing data," where the vertical axis gives the time in the program and the horizontal axis gives the inches lost. Then find the correlation between time and inches lost, using a method of your choice.

9 Find the error variance for the number of push-ups in the data given in Exercise of this section using Table 8.41. Use the error variance to find the correlation between age and number of push-ups. Is the correlation positive or negative? Why?

10 Find the correlation between length and body weight of mice using the slope method and Table 8.35. Compare with your answer obtained in Exercise 5 of this section.

11 Find the error variance when estimating temperature from cricket chirps using Table 8.34. Then find the correlation between cricket chirps and temperature using the error variance method and compare with the value you obtained in Exercise 6 of this section.

SPECIAL INTEREST FEATURE

Statistics and Health

Several studies over recent years have related sugar consumption to poor health. A recent book,[4] however, has raised questions about the claims concerning harmful effects of sugar.

Sugar and Heart Disease

A widely publicized report by an English physician in the 1960s charged that sugar can cause heart disease. The doctor's research involved comparing the sugar intake of 45 heart patients with 13 hospitalized accident victims and 12 men residing in the area. It was found that the heart patients consumed substantially more sugar than the other persons studied. But when other scientists repeated the study in other hospitals, using as a comparison group only nonhospitalized persons, they found no significant difference between sugar consumption of the heart patients and the others.

The authors of the book state:

> *A quick look at some simple statistics illustrates why the sugar–heart-disease theory makes so little sense. If the relationship were positive, countries with high sugar consumption should have a higher rate of heart disease. In fact, some – such as Venezuela, Cuba, Brazil, and Costa Rica – have a rather low incidence.*

Note to student: Do you believe this claim?

Sugar and Tooth Decay

It is well known that sugar contributes to cavities. There is a strong positive relationship between exposure time to sweets and tooth decay. The longer sugar remains in contact with your teeth, the more damage will be done.

The only way to reduce exposure time is to eliminate sticky sweets between meals, and to brush your teeth often. As the authors of the book point out:

> *The bacteria in the mouth that cause cavities aren't the least bit particular about the type of sugar they feed on. In fact, some "health" foods, such as raisins, can nestle among the teeth all day, providing hours of tasty fare for bacteria.*

8.7   RELATION AND CAUSATION

In this chapter, we have seen many examples of data in which there was found to be a strong relationship between the two variables. For example, in the age and push-up data, we saw a negative relationship. It seems reasonable in this case, as well, to use the data to conclude that age *causes* a person to do physical tasks such as running, jogging or playing football less well or less efficiently.

However, the existence of a relationship between two variables does not prove that one *causes* the other. Suppose, for example, we were to collect data on the shoe size and score on a twelfth grade mathematics test on students from seventh through twelfth grades. We would expect the data to look something like that in Figure 8.20. That is, we would expect that people who get better scores on the test will wear bigger shoes. But this is far from saying that a knowledge of mathematics *causes* big feet (or vice versa)! In this, there is another factor to take into account – the age of the person, which causes *both* feet to get bigger and mathematics scores to improve.

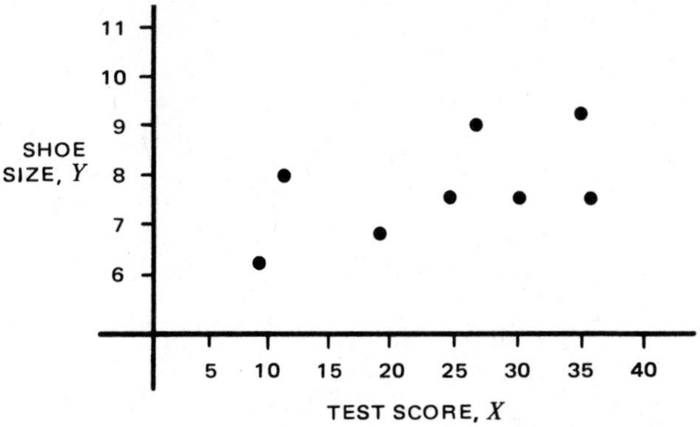

As another example, think of what we see around us as the season of autumn approaches. The leaves change color and fall and snow falls. Leaves changing and snow falling are related – very strongly. But falling leaves do not *cause* snow to fall.

There are many examples of the perils of interpreting correlations between variables. The radiation and cancer data in the key problem of Chapter Eight provides an illustration between the amount of radiation in the environment and the rate of cancer. But the data do not *prove* (though they strongly suggest) that the radiation is the cause of the cancer. It could be that there are other, yet unknown, factors that enter into the picture, such as minerals or pollutants in the food supply.

One of the best-known examples of the confusion about correlation and cause is that of smoking and cancer. For the past 50 years or more, research projects have produced data showing a correlation (positive) between the amount of smoking that a person does and the person's chances of getting cancer. On the one side of the debate, medical associations have claimed smoking as the cause of cancer. On the other side – notably by the tobacco industry – it has been argued that although smoking and cancer are statistically related, smoking has not been established as the cause of cancer. As time goes on and statistical evidence continues to grow, it seems more evident that smoking can indeed be declared as a significant factor in causing cancer.

FIGURE 8.20

EXERCISES

1   The data in Table 8.47 were reported by Dr. Al Shulte of the Pontiac, Michigan, schools.[5] The data are the number of personal fouls committed (X) and the total number of points scored (Y) by the junior varsity basketball team at Waterford-Kettering High School in Michigan.

TABLE 8.47

PLAYER	PERSONAL FOULS COMMITTED (X)	TOTAL POINTS SCORED (Y)
Bogert	0	1
Bone	0	0
Campbell	0	0
Forbes	0	2
Godoshian	6	6
Graham	17	48
Madill	7	21
Manning	31	75
McGrath	9	18
Nutter	4	2
Nyberg	24	42
Shipman	15	60
Spencer	16	37
Watson	0	3

(a)   Draw a graph of these data.

(b)   Suppose the slope of the best-fitting line for the points scored – fouls committed data is 2.5. Assuming the regression line goes through the mean data point, what is the equation of this line?

(c)   Find the covariance between X and Y and the standard deviation of X and Y. What is the correlation between X and Y?

(d)   If these data are interpreted causally, what are the implications for coaching?

2   The article at the right is a news clipping from *Time*.[6] What variables are reported to be related? What is the causal relationship implied? Comment on the soundness of the interpretation of this causal relationship.

CAPSULES

NEXT, AN UNFATTENING FAT?

In diet-conscious America, when one fad falls, the next amazing shrinking cure is always on its way. And at Procter & Gamble right now they are talking about a new substance that can create a creamy-rich milkshake or a buttery spread that is not the least bit fattening. It is a zero-calorie dead ringer for dietary fat called sucrose polyester (SPE), and last week researchers at the University of Cincinnati Medical Center reported SPE's first successful test. The compound contains eight fatty acids instead of the three that make up ordinary fats. As a result, digestive enzymes cannot break it down, and it passes unaltered out of the body. In the Cincinnati study, ten chronically obese patients were fed a diet much like their usual one for 20 days and, for another 20-day period, the same diet with SPE replacing about 540 daily calories of fat. Subjects were unable to detect any difference in the taste of the food, and did not attempt to compensate for lost calories by snacking more. They lost an average of 8 lbs. during their time on the SPE diet, and, as a bonus, their serum cholesterol levels dropped 10%. How soon will SPE be on your grocer's shelf? Not for a long time, if ever. Though no serious side effects were reported, SPE is "potent stuff," says Dr. Charles Glueck, who headed the study. The compound has been classified by the FDA as a drug. When it becomes available, several years from now, it will be sold by prescription only.

3   Find a newspaper clipping or quote an item from radio or television news which uses data in two variables in such a way as to imply a causal relationship between them.

4   What correlations are evident from these graphs shown in Figure 8.21? Is there a causal relationship, do you think, between one variable and another? Explain.
[*Note:* Relationships between variables need not be linear (a straight line).]

FIGURE 8.21
CORRELATION BETWEEN THE INCREASE IN HUMAN POPULATION AND THE EXTINCTION OF MAMMALS AND BIRDS SINCE 1600

BILLIONS

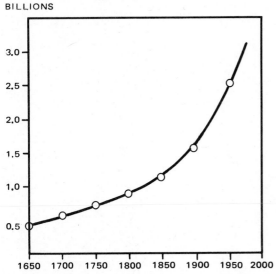

(a) THE INCREASE IN HUMAN POPULATION OVER THE LAST THREE HUNDRED YEARS.

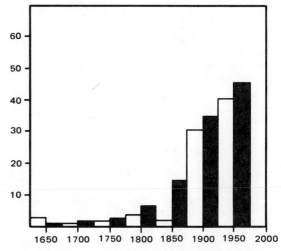

(b) THE NUMBER OF EXTERMINATED MAMMAL FORMS (WHITE BARS) AND BIRD FORMS (BLACK BARS) ELIMINATED OVER THE LAST THREE HUNDRED YEARS. EACH BAR REPRESENTS A FIFTY-YEAR PERIOD.[7]

Working with Data

KEY PROBLEM

How Much Does a 40-Pound Box of Bananas Weigh? [1]

Bananas boxed and shipped from the tropics (usually Central America) are supposed to weigh 40 pounds net when they arrive at their destination in North America, Europe, or Japan. But bananas lose weight due to evaporation and ripening. Also, different boxes weigh different amounts, since no two bunches of bananas are exactly the same.

A packing plant must adopt certain rules for weights of boxes when shipping so that the boxes are not less than 40 pounds when they are sold. In fact, a government-inspired regulation states that a box containing less than 40 pounds (no matter how close to 40 pounds) cannot be sold.

Ideas of this chapter are used to help make such rules for shipping.

233

9.1   SLIDING, STRETCHING AND SHRINKING

SLIDES.   It can be very useful, at times, to transform, or change, a set of scores from their original values to new values.

The simplest transformation to do on a set of scores is to slide all of them either to the right (which makes each score larger by the same amount) or to the left (which makes each score smaller by the same amount).

■ **Example 1:**   Six students get these scores on a quiz:

$$0, \quad 1, \quad 2, \quad 2, \quad 3, \quad 7$$

The scores are shown on the number line as in Figure 9.1(a).

FIGURE 9.1

Now, let's say the instructor decides to add 3 bonus points to each score (Table 9.1).

TABLE 9.1

OLD SCORE	ADD 3	NEW SCORE
0	+3	3
1	+3	4
2	+3	5
2	+3	5
3	+3	6
7	+3	10

Old total = 15   18   New total = 33

Old mean = $\dfrac{15}{6}$ = 2.5   New mean = $\dfrac{33}{6}$ = 5.5

As you can see in Figure 9.1 (b), adding 3 to each score moves each score 3 points to the right. This effect on the data is called a *slide*. Also note that the mean of the score is increased by 3, so it slides 3 points to the right as well. We call this increase a *slide factor of +3.*

Rule 9.1: If the same (positive) number is added to each score, the mean is *increased* by that number.

New mean = 5.5
New mean = old mean + 3.0

Note that a slide does not affect the *range* of a set of scores. For the *old scores,* the range was from 0 to 7.

Old range: $7 - 0 = 7$ points

For the *new scores,* the range is from 3 to 10.

New range: $10 - 3 = 7$ points ■

■ **Example 2:**   Suppose that the price of a can of corn is surveyed in five supermarkets:

53¢   62¢   63¢   71¢   77¢

Old mean = $\dfrac{326}{5}$ = 65.2 ¢

If the price were now lowered 10¢ in each store, the new prices would be shown on a number line like Figure 9.2. The slide factor here is −10.

FIGURE 9.2

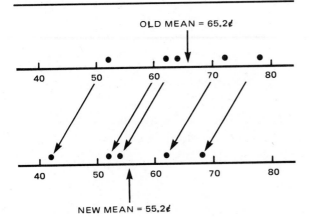

We see that the prices have moved to the left by 10 points. Note also that the mean has moved 10 points to the left.

New mean = 55.2¢ = old mean − 10¢

 Rule 9.2: If the same number is subtracted from each score, the mean score is *decreased* by that number.

Note, again, that we have, for the old scores:

Range = 77¢ − 53¢ = 24¢

and for the new scores:

Range = 67¢ − 43¢ = 24¢

So, the range is not affected by a slide to the left.

 Rule 9.3: Sliding a set of scores does not affect the range of the scores.

STRETCHING AND SHRINKING

■ **Example 3:** For the set of scores in Table 9.2, suppose we multiply each by 2.

TABLE 9.2

OLD SCORE	× 2	NEW SCORE
0	0 × 2	0
1	1 × 2	2
2	2 × 2	4
5	5 × 2	10
7	7 × 2	14
15		30

Old mean = $\frac{15}{6}$ = 2.5   New mean = $\frac{30}{6}$ = 5.0

The results are shown in Figure 9.3.

New mean = 2 × old mean
5 = 2 × 2.5

FIGURE 9.3

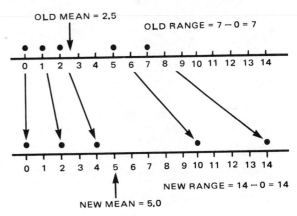

THE RANGE IS STRETCHED BY A FACTOR OF 2

The range is also stretched by that amount, which we call the *stretch factor*. Here the stretch factor is 2.

$$\text{New range} = 2 \times \text{old range}$$
$$= 2 \times 7$$
$$= 14$$

We have seen that stretching involves multiplying by some number greater than 1, such as 3 or 7. But suppose we multiply by a number less than 1, like $\frac{2}{3}$, as in the following example.

■ **Example 4**: A store has a sale and advertises that everything is one-third off. That is, the sale price is two-thirds of the regular price (Table 9.3).

TABLE 9.3

ITEM	REGULAR PRICE	SALE PRICE
Dress	$30.00	$\frac{2}{3}$ of 30 = $20.00
Shoes	$24.00	$\frac{2}{3}$ of 24 = $16.00
Hat	$21.00	$\frac{2}{3}$ of 21 = $14.00

We show these prices in Figure 9.4. The effect of multiplying by a number between 0 and 1 is to *shrink* a set of scores.

$$\text{Old range} = \$30 - \$21 = \$9$$
$$\text{New range} = \$20 - \$14 = \$6$$
$$\text{New range} = \frac{2}{3} \times \text{old range} = \frac{2}{3} \times \$9 = \$6$$

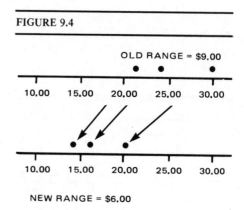

FIGURE 9.4

OLD RANGE = $9.00

NEW RANGE = $6.00

■ **Example 5**: The scores 4, 8, 5, 7, and 10 are quiz scores. Multiply each score by one-half.

Solution:

TABLE 9.4

OLD SCORE	$\times \frac{1}{2}$	NEW SCORE
4	$\frac{1}{2} \times 4$	2
8	$\frac{1}{2} \times 8$	4
5	$\frac{1}{2} \times 5$	2.5
7	$\frac{1}{2} \times 7$	3.5
10	$\frac{1}{2} \times 10$	5

$$\text{Old mean} = \frac{34}{5} = 6.8$$

$$\text{New mean} = \frac{17}{5} = 3.4$$

$$\text{New mean} = \frac{1}{2} \times \text{old mean} = \frac{1}{2} \times 6.8 = 3.4 \quad ■$$

INVERSE OPERATIONS

Example 5 involved multiplying by one-half. The same effect on the scores can be produced by *dividing* each score by 2. This is because multiplication and division are called *inverse operations*. One undoes the work of the other. So, we can think of multiplying by a number greater than 1 as stretching and dividing by a number greater than 1 as shrinking.

 Stretching a set of scores by 2 (multiplying each score by 2) is *undone* by *shrinking* a set of scores by 2 (dividing each score by 2).

We will think of shrinking in terms of multiplying by a fraction between 0 and 1. But we could also think of shrinking as dividing by numbers greater than one. We call such numbers *shrink factors.*

■ **Example 6:**   Here is a set of data: 6, 9, 4, 12. We *multiply* each by one-third, as shown in Table 9.5.

Solution:

TABLE 9.5

OLD SCORE	$\times \frac{1}{3}$	NEW SCORE
6	$\frac{1}{3} \times 6$	2
9	$\frac{1}{3} \times 9$	6
4	$\frac{1}{3} \times 4$	$1\frac{1}{3}$
12	$\frac{1}{3} \times 12$	4

We can get the same shrinking effect by *dividing* each score by 3. (Shrink factor is 3.)

Old Score	Shrink Factor: ÷3	New Score
6	$\frac{6}{3}$	2
9	$\frac{9}{3}$	3
4	$\frac{4}{3}$	$1\frac{1}{3}$
12	$\frac{12}{3}$	4

■

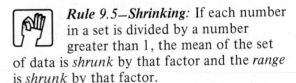 *Rule 9.4—Stretching:* If each score in a set is multiplied by the same number greater than 1, the mean of the set of data is *stretched* by that number and that *range* is *stretched* by that number.

Rule 9.5—Shrinking: If each number in a set is divided by a number greater than 1, the mean of the set of data is *shrunk* by that factor and the *range* is *shrunk* by that factor.

We can use these rules to find the mean of a set of scores in two ways. Suppose we want to find the mean of this set of data:

$$150, \quad 260, \quad 480, \quad 810$$

Method 1:   There are four scores. Add them and divide by 4.

$$\text{Mean} = \frac{150 + 260 + 480 + 810}{4} = \frac{1700}{4} = 425$$

Method 2:   Divide each of the scores by 10. Find the mean of the shrunken set. Multiply the shrunken mean by 10 to get the old mean (Table 9.6).

TABLE 9.6

OLD SCORE	SHRINK FACTOR: ÷10	NEW SCORE
150	150 ÷ 10	15
260	260 ÷ 10	26
480	480 ÷ 10	48
810	810 ÷ 10	81

$$\text{New mean} = \frac{15 + 26 + 48 + 81}{4} = \frac{170}{4} = 42.5$$

$$\text{Old mean} = 10 \times 42.5 = 425$$

EXERCISES

1   For this set of quiz scores, add 5 to each score (that is, use a slide factor of +5).

Old Score	Slide Factor	New Score
3		
6		
7		
6		
8		
12		

Then draw a graph to show the slide of +5.

2   For the scores in Exercise 1 of this section, find the old mean, the new mean, the old range, and the new range.

3   A survey shows these prices for a quart of milk in eight stores:

48¢ 52¢ 49¢ 58¢ 52¢ 61¢ 59¢ 55¢

Suppose that the dairy decides that the average (mean) price of a quart of milk can be allowed to rise by 5¢. If the price in each store is to be increased by the same amount, how much will a quart of milk in each store be after the increase? What is the range in the cost of a quart of milk in the eight stores before and after the price increase?

4   The prices of the same model of calculator in six cities are reported as follows:

$15.50 $16.25 $14.95 $15.80 $16.50 $15.80

Competition from other manufacturers forces the average (mean) price *down* by $1.65. If the price in each city goes down by the same amount, what will each calculator cost? What will be the range in prices across the six cities before and after the price increase?

Old Price	Slide Factor	New Price
$15.50		
16.25		
14.95		
15.80		
16.50		
15.80		

5   This set of data is to be stretched by a factor of 3. Find each new score.

Old Score	Stretch Factor	New Score
6		
7		
9		
10		
6		
7		

6   Find the mean of the new scores in Exercise 5 of this section by the two methods given in this section.

7   Draw a graph of the old scores and the new scores in Exercise 5 of this section. Show the old range and the new range on your graph.

8   This set of data is to be shrunk by a factor of 3. Find each new score.

Old Score	Shrink Factor	New Score
24		
18		
15		

9   Find the mean of the new scores in Exercise 8 using the two methods of this section.

10   For the data in Exercise 8 of this section, draw graphs of the old scores and the new scores and show the range of each set.

11   A record store advertises a half-price sale. Find the sale price of these albums (to nearest cent).

Regular Price	?	Sale Price
$6.50		
6.25		
7.00		
5.95		
5.50		

12   What is the average (mean) price of the albums in Exercise 11 of this section before the sale and after the sale?

13   Find the ranges of the regular prices and the sale prices in Exercise 11 of this section. Check the rule that shows how the range is affected by a shrink factor.

9.2   TAILOR-MADE SCORES

Let's review the effects of slides on a set of scores.

1.   A slide to the right (adding the same positive number to each score) *has these effects:*

- □ The mean of the scores is increased.
- □ The range of the scores is unchanged.

From Figure 9.5, it appears that the amount of variation in a set of scores is *not* changed by a slide. For example, it can be shown in more advanced courses that the standard deviation is not changed by a slide either. So we have another statement about slides on a set of scores:

- □ The standard deviation is unchanged.

FIGURE 9.5

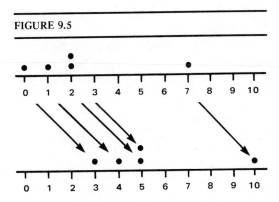

2.   A slide to the left (subtracting the same positive number from each score) *has these effects:*

- □ The mean of the scores is decreased.
- □ The range of the scores is unchanged.
- □ The standard deviation of the scores is unchanged.

Now recall the effects of stretching and shrinking on a set of scores. Specifically, remember that both the range and the mean were affected.

3.   Stretching (multiplying by a number such as 10) *has these effects:*

  □ The mean is multiplied by the stretch factor.
  □ The range is multiplied by the stretch factor.

It can also be shown that:

  □ The standard deviation is multiplied by the stretch factor.

Similarly, we have:

4.   Shrinking (dividing by a number like 3) *has these effects:*

  □ The mean is divided by the shrink factor.
  □ The range is divided by the shrink factor.
  □ The standard deviation is divided by the shrink factor.

COMBINING SLIDES AND STRETCHES

■ **Example 1:**   Suppose we first multiply each score by a stretch factor of, for example, 5 and then add 3 to each score. That is,

$$\text{New score} = 5 \times \text{old score} + 3$$

What is the effect on the mean, range, and standard deviation of a set of data with mean of 10 and standard deviation of 6?

Solution:

We first consider the mean.

1.   Multiplying each score by 5 multiplies the mean by 5 (stretch factor is 5).

$$\text{New mean} = 5 \times \text{old mean}$$
$$= 5 \times 10 = 50$$

2.   Adding 3 to each score increases the mean by 3. The mean was 50; now it is

$$50 + 3 = 53$$

Now for the standard deviation:

1.   Multiplying each score by 5 multiplies the standard deviation by 5.

$$\text{New standard deviation} = 5 \times \text{old standard deviation}$$
$$= 5 \times 6 = 30$$

2.   Adding 3 to each score has no effect on the standard deviation. Standard deviation remains as 30.

So, the change rule

$$\text{New score} = 5 \times (\text{old score}) + 3$$

produces a new set of data with

$$\text{New mean} = 5 \times (\text{old mean}) + 3$$
$$= 5 \times 10 + 3$$
$$= 53$$

$$\text{New standard deviation} = 5 \times \text{old standard deviation}$$
$$= 5 \times 6$$
$$= 30 \quad ■$$

CHANGING SCORES

Table 9.7 shows a set of scores obtained by eight students on a quiz.

TABLE 9.7

STUDENT	SCORE
Dean	8
Mary	6
Pete	5
Carlos	6
Stacey	9
Wendy	8
Kelly	7
Fred	3
Total	52

Using the methods learned in Chapter Two, we find the mean and standard deviation of these scores (Table 9.8).

TABLE 9.8

$$\text{Mean} = \frac{\text{total}}{N} = \frac{52}{6} = 6.5$$

STUDENT	DEVIATION FROM MEAN	(DEVIATION)^2
Dean	$8 - 6.5 = 1.5$	2.25
Mary	$6 - 6.5 = -.5$	.25
Pete	$5 - 6.5 = -1.5$	2.25
Carlos	$6 - 6.5 = -.5$	.25
Stacey	$9 - 6.5 = 2.5$	6.25
Wendy	$8 - 6.5 = 1.5$	2.25
Kelly	$7 - 6.5 = .5$	.25
Fred	$3 - 6.5 = -3.5$	12.25
Totals	0	26.00

$$\text{Variance} = \frac{26.00}{8} = 3.25$$

$$\text{Standard deviation} = \sqrt{3.25} \doteq 1.80$$

CHANGING THE MEAN

Although the mean and standard deviation for this set of data are perfectly respectable (and correct!), there are times in statistics when we want to report the scores in terms of some other mean or standard deviation. (See, for example, Section 9.3.)

Suppose, for the sake of argument, we would prefer our set of scores to have a mean of 10. We now know how this can be handled, for we have the rule:

New mean = old mean + slide factor

So, the needed slide factor is:

Slide factor = new mean − old mean
= 10.0 − 6.5 = 3.5

Thus we can have a set of scores with a mean of 10 by adding 3.5 to each score! (See Table 9.9.)

TABLE 9.9

STUDENT	OLD SCORE	SLIDE FACTOR	NEW SCORE
Dean	8	+ 3.5	11.5
Mary	6	+ 3.5	9.5
Pete	5	+ 3.5	8.5
Carlos	6	+ 3.5	9.5
Stacey	9	+ 3.5	12.5
Wendy	8	+ 3.5	11.5
Kelly	7	+ 3.5	10.5
Fred	3	+ 3.5	6.5
Totals	52	+ 28.0	80.0

$$\text{Old mean} = \frac{52}{8} = 6.5$$

$$\text{New mean} = \frac{80}{8} = 10.0$$

You should check for yourself that the mean of the new set of scores is indeed 10. Suppose, on the other hand, we wanted a mean smaller than 6.5—for example, 5 (a nice round number!). We can use a slide factor of −1.5 on each score (since 6.5 − 1.5 = 5.0). See Table 9.10.

TABLE 9.10

STUDENT	OLD SCORE	SLIDE FACTOR	NEW SCORE
Dean	8	− 1.5	6.5
Mary	6	− 1.5	4.5
Pete	5	− 1.5	3.5
Carlos	6	− 1.5	4.5
Stacey	9	− 1.5	7.5
Wendy	8	− 1.5	6.5
Kelly	7	− 1.5	5.5
Fred	3	− 1.5	1.5
Totals	52	− 12.0	40.0

$$\text{Old mean} = \frac{52}{8} = 6.5$$

$$\text{New mean} = \frac{40}{8} = 5.0$$

CHANGING THE STANDARD DEVIATION

First, recall that slides do not change the range or the standard deviation of a set of data. So, if you want to do the work, you can see for yourself that the slides just used on the above scores do not affect their standard deviation. The standard deviation of the new scores remains at 1.80, the same as the old standard deviation.

Stretches and shrinks, however, *do* change the range and standard deviation of a set of scores. For example, if we want a set of data to have a standard deviation twice as big, we can use a stretch factor of 2 (multiply each score by 2). We will not show how this is done. We will, however, look at the case of reducing the standard deviation. We have found that the quiz scores have a standard deviation of 1.80. Suppose we want to reduce this by one-half. We can use a shrink factor of 2 (Table 9.11).

TABLE 9.11

STUDENT	OLD SCORE	SHRINK FACTOR: 2	NEW SCORE	DEVIATION	(DEVIATION)^2
Dean	8	8 ÷ 2	4	4 − 3.25 = .75	.5625
Mary	6	6 ÷ 2	3	3 − 3.25 = − .25	.0625
Pete	5	5 ÷ 2	2.5	2.5 − 3.25 = − .75	.5625
Carlos	6	6 ÷ 2	3	3 − 3.25 = − .25	.0625
Stacey	9	9 ÷ 2	4.5	4.5 − 3.25 = 1.25	1.5625
Wendy	8	8 ÷ 2	4	4 − 3.25 = .75	.5625
Kelly	7	7 ÷ 2	3.5	3.5 − 3.25 = .25	.0625
Fred	3	3 ÷ 2	1.5	1.5 − 3.25 = − 1.75	3.0625
Totals	52		26.0		6.5

$$\text{New variance} = \frac{6.5}{8} = .8125$$

$$\text{New standard deviation} = \sqrt{.8125} \doteq .90$$

Notice, first, that the means are affected as we expected them to be.

$$\text{Old mean} = \frac{52}{8} = 6.5$$

$$\text{New mean} = \frac{26}{8} = 3.25$$

But, the shrink factor has brought the scores closer together, as the standard deviations of the old and new scores show.

$$\text{Old standard deviation} = 1.80$$

$$\text{New standard deviation} =$$

$$\frac{\text{old standard deviation}}{2} = \frac{1.80}{2} = .90$$

The order in which slides and stretches are done makes a difference in the effect on the mean.

■ **Example 2:**   A set of data has a mean of 10. Use a stretch of 2 and slide of 3.

Solution:

1.   Do the stretch of 2 and then the slide of 3.

$$\text{New mean} = 2 \times (\text{old mean}) + 3$$
$$= 2 \times 10 + 3$$
$$= 23$$

2.   Do the slide of 3 and then the stretch of 2.

$$\text{New mean} = (\text{old mean} + 3) \times 2$$
$$= (10 + 3) \times 2$$
$$= 13 \times 2$$
$$= 26 \ ■$$

Therefore, when slides or stretches (shrinks) are to be done, the *order* in which they are to be done must be made clear.

EXERCISES

1   The scores received by six students on a quiz were:

$$8, \ 5, \ 7, \ 5, \ 8, \ 9$$

(a)   Find the mean, range, and standard deviation of these data.

(b)   Add 10 to each of the scores (slide factor of +10). Find the mean, range, and standard deviation of the new scores.

2   Multiply each of the quiz scores in the above exercise by 5. What are the mean, range, and standard deviation of the new scores?

3   Change the quiz scores in Exercise 1 of this section to a new set of scores using the rule: First multiply each score by 5, then add 10 to the result. That is,

$$\text{New score} = 5 \times (\text{old score}) + 10$$

For example, the first quiz score is 8:

$$\text{New score} = 5 \times 8 + 10 = 40 + 10 = 50$$

What are the mean, range, and standard deviation of the new scores?

4   Explore the effect of interchanging the order of the transformations given in Exercise 3 above. Now, for the data of Exercise 1 of this section, do a slide of +10 and then a stretch of 5. Compare the *mean* of this new set of data with the results obtained in Exercise 3 of this section.

5   Change the quiz scores in Exercise 1 of this section by subtracting 5 from each (a slide factor of −5).

(a)   Complete this table. The first line is done for your.

Old Score	Slide Factor	New Score
8	8 − 5	3
5		
7		
5		
8		
9		

(b)   Find the mean, range, and standard deviation of the new scores.

6   Change the quiz scores in Exercise 1 of this section by dividing each by 2.

(a)   Complete this table. The first line is done for you.

Old Score	Shrink Factor: ÷2	New Score
8	8 ÷ 2	4
5		
7		
5		
8		
9		

(b)   Find the mean, range, and standard deviation of the new scores found in (a).

7   Change the quiz scores in Exercise 1 of this section to a new set by using the rule: First subtract 5 from each score, then divide the results by 2. That is,

$$\text{New score} = \frac{(\text{old score} - 5)}{2}$$

For example, if the old score is 8,

$$\text{New score} = \frac{(8 - 5)}{2}$$

$$= \frac{3}{2} = 1.5$$

What are the mean and standard deviation of the new scores?

8   A set of scores has mean of 10 and standard deviation of 2. Each score is multiplied by 3 and then 5 is added to the product. That is, the rule is:

$$\text{New score} = 3 \times (\text{old score}) + 5$$

What are the mean and standard deviation of the new set of scores?

9   A set of scores has a mean of 50 and a standard deviation of 10. The rule is:

$$\text{New score} = 2 \times (\text{old score}) + 10$$

Find the mean and standard deviation of the new scores.

10   Here is a set of scores obtained by five students on a quiz:

Student	Score
Sue	6
Mark	5
Scott	8
Stacey	8
Karyn	7

Find the mean of the scores. Then, use a slide factor of 10 on each score and find the new mean using the new scores.

11   Find the mean and standard deviation of the quiz scores in Exercise 1 of this section before and after the slide of +10. Compare your answers.

12 What slide factor would be needed so that the quiz scores in Exercise 1 of this section would have a new mean of 12? Show how you obtained your answer. What stretch (or shrink) factor would be needed so that the quiz scores would have a new standard deviation of 5?

13 Suppose we want the quiz scores in Exercise 1 of this section to have a standard deviation three times as large as it is now. Would you use a slide or a stretch (or shrink) factor on the scores? What would this factor be?

14 Carry out the transformation that you found to be needed in Exercise 12 of this section and calculate the new standard deviation. What is the new mean?

9.3 STANDARD SCORES

It is very useful in statistics to be able to produce a set of scores with a specified mean and standard deviation. And since 0 and 1 have nice properties, they are often chosen as the desired values for the mean and standard deviation, respectively. For example, suppose we want to change the quiz scores of the previous section (repeated in Table 9.12) so that they have a mean of 0 and a standard deviation of 1.

TABLE 9.12

STUDENT	OLD SCORE
Dean	8
Mary	6
Pete	5
Carlos	6
Stacey	9
Wendy	8
Kelly	7
Fred	3

CHANGING THE MEAN TO ZERO

Since the mean of these scores is, at present, 6.5, we need a slide factor to lower the mean.

Slide factor = new mean − old mean = 0 − 6.5 = − 6.5

So, we slide each score 6.5 units to the left (Table 9.13).

TABLE 9.13

STUDENT	OLD SCORE	SLIDE FACTOR: −6.5	NEW SCORE
Dean	8	8 − 6.5	1.5
Mary	6	6 − 6.5	− .5
Pete	5	5 − 6.5	−1.5
Carlos	6	6 − 6.5	− .5
Stacey	9	9 − 6.5	2.5
Wendy	8	8 − 6.5	1.5
Kelly	7	7 − 6.5	.5
Fred	3	3 − 6.5	−3.5
			0

We find the new mean to be 0.

But notice also that we have seen this "new score" before. Since the slide factor is the old mean, the new score is what we called the deviation score in Chapter Two! So, we have a rule.

Rule 9.6: To change a set of data so that their mean is 0, do a slide transformation, which involves subtracting the old mean from each score. (The slide factor is the old mean.)

CHANGING THE STANDARD DEVIATION TO ONE

We have already seen that slides do not change the standard deviation of a set of scores. So, the standard deviation of the new scores in Table 9.12 is 1.80.

We need to make the standard deviation smaller, so we will shrink the set. But what should the shrink factor be?

Let's try 2. A shrink factor of 2 would cut the standard deviation in half—that is, the new standard deviation would be

$$\frac{1.80}{2} = .90$$

So, 2 is too big. We need a shrink factor *smaller* than 2 to produce a standard deviation of 1.0.

The necessary shrink factor is, as perhaps you have already guessed, the standard deviation itself! So, the needed shrink factor is 1.80 (Table 9.14).

Notice, too, that the mean of the scores after the slide is still 0, as we required; when 0 is divided by the shrink factor, the result is again 0.

 Rule 9.7: To change a set of data so that their mean is 0 and their standard deviation is 1: (a) Do a slide transformation, by subtracting the old mean from each score. (b) Do a shrink transformation, by dividing the result of (a) by the old standard deviation (the shrink factor is the old standard deviation). *Note:* If the shrink factor is less than 1, the effect of this transformation is a stretch!

TABLE 9.14

STUDENT	OLD SCORE	SLIDE FACTOR: −6.5	SHRINK FACTOR: ÷1.80	NEW SCORE (z SCORE)
Dean	8	8 − 6.5	$\dfrac{(8 - 6.5)}{1.80}$	.83
Mary	6	6 − 6.5	$\dfrac{(6 - 6.5)}{1.80}$	− .28
Pete	5	5 − 6.5	$\dfrac{(5 - 6.5)}{1.80}$	− .83
Carlos	6	6 − 6.5	$\dfrac{(6 - 6.5)}{1.80}$	− .28
Stacey	9	9 − 6.5	$\dfrac{(9 - 6.5)}{1.80}$	1.39
Wendy	8	8 − 6.5	$\dfrac{(8 - 6.5)}{1.80}$	.83
Kelly	7	7 − 6.5	$\dfrac{(7 - 6.5)}{1.80}$	.28
Fred	3	3 − 6.5	$\dfrac{(3 - 6.5)}{1.80}$	−1.94

The resulting score, called a *z-score*, is thus obtained by the following transformation.

PROPERTIES OF z SCORES

If you have a set of scores and change them to z scores, then the following are true:

1. The mean of the z scores is 0.
2. The standard deviation of the z scores is 1.

Why would we ever want to change means and standard deviations of scores at all? An important reason for wanting to change the mean and standard deviation of a set of scores is to help us make meaningful comparisons between different sets of scores.

Suppose, for example, that Tom takes two tests in statistics. The scores on the tests are:

Test I	Test II
41 47 62 48 51	33 28 39 22 37
56 47 43 51 54	33 33 30 33 33

On the first test, he gets 54 and on the second he gets 37. How can we compare how he did on the two tests? We need more information. If the second test were much more difficult than the first, a score of 37 may be much better than a score of 54, for example.

One way of determining the difficulty of a test is to find the average, or mean score, on each test. Then, by comparing Tom's score with the mean, we can compare his performance on the test with the average score. But this score is just the deviation score:

Deviation score for Tom = Tom's score − mean score

We have these statistics on the two tests:

Mean score for test I: 50.0
Mean score for test II: 32.1

Now for Tom's score of 53 on the first test, we have

Tom's deviation score for test I = 54 − 50.0 = 4.0
Tom's deviation score for test II = 37 − 32.1 = 4.9

So, now we know that Tom was 4 points above the mean on test I and 4.9 points above the mean on test II. We now are in a better position to comment on Tom's performance on the two tests. Even though 37 is much less than 54, both scores were good in that they were above the mean score for the class in each case.

But now we have more information about Tom's score. Figure 9.6 shows a graph of the scores of the class for each test. Notice that in test I, the scores were spread out more than in test II.

FIGURE 9.6

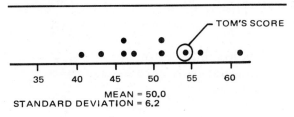

MEAN = 50.0
STANDARD DEVIATION = 6.2

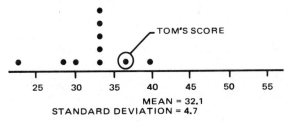

MEAN = 32.1
STANDARD DEVIATION = 4.7

We can describe the amount of spread by calculating the standard deviation of each set of scores:

Standard deviation of test I scores = 6.2
Standard deviation of test II scores = 4.7

Note: This may be a good place to use a calculator that finds standard deviation.

A good way of describing how well Tom did on the test is to tell how many standard deviations his score is either above or below the mean of all of the scores on the test. We can find this out by dividing Tom's deviation score by the standard deviation of the set of scores.

$$\text{Tom's } z \text{ score for test I} = \frac{54 - 50.0}{6.2} = \frac{4}{6.2} \doteq 0.65$$

$$\text{Tom's } z \text{ score for test II} = \frac{37 - 32.1}{4.7} = \frac{4.9}{4.7} \doteq 1.04$$

We see that Tom's score of 37 on test II was a little more than one standard deviation above the mean (his z score is 1.04), while his score of 54 on test I was less than one standard deviation above the mean (his z score is .65). So, relative to the other scores, he did better (got a higher score) on test II than on test I.

Now note another useful property of z scores. Suppose that Harry got a score of 41 on test I. Harry's z score is

$$\frac{41 - 50}{6.2} = \frac{-9}{6.2} = -1.45$$

This score is negative. So we know right away that Harry's score is below the mean. In fact, the z score of −1.45 tells us that his score was about one-and-one-half standard deviations *below* the mean on the test.

TRANSFORMED SCORES

If we have a set of z scores, we can change them to a new set of scores having *any* mean and standard deviation we choose. Using the transformation rules of this chapter, let's change this set of scores to a set with a mean of 100 and a standard deviation of 15 (Table 9.15).

The transformation rule used in Table 9.15 is:

$$T \text{ score (transformed score)} = (z \text{ score}) \times 15 + 100$$

Let's talk a bit about this rule to see how it works. The old scores were first changed to z scores. The z scores have a mean of 0 and a standard deviation of 1.0. We multiply each z score by 15, the desired standard deviation of the new set of scores (the transformed scores). This factor of 15 stretches the standard deviation from 1.0 to 15.0. But what happens to the mean? The old mean was 0. The new mean, after a stretch of 15, is 15 times 0, which is still 0. So, the mean has not been affected by this stretch. That is, the result of the rule

$$z \times 15$$

is a set of scores with a mean of 0 and a standard deviation of 15.

Now, we add 100 to each score. This is a slide to the right of 100 points. So, the mean of the set of scores has also been slid to the right by 100 points.

$$\text{New mean} = \text{old mean} + 100$$

The mean was 0. Now it is 100. What about the standard deviation? Remember that a slide does not affect the standard deviation, or spread, of a set of scores. So the standard deviation is still 15.

TABLE 9.15

STUDENT	OLD SCORE	z SCORE	STRETCH FACTOR: ×15	SLIDE FACTOR: +100	TRANSFORMED SCORE (ROUNDED)
Dean	8	.83	.83 × 15 = 12.45	12.45 + 100 = 112.45	112
Mary	6	− .28	− .27 × 15 = 4.20	− 4.20 + 100 = 95.80	96
Pete	5	− .83	− .83 × 15 = −12.45	−12.45 + 100 = 87.55	88
Carlos	6	− .28	− .28 × 15 = 4.20	− 4.20 + 100 = 95.80	96
Stacey	9	1.39	1.39 × 15 = 20.85	20.85 + 100 = 120.85	121
Wendy	.8	.83	.83 × 15 = 12.45	12.45 + 100 = 112.45	112
Kelly	7	.28	.28 × 15 = 4.20	4.20 + 100 = 104.20	104
Fred	3	−1.94	−1.94 × 15 = 29.10	−29.10 + 100 = 70.90	71
Total		0			800

$$\text{Mean of } z \text{ scores} = \frac{0}{8} = 0$$

$$\text{Mean of } T \text{ scores} = \frac{800}{8} = 100$$

The transformation rule

$$T \text{ score} = z \times 15 + 100$$

gives a set of scores with a mean of 100 and a standard deviation of 15.

Let's use transformed scores to compare how well Tom did on test I and test II. We will transform his scores to new scores having a mean of 100 and a standard deviation of 15.

Test	Old Score	z Score	Transformed Score
I	54	.65	.65 × 15 + 100 = 9.75 + 100 ≐ 110
II	37	1.04	1.04 × 15 + 100 = 15.60 + 100 ≐ 116

The transformed scores tell us that Tom did, indeed, do better on test II than test I. (His transformed score was 116 on test II and 110 on test I.)

EXERCISES

1   Here are the scores obtained by a class on a quiz:

$$5, \ 7, \ 6, \ 5, \ 7$$

Change each score to a z score.

2   A set of data has a mean of 18.1 and a standard deviation of 4.2. How can the data be changed so that they have a mean of 0 and a standard deviation of 1.0?

3   A set of scores has a mean of 20.3 and a standard deviation of 5.3. How can each score be changed so that the scores have a mean of 0 and a standard deviation of 1.0?

4   How can the data in Exercise 2 of this section be changed so that they have a mean of 50 and a standard deviation of 20?

5   How can the data in Exercise 2 of this section be changed so they have a mean of 500 and a standard deviation of 50?

6   A set of test scores has a mean of 45.3 and a standard deviation of 4.8. Polly got a score of 52 on the test and Ron got a score of 40. Change their scores to z scores. Now change their scores so the scores have a mean of 100 and a standard deviation of 10.

7   Barb took two tests. The mean and standard deviation of each test were as shown below:

Test	Mean	Standard Deviation
I	34	5.3
II	48.3	6.8

Barb got scores of 40 on test I and 44 on test II. Change her scores to z scores and use them to decide on which test she did better.

8   Change the scores in Exercise 7 above so that for both tests the mean is 200 and the standard deviation is 100. What is Barb's score on each test when changed to these new values? On which test did Barb do better?

9   Here is a set of scores on a quiz. Change them so that they have a mean of 50 and a standard deviation of 10.

$$5, \ 7, \ 3, \ 4, \ 5, \ 5, \ 6, \ 4, \ 8, \ 6$$

10   Here is a set of scores on a final exam. Change each so that the scores have a mean of 100 and a standard deviation of 15.

$$32, \ 45, \ 43, \ 48, \ 51, \ 50, \ 43, \ 44, \ 37, \ 55$$

9.4   THE UNIFORM (FLAT) CURVE

So far in this chapter, we have learned some ways of changing the mean and standard deviation of a set of data. Now we will consider another important aspect of working with data, that of *smoothing*. (We talked about this idea in Chapter Six.)

We begin with an example.

■ **Example 1 — The Fair Die:**   Suppose we throw a die 300 times and obtain the results in Table 9.16.

TABLE 9.16

OUTCOME	f	RELATIVE FREQUENCY (P)
1	42	.14
2	58	.19
3	61	.20
4	36	.12
5	38	.13
6	65	.22
Totals	300	1.00

We have already seen how we can draw graphs of these data. A histogram and frequency polygon would be as given in Figure 9.7.

The *relative* frequency polygon for these data looks like Figure 9.8.

Remember the only difference between a frequency polygon and a *relative* frequency polygon—it is in the vertical axis. A frequency polygon tells us how many of each outcome were obtained. For example, a 1 was obtained 42 times in the 300 tosses. A relative frequency polygon tells us *what fraction*, or proportion, of the total number of 300 tosses was obtained for each outcome. So, for example, we know that

$$\frac{42}{300} \text{ or } .14$$

of all of the tosses turned out to be 1s.

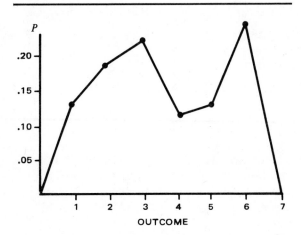

FIGURE 9.8

We will now see how a graph of data can be used to obtain a probability model. We throw our die a total of 3000 times, instead of 300 times, and get the results in Table 9.17.

FIGURE 9.7

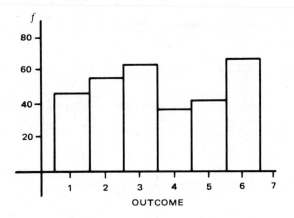
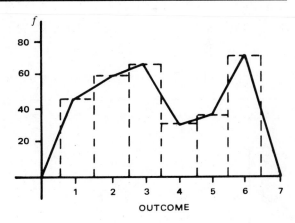

TABLE 9.17

OUTCOME	f	RELATIVE FREQUENCY
1	514	.17
2	498	.17
3	522	.17
4	489	.16
5	504	.17
6	473	.16
Totals	3000	1.00

Notice that while the *frequency* polygon is still bumpy, the *relative* frequency polygon has smoothed out quite a bit for this very large number of tosses (3000). See Figure 9.9.

If we continue tossing our die for thousands and thousands of times, what would we expect to happen? Well, if our die is *not* loaded (or biased), we would expect that we would get each of the 6 outcomes about one-sixth of the time. Remember,

$$\frac{1}{6} \doteq .17$$

We can come to this same conclusion by referring to the relative frequency polygon in Figure 9.9. Notice that for 3000 throws of the die, the relative frequency polygon is almost a straight, smooth, horizontal line. The horizontal line shows that we expect each outcome to occur one-sixth of the time for a fair (not loaded) die. Such a flat curve is said to describe a *uniform distribution* (Figure 9.10).

FIGURE 9.10

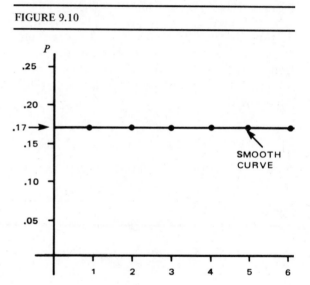

FIGURE 9.9

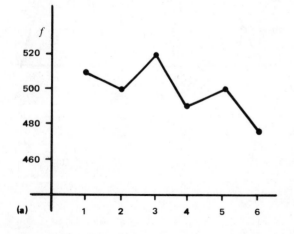

(a)

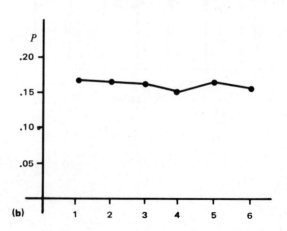

(b)

Drawing a straight, horizontal line amounts to stating that, on the basis of 3000 rolls, we assume the die to be fair. That is, we have used our data as a basis for making the assumption of equally likely outcomes. Thus,

$$p(1) = \frac{1}{6}, \quad p(2) = \frac{1}{6},$$

and so on.

Therefore, we should think of smoothed curves as providing theoretical probabilities. This is a very important idea in statistics.

We stated earlier that estimated probabilities get closer and closer to theoretical probabilities as more and more trials are conducted. For example, the estimated probability of getting a 1 on a die is

$$P(1) = .14$$

based on 300 trials and

$$P(1) = .171$$

based on 3000 trials. These estimates are getting closer and closer to

$$p(1) = \frac{1}{6} \doteq .17$$

the theoretical probability of getting a 1. The same is true for the other five sides of the die.

So, we can think of smoothing a graph as a way of removing the estimation error from our real-world data and having as a result theoretical probabilities. For example,

$$P(1) = .14 \text{ becomes } p(1) = .17$$

as a result of smoothing, and

$$P(2) = .19 \text{ becomes } p(2) = .17$$

and so on. As we have noted several times in this book, the experimental probabilities are estimates of theoretical probabilities, which are based on some probabilistic model (such as a fair die).

In Chapters Five and Six, we studied ways in which we could decide if a certain probabilistic model is an appropriate one to assume as a source of the data that have been obtained. In particular, the chi-square test can tell us whether a horizontal line is an appropriate smooth curve for a set of data. Remember, in Chapter Six, we were deciding whether or not a fair die was an appropriate model for our data (see especially Section 6.5). ∎

■ **Example 2–The Well-Shuffled Deck of Cards:**   A regular deck of playing cards has 52 cards, 13 of each of four suits: hearts; diamonds; clubs; spades. Suppose we shuffle the deck well, deal a card, and determine its suit. We make a tally in Table 9.18, replace the card in the deck, shuffle the deck, and repeat 100 times.

TABLE 9.18

OUTCOME	f
Hearts	24
Diamonds	21
Clubs	24
Spades	31
Total	100

The frequency polygon and relative frequency polygon in Figure 9.11 are graphs of the data. Because the relative frequency polygon is somewhat smooth, and flat, we draw in a horizontal line. Thus we assume our model is that of four equally-likely outcomes. Each suit has an equal chance of being dealt. So, for example, out of 100 deals from the deck (in our game, the card is always put back), we expect hearts to be dealt

$$\frac{1}{4} \times 100$$

or 25 times.

From our table and from the graphs, we have these estimated probabilities:

$$P(\text{hearts}) = .24$$
$$P(\text{diamonds}) = .21$$
$$P(\text{clubs}) = .24$$
$$P(\text{spades}) = .31$$

The smooth curve suggests that each of the four suits is expected to appear an equal number of times—25. We therefore estimate these theoretical probabilities to be:

$$p(\text{hearts}) = \frac{1}{4}$$

$$p(\text{diamonds}) = \frac{1}{4}$$

$$p(\text{clubs}) = \frac{1}{4}$$

$$p(\text{spades}) = \frac{1}{4}$$

We will apply the chi-square test to these data to see if the cards seem to have been fairly dealt from a well-shuffled deck. Table 9.19, which you should recognize from Chapter Six, shows the expected number of cards dealt from each suit, assuming a uniform distribution.

FIGURE 9.11

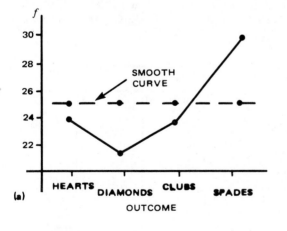

(a)

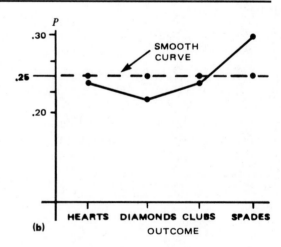

(b)

TABLE 9.19

OUTCOME	EXPECTED $f(E)$	OBTAINED $f(O)$	$E - O$	$(E - O)^2$
Hearts	25	24	1	1
Diamonds	25	21	4	16
Clubs	25	24	1	1
Spades	25	31	-6	36
				Total 54

$$\chi^2 = \frac{1}{25} + \frac{16}{25} + \frac{1}{25} + \frac{36}{25} = \frac{54}{25} = 2.16$$

This chi-square has 3 degrees of freedom, since there are four rows in Table 9.19. We see from Appendix C that a chi-square this large, or larger, occurs frequently (more than .10 of the time) by chance. So, our data are consistent with our assumption that we were dealing from a well-shuffled deck. (You may wish to refer back to Chapter Six for a review of the chi-square test.) ∎

These examples illustrate a central idea of statistics. That is, we often start with data from the real world—number of cases of lung cancer, yield of a crop when fertilizer has been applied, number of accidents occurring in various shifts of a factory, and so on—and search for a theoretical probability model that can be thought of as generating such data.

EXERCISES

1   A six-sided die was rolled 120 times and the outcomes in Table 9.20 were obtained. Draw a frequency polygon and relative frequency polygon of these data and draw in the smooth curve to fit these data, assuming the die was fair. Through what frequency on the vertical axis does the smooth curve pass?

TABLE 9.20

OUTCOME	f
1	23
2	18
3	17
4	25
5	21
6	16
Total	120

Interpret the smooth curve you have drawn.

2   Toss a fair (not loaded) six-sided die 60 times and record the data as in Exercise 1 above. Draw a relative frequency polygon of the data and then draw in the smooth curve for your data. Through what value does the smooth curve pass?

3   The last digit was recorded from each of 1000 telephone numbers in a phone book, as shown in Table 9.21.

TABLE 9.21	
DIGIT	FREQUENCY
1	77
2	93
3	106
4	123
5	94
6	88
7	117
8	104
9	92
0	106
Total	1000

Draw a relative frequency polygon of these data. Draw the smooth curve to fit these data. Through what frequency on the vertical axis does the smooth curve pass? From your graph, what is $P(1)$? What is $P(5)$? What is P(even number)? Also from your graph, what do you estimate are these theoretical probabilities to be, assuming a uniform (flat) probability distribution: $p(1)$; $p(6)$; $p(0)$; p(even number)?

4   From your own telephone book, record the last digit of 100 numbers. Answer the questions for your data that are asked in Exercise 3.

5   In Exercise 6 of Section 1.4, it is stated that the United Nations Demographic Yearbook for 1977 reports the world's population for 1975 as 1,987,049,000 women and 1,979,956,000 men. Draw a relative frequency polygon of these data. Draw in the smooth curve to fit these data, assuming a uniform distribution. Find P(woman) from your graph. What do you estimate p(woman) to be?

6   A coin was tossed 1000 times. Draw a relative frequency polygon for the following data.

Outcome	f
H	457
T	543
	1000

Then draw in the smooth curve assuming the coin to be fair. Find these values:

(a) P(heads)
(b) P(tails)
(c) p(heads)
(d) p(tails)

7   Toss a coin and roll a six-sided die. Record the outcome using a pair of numbers. Let heads be 0 and tails be 1. Then, for example, if you get heads on the coin and a 5 on the die, you would write $(0,5)$. Repeat the tosses of the coin and die for a total of 60 tosses. Draw a relative frequency polygon of your data. Then draw in the smooth curve assuming the coin and die to be fair.

(a)   What are these values?
    $P(0,5)$; $P(1,6)$
(b)   Estimate these theoretical probabilities:
    $p(0,5)$; $p(1,6)$

8   Take a deck of playing cards. Shuffle it well and deal a card. Make a record of the suit of the card; then put the card back in the deck. Repeat for a total of 40 times. Draw a relative frequency polygon of your results, then draw in the smooth curve. What are these values?

(a) P(hearts)   (b) P(clubs)
(c) P(red card)   (d) p(hearts)
(e) p(clubs)   (f) p(red card)

9   A cork can fall one of three ways when tossed: on its top, its bottom, or its side, as shown in Figure 9.12. Suppose a cork is tossed 1000 times with these results:

Outcome	Frequency
Top	109
Bottom	653
Side	238
Total	1000

Draw a relative frequency polygon of these data and find these estimated probabilities: $P(\text{top})$; $P(\text{bottom})$; $P(\text{side})$. Does it appear that these data smooth to a straight (horizontal) line?

10   Three coins were tossed 100 times and the number of heads was recorded as follows:

N Heads	Frequency
0	10
1	29
2	41
3	20
Total	100

Draw a relative frequency polygon of these data and find: $P(0 \text{ heads})$; $P(1 \text{ head})$; $P(3 \text{ heads})$. Do you think that a horizontal line is an appropriate smooth curve for these data?

FIGURE 9.12

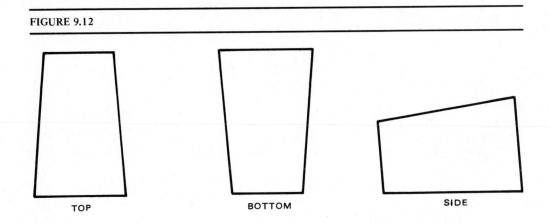

TOP   BOTTOM   SIDE

11 The final fraction in the closing stock price of each 317 low-priced stocks on the American Stock Exchange for a certain day is given in Table 9.22. Draw a relative frequency polygon of these data. Under the assumption that final stock market fractions have a uniform distribution, what is the expected number of stocks for each fraction?

TABLE 9.22[2]
FREQUENCY DISTRIBUTION OF FINAL FRACTION ON STOCKS SELLING AT OR BELOW 10 ON THE AMERICAN STOCK EXCHANGE

FRACTION	EIGHTHS	FREQUENCY
0	0	60
$\frac{1}{8}$	1	30
$\frac{1}{4}$	2	29
$\frac{3}{8}$	3	27
$\frac{1}{2}$	4	47
$\frac{5}{8}$	5	49
$\frac{3}{4}$	6	37
$\frac{7}{8}$	7	38
		317

12 In Exercise 11 above, remove from the table the zero cases (the 60 stocks whose final fraction was 0). Now do a chi-square test to see if the remaining data can be assumed to come from a uniform probability model.[3]

13 The statistician Carlson reports 10,000 digits obtained from a telephone book as shown in Table 9.23.[4] Carlson states, "It may not be obvious, but there are too few 9's here" if the digits are to be assumed to be truly random (to be generated by a uniform probability model such as a fair die). Do a test and make a decision as to whether Carlson's assertion should be accepted as correct.

TABLE 9.23

DIGIT	f
0	1,026
1	1,107
2	997
3	966
4	1,075
5	933
6	1,107
7	972
8	964
9	853
Total	10,000

9.5 NORMAL (BELL-SHAPED) CURVE

We begin with data on the weight of the 50 girls from Central High. They are reported in the stem-and-leaf table (Table 9.24).

TABLE 9.24

STEM	LEAF	FREQUENCY
15	3, 2	2
14	8, 6	2
13	2, 8, 5, 7, 9, 3, 4, 1, 7, 5, 2	11
12	2, 0, 9, 7, 4, 8, 1, 5	8
11	7, 4, 5, 4, 8, 4, 0, 9, 0, 2	10
10	3, 5, 6, 8, 3, 0, 7, 1, 0	9
9	2, 9, 8, 2	4
8	8, 9, 9, 2	4
	Total	50

The mean of this set of data is

$$\overline{X} = \frac{\text{sum of weights}}{\text{number of girls}} = \frac{5883}{50} = 117.7 \text{ pounds}$$

The standard deviation of the data is

$$S = \sqrt{\frac{\text{sum of (deviation)}^2}{\text{number of girls}}} = \sqrt{\frac{15,793.22}{50}} \doteq 17.7 \text{ pounds}$$

Let's use a stem-and-leaf table to ask another question about the data on the girls at Central High. Suppose that we go to the school and select one of these girls at random. That is, each of the 50 girls is as likely to be chosen as any other. What is the probability that the girl we pick has a weight between 100 pounds and 108 pounds, inclusive?

Table 9.24 tells us that of the 50 girls, 9 weigh between 100 pounds and 108 pounds. So we can say that

$$p(\text{girl picked at random weighs between 100 and 108 pounds}) = \frac{9}{50} = .18$$

Other probability statements about the weights of the girls at Central are:

$$p(131 \leqslant \text{weight of girl picked at random} \leqslant 139) = \frac{11}{50} = .22$$

$$p(\text{weight of girl picked at random} \leqslant 100) = \frac{10}{50} = .20$$

Figure 9.13(a) shows a frequency polygon of the weights of the 50 girls at Central High School. Suppose we now go out and pick another 550 girls at random from the same school, so that we have the weights for a total of 600 girls, as shown in Figure 9.13(b).

FIGURE 9.13

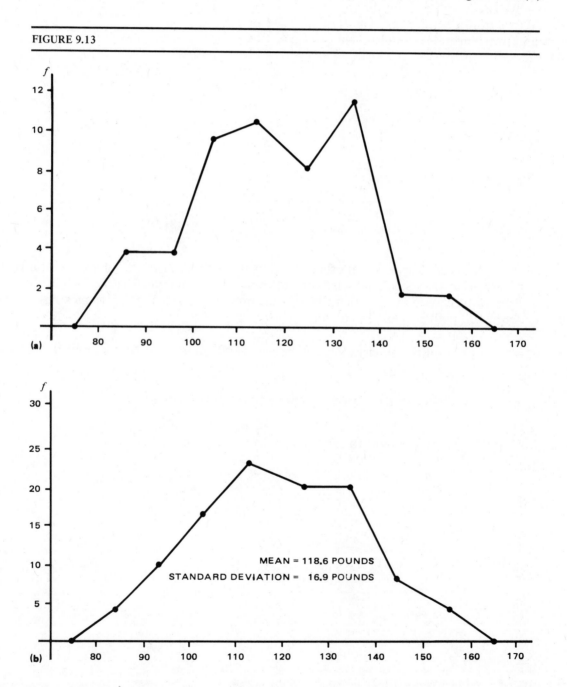

Now let's smooth the polygon for the 600 weights. We get a curve like the one in Figure 9.14. Similarly, if we draw a relative frequency polygon and smooth it, we get the graph in Figure 9.15. Notice that the graph is rather bell-shaped.

FIGURE 9.14

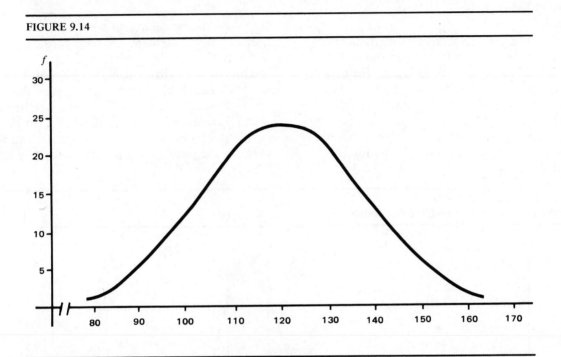

FIGURE 9.15

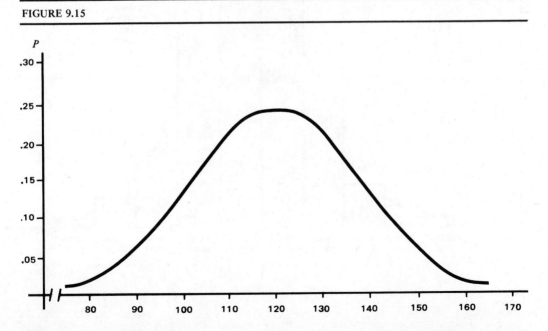

Now let's consider the results of 102 laboratories that tested the strength of the same batch of concrete at a construction site.[5] We get the frequency polygon and smoothed curve in Figure 9.16.

We have obtained another bell-shaped curve. It is a very important fact about the real world that the frequency polygons for many sets of data can be smoothed to bell-shaped curves. Statisticians have a name for such curves. They call these curves *normal*. A good reason for the name *normal* is that in many normal, or natural cases in daily living, the frequency polygon (and the relative frequency polygon) of data which is obtained can be smoothed to a curve having this bell-shaped appearance.

One statistical property of a normal curve is that the mean, median, and mode are all the same. Let's take the example of the weights of the high school girls. From the graph in Figure 9.17 and Table 9.23, it can be seen that the smoothed curve is highest for a weight of about 116 pounds. This means that a weight of about 116 pounds is most frequent, and we therefore choose 116 as the mode for the data.

FIGURE 9.16
HISTOGRAM FOR CEMENT TESTS REPORTED BY 102 LABORATORIES

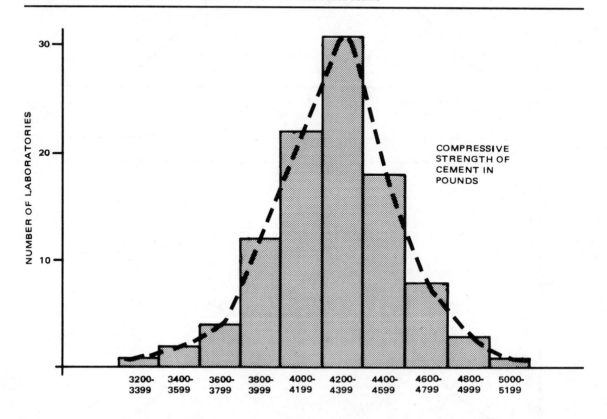

FIGURE 9.17

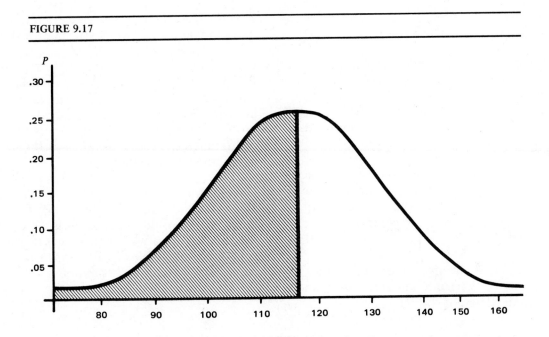

The median can be found from the graph. Since one-half of the scores are below the median and one-half are above, we find the point that divides the area under the smooth curve into two equal parts. The line marked in the graph does this. Since we can think of this line as marking off equal proportions of the weights, this line then determines the median weight, which is about 116 pounds.

An important property of normal curves has to do with the areas, and, therefore, with probabilities. As was stated in Chapter One, the bars of histograms and *therefore* the areas enclosed by polygons can be interpreted as probabilities. (Also recall that in Chapter Six, we used areas under curves to give us probabilities for chi-square statistics.) These probabilities are so useful that special tables have been prepared to tell what these theoretical probabilities are for various points under the normal curve. We will see how to use these tables in Section 9.6.

SOME PROPERTIES
OF THE NORMAL CURVE

Each normal curve has a mean and a standard deviation. Look at three normal curves, shown in Figure 9.18. They each have the same standard deviation (it happens to be 5.0) but different means. The mean tells the location of the center of the curve.

Figure 9.19 shows three normal curves with the same mean, 0, but different standard deviations. The graphs are drawn one on top of the other so that you can compare their standard deviations.

If we are finding the area under a part of a normal curve, we need to know the mean and standard deviation of the curve.

Suppose, for example, we have a curve with a mean of 0 and standard deviation of 1. This curve is shown in Figure 9.20.

It can be shown mathematically that the area under a normal curve is divided up as shown in Figure 9.21. That is, one standard deviation on either side of the mean represents 68%, or approximately two-thirds, of all of the scores in a normal distribution.

Let's look at this in terms of the weights of 100 twelfth grade girls (Table 9.25). Theoretically, we would expect two-thirds of all of the girls to have a weight between one standard deviation above the mean, *if their weights are normally distributed.*

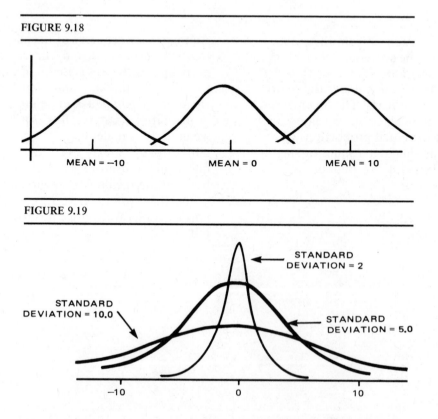

FIGURE 9.18

MEAN = −10   MEAN = 0   MEAN = 10

FIGURE 9.19

STANDARD
DEVIATION = 2

STANDARD
DEVIATION = 10.0

STANDARD
DEVIATION = 5.0

−10   0   10

FIGURE 9.20

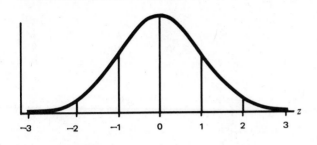

FIGURE 9.21

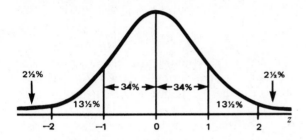

For this sample of 100 girls, the mean is 118.67 pounds and the standard deviation is 16.91 pounds.

One standard deviation below the mean:

$$118.67 - 16.91 = 101.76 \text{ pounds}$$

One standard deviation above the mean:

$$118.67 + 16.91 = 135.58 \text{ pounds}$$

So, we estimate that 68% of the girls weigh between 101.8 pounds and 135.6 pounds.

Since we know the actual weights of the 100 girls, we can find how many are between 101.8 and 135.6 pounds by referring to Table 9.25.

We count all the weights for the 11- and 12-stems (that is, $22 + 16 = 38$). Then, from the 10-stem, we include 12 weights and from the 13-stem, we include 13 weights. Therefore,

$$38 + 12 + 13 = 63$$

girls are in the interval 101-135 pounds. It appears that there is some evidence that the distribution of the weights of the girls is close to normal. Therefore, our theoretical probability model, a normal probability distribution, is plausible as a model for our real-world data, weights of high school girls.

TABLE 9.25
STEM-AND-LEAF TABLE OF WEIGHTS OF 100 HIGH SCHOOL GIRLS

STEM	LEAF	FREQUENCY
15	3, 2, 3	3
14	8, 6, 2, 5, 4, 7, 9, 2, 3	9
13	2, 8, 5, 7, 9, 3, 4, 1, 7, 5, 2, 7, 0, 2, 0, 3, 5, 3	18
12	2, 0, 9, 7, 4, 8, 1, 5, 5, 7, 4, 1, 4, 2, 2, 4	16
11	7, 4, 5, 4, 8, 4, 0, 9, 0, 2, 6, 5, 8, 1, 0, 0, 0, 6, 2, 0, 0, 3	22
10	3, 5, 6, 8, 3, 0, 7, 1, 0, 6, 2, 0, 2, 5, 8, 2	16
9	2, 9, 8, 2, 9, 7, 4, 7, 4, 8, 7	11
8	8, 9, 9, 2, 9	5
		Total 100

EXERCISES

1   The chest measurements of 1516 soldiers (from the Army of the Potomac) are given in Table 9.26.[6]

TABLE 9.26

CHEST CIRCUMFERENCE (DISTANCE AROUND, IN INCHES)	NUMBER OF SOLDIERS
28	2
29	4
30	17
31	55
32	102
33	180
34	242
35	310
36	251
37	181
38	103
39	42
40	19
41	6
42	2

Draw a frequency polygon of these data. Do the data appear to be normally distributed? (Use your judgment.)

2   Construct a stem-and-leaf table for these data, the heights (in inches) of 30 cub scouts at a summer camp.

```
48  52  43  47  51  40  46  62  53  50
45  45  45  40  55  53  44  50  50  58
53  50  51  53  50  44  49  49  40  51
```

3   Find the mean and standard deviation of the data in Exercise 2 of this section.

4   Here are the weights in grams of 50 washers. Construct a stem-and-leaf table of the data. Draw a histogram, frequency polygon, and relative frequency polygon of the data.

```
3.17  3.13  3.2   3.19  2.93  3.14  3.02  3.11
3.20  2.96  3.09  3.13  3.00  3.09  3.03  3.23
3.00  3.17  3.14  3.12  3.18  3.04  3.20  3.07
3.05  3.17  2.96  3.07  3.12  3.06  3.06  3.18
3.12  2.96  3.06  3.25  3.09  3.16  2.98  3.13
3.08  3.10  3.08  3.20  3.02  3.00  3.12  3.09
3.12  3.06
```

5   A sample of 40 newly minted pennies is weighed on a very delicate scale to the nearest .001 gram. Here are the weights.

```
3.164  3.164  3.081  3.07   3.157  3.063  3.154
3.024  3.125  3.125  3.128  3.037  3.087  3.066
3.034  3.16   3.099  3.033  3.042  3.111  3.073
3.001  3.055  3.066  3.174  3.092  3.041  3.073
3.061  3.127  3.102  3.059  3.171  3.176  3.112
3.039  3.12   3.112  3.042  3.125
```

Construct a stem-and-leaf table and relative frequency polygon of the data. (*Hint:* The stems for these data are 300, 301, 302, ...)

6   For the washer data in Exercise 4 of this section, suppose a washer is picked at random from the sample. What is the probability that the washer weighs between 3.00 and 3.20 grams? Less than 3.00 grams?

7   For the data in Exercise 5 of this section, suppose a penny is picked at random from the sample of 40 pennies. What is the probability that the penny weighs more than 3.10 grams?

8   Here is a frequency table giving weights in pounds of 50 male recruits for a summer work program. Draw a relative frequency polygon of these data. Then draw a smooth curve for the data. Does the curve appear to be normal?

```
110  141  135  113  121  114  131  128  157
146  140  147  126  146  104  119  147  135
140  144  169  169  113  136  138  157  163
139  115  138  106  117  137  131  123  143
166  148  127  124  158  120  105  117  161
119  148  139  154  140
```

9 The mean and standard deviation of the data in Exercise 8, above, are as follows:

Mean = 135.3 pounds
Standard deviation = 17.3 pounds

Using this information and assuming the distribution of the weights to be normal, find an interval around the mean in which two-thirds of the recruits could be expected to weigh. Then check to see how many recruits are actually in the interval.

10 Refer to the penny data in Exercise 5 of this section. The mean and standard deviation of the weights of the 100 pennies are:

Mean = 3.09 grams
Standard deviation = .05 gram

Assume these weights can be smoothed to a normal curve. Then, 68% of the pennies would be expected to have weights in what interval? How many of the pennies were *actually* in each interval?

11 The length in centimeters of 50 laboratory mice used in a psychological experiment were as follows:

12.2	14.3	13.5	12.6	11.6	15.4	16.0	17.2
12.7	12.4	10.7	14.5	17.0	14.2	13.4	11.4
13.3	14.0	13.6	14.2	12.4	16.0	13.4	16.0
13.1	10.5	11.2	15.4	13.2	12.8	12.8	14.3
12.8	13.5	13.4	13.8	13.6	12.0	11.6	13.1
15.3	13.5	14.6	14.1	13.0	14.2	14.5	15.9
12.3	13.6

Construct a frequency distribution of these data. Does it appear that the data are normally distributed? Why or why not?

12 Roll two six-sided dice and record the sum of the dots on the top of the dice. Repeat for 30 rolls of the two dice. Record your results. Do the data appear to be normally distributed? Why or why not?

SPECIAL INTEREST FEATURE

Statistics and Individual Differences[7]

The French scientist, A. Quetelet (1796–1874), first noted that if you take a large group of people and measure their heights, the frequency distribution of these heights will be nearly a normal distribution. He also studied the distributions of other characteristics such as weight, chest girth, and arm length, and found that all of these follow nearly the same kind of distribution. This property of the distribution of physical characteristics occurred so frequently that the British scientist Sir Francis Galton (1822–1911) coined the term "normal" to describe these distributions.

This property of most physical characteristics has many practical applications. For example, the designer of an airplane cockpit must arrange it so that most pilots are comfortable and can reach all of the controls. Clearly, this requires knowledge of average heights, average arm lengths, and so on, as well as knowledge of the variability around these averages so that *most* pilots will be accommodated.

9.6 NORMAL-CURVE TABLES

Since the normal curve is so useful, tables have been prepared that provide the area under the curve corresponding to given values of z. Remember from the early parts of this chapter how you can change any set of scores to z scores and change z scores to those with *any* desired mean and standard deviation. If you can also assume that the data with which you are working can be smoothed to a normal curve, then a table of normal-curve areas is very useful in finding probabilities. But first, let's see how to use a table of normal-curve areas.

Figure 9.22 shows a theoretical normal curve with a mean of 0 and a standard deviation of 1.0. This is sometimes called a *standard normal curve*, since the scale of the horizontal axis is z scores (or standard scores).

Our normal-curve table deals with z scores and areas (probabilities). Only parts of the normal-curve table are given in this chapter. A more complete table is given in Appendix C.

1. **z-scores:** Along the left column and the top row of the table (see Table 9.27).

2. **Areas:** Within the body of the table. Since the total area under the normal curve for z scores is 1.0, all areas given in the table are less than 1.0. (Look at Table 9.27 and note that this is the case.)

Before we use a normal-curve table, we have to make a decision: Are we going to the table with a z score or with a probability? If we go to the table with a z score, then we usually want a probability from the table. If we go to the table with a probability (area), then we usually want to read a z score from the table.

FIGURE 9.22

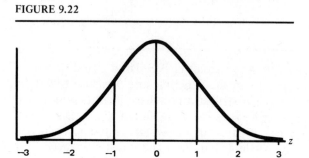

TABLE 9.27[8]

z	0	1	2	3	4	5	6	7	8	9
.0	.5000	.5040	.5080	.5120	.5160	.5199	.5239	.5279	.5319	.5359
.1	.5398	.5438	.5478	.5517	.5557	.5596	.5636	.5675	.5714	.5733
.2	.5793	.5832	.5871	.5910	.5948	.5987	.6026	.6064	.6103	.6141
.3	.6179	.6217	.6255	.6293	.6331	.6368	.6406	.6443	.6480	.6517
.4	.6554	.6591	.6628	.6664	.6700	.6736	.6772	.6808	.6844	.6879
.5	.6915	.6950	.6985	.7019	.7054	.7088	.7123	.7157	.7190	.7224
.6	.7257	.7291	.7324	.7357	.7389	.7422	.7454	.7486	.7517	.7549
.7	.7580	.7611	.7642	.7673	.7704	.7734	.7764	.7794	.7823	.7852
.8	.7881	.7910	.7939	.7967	.7995	.8023	.8051	.8078	.8106	.8133
.9	.8159	.8186	.8212	.8238	.8264	.8289	.8315	.8340	.8365	.8389
1.0	.8413	.8438	.8461	.8485	.8508	.8531	.8554	.8577	.8599	.8621
1.1	.8643	.8665	.8686	.8708	.8729	.8749	.8770	.8790	.8810	.8830
1.2	.8849	.8869	.8888	.8907	.8925	.8944	.8962	.8980	.8997	.9015
1.3	.9032	.9049	.9066	.9082	.9099	.9115	.9131	.9147	.9162	.9177
1.4	.9192	.9207	.9222	.9236	.9251	.9265	.9279	.9292	.9306	.9319
1.5	.9332	.9345	.9357	.9370	.9382	.9394	.9406	.9418	.9429	.9441
1.6	.9452	.9463	.9474	.9484	.9495	.9505	.9515	.9525	.9535	.9545
1.7	.9554	.9564	.9573	.9582	.9591	.9599	.9608	.9616	.9625	.9633
1.8	.9641	.9649	.9656	.9664	.9671	.9678	.9686	.9693	.9699	.9706
1.9	.9713	.9719	.9726	.9732	.9738	.9744	.9750	.9756	.9761	.9767
2.0	.9772	.9778	.9783	.9788	.9793	.9798	.9803	.9808	.9812	.9817

■ **Example 1:**   Using the table of normal-curve areas, find

$$p(z < 1.65)*$$

Solution:

We are asked to find the (theoretical) probability that a z score picked at random from a normally distributed population is less than 1.65.

Here we are going to the table (Table 9.28) with a z score (1.65) and want to get an area from the table.

Figure 9.23 is a picture of what is going on. A z score of 1.65 is shown on the graph. The shaded area to the *left* of $z = 1.65$ represents that *probability* that a random z is less than 1.65.

We go to the table with $z = 1.65$ and find the value of 1.6 in the left column. Then we look along the top row of the table for the column headed 5. These column headings are the

FIGURE 9.23

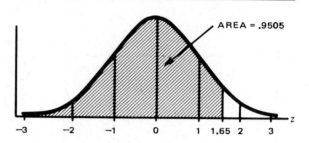

AREA = .9505

hundredths digit for a z score. So, for $z = 1.65$, we need column 5. Look in the body of the table for the number at which the row for $z = 1.65$ and column 5 intersect. We see the number is .9505. This is an area. (Remember? We are going to the table with a z score and looking for an area.)

The area of .9505 is for the shaded region to the left in Figure 9.23. So,

$$p(z < 1.65) = .9505 ■$$

TABLE 9.28

z	0	1	2	3	4	5	6	7	8	9
1.3	.9032	.9049	.9066	.9082	.9099	.9115	.9131	.9147	.9162	.9177
1.4	.9192	.9207	.9222	.9236	.9251	.9265	.9279	.9292	.9306	.9319
1.5	.9332	.9345	.9357	.9370	.9382	.9394	.9406	.9418	.9429	.9441
1.6	.9452	.9463	.9474	.9484	.9495	.9505	.9515	.9525	.9535	.9545
1.7	.9554	.9564	.9573	.9582	.9591	.9599	.9608	.9616	.9625	.9633
1.8	.9641	.9649	.9656	.9664	.9671	.9678	.9686	.9693	.9699	.9706
1.9	.9713	.9719	.9726	.9732	.9738	.9744	.9750	.9756	.9761	.9767
2.0	.9772	.9778	.9783	.9788	.9793	.9798	.9803	.9808	.9812	.9817
2.1	.9821	.9826	.9830	.9834	.9838	.9842	.9846	.9850	.9854	.9857
2.2	.9861	.9864	.9868	.9871	.9875	.9878	.9881	.9884	.9887	.9890

*Since z is a *continuous* variable (for example, it may assume all values in the interval 1.0–2.0), it can be shown that $p(z < 1.65) = p(z \leq 1.65)$. See Appendix B.
  From now on in this book, we will usually use "less than" instead of "less than or equal to" (and "greater than" instead of "greater than or equal to") when we are dealing with continuous variables such as z or χ^2.

■ **Example 2:**   Using the normal-curve table, find

$$p(z < -1.96)$$

Solution:

We are going to the table (Table 9.29) with a z of -1.96 and want to find an area.

Look for -1.9 on the left and for the column headed by 6. We find an *area* of .0250 in the body of the table. This is the area of the shaded region shown in Figure 9.24. So,

$$p(z < -1.96) = .0250 \; ■$$

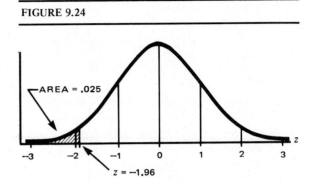

FIGURE 9.24

AREA = .025

$z = -1.96$

TABLE 9.29

z	0	1	2	3	4	5	6	7	8	9
−2.2	.0139	.0136	.0132	.0129	.0125	.0122	.0119	.0116	.0113	.0110
−2.1	.0179	.0174	.0170	.0166	.0162	.0158	.0154	.0150	.0146	.0143
−2.0	.0228	.0222	.0217	.0212	.0207	.0202	.0197	.0192	.0188	.0183
−1.9	.0287	.0281	.0274	.0268	.0262	.0256	.0250	.0244	.0239	.0233
−1.8	.0359	.0351	.0344	.0336	.0329	.0322	.0314	.0307	.0301	.0294
−1.7	.0446	.0436	.0427	.0418	.0409	.0401	.0392	.0384	.0375	.0367
−1.6	.0548	.0537	.0526	.0516	.0505	.0495	.0485	.0475	.0465	.0455
−1.5	.0668	.0655	.0643	.0630	.0618	.0606	.0594	.0582	.0571	.0559
−1.4	.0808	.0793	.0778	.0764	.0749	.0735	.0721	.0708	.0694	.0681
−1.3	.0968	.0951	.0934	.0918	.0901	.0885	.0869	.0853	.0838	.0823
−1.2	.1151	.1131	.1112	.1093	.1075	.1056	.1038	.1020	.1003	.0985

■ **Example 3:**   Using the normal-curve table, find

$$p(z > 1.65)$$

Solution:

In this example, we are to find the probability of getting a random *z greater than* a given value of 1.65. But our table gives areas (probabilities) for *z less than* (to the left of) given *z* values. Figure 9.25 shows a graph of what we are to find.

FIGURE 9.25

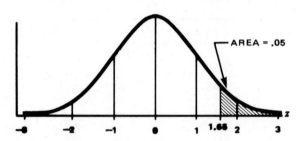

For *z* values *greater than* a given value, we need the area to the right of the given *z*. How can we find this area? Conveniently, the total area under a probability distribution for *z* scores is 1.0. So, we know that the area to the *right* of *z* = 1.65 is 1 minus the area under the curve to the *left* of *z* = 1.65. Therefore,

$$p(z > 1.65) = 1 - .9595 = .0495$$

or, .05 when rounded to nearest hundredth. ■

■ **Example 4:**   Suppose we want to draw a *z* at random from a normal population, such that the *z* is less than some particular value. Suppose also that the probability of getting such a *z* is .20. Below what particular value of *z* is the random *z* to be drawn?

Mathematically, we can state this problem as:

$$p(z < ?) = .20$$

Solution:

The picture of the problem is shown in Figure 9.26. Here we are given an *area* (a probability) and need to find a *z*. So, we go to the table with an area of .20 and will return with a corresponding *z*.

FIGURE 9.26

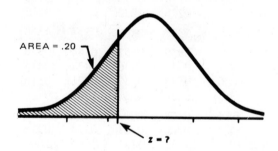

We look through the table (Table 9.30) and find .2005, which when rounded is closest to our given area of .20. The corresponding *z* for an area of .20 is −.84. We find a *z* of −.84 as follows. We look along the row in which .2005 appears and find, in the far left column, a *z* of −.8. Then we look at the top of the column in which .2005 appears to find the digit in the hundredths place for *z*. There we find a 4. So we read from the table a *z* of −.84 and have

$$p(z < -.84) = .20 ■$$

TABLE 9.30

z	0	1	2	3	4	5	6	7	8	9
−1.0	.1587	.1562	.1539	.1515	.1492	.1469	.1446	.1423	.1401	.1379
− .9	.1841	.1814	.1788	.1762	.1736	.1711	.1685	.1660	.1635	.1611
− .8	.2119	.2090	.2061	.2033	.2005	.1977	.1949	.1922	.1894	.1867
− .7	.2420	.2389	.2358	.2327	.2296	.2266	.2236	.2206	.2177	.2148
− .6	.2743	.2709	.2676	.2643	.2611	.2578	.2546	.2514	.2483	.2451
− .5	.3085	.3050	.3015	.2981	.2946	.2912	.2877	.2843	.2810	.2776
− .4	.3446	.3409	.3372	.3336	.3300	.3264	.3228	.3192	.3516	.3121
− .3	.3821	.3783	.3745	.3707	.3669	.3632	.3594	.3557	.3520	.3483

■ **Example 5:**   Find

$$p(z > ?) = .20$$

Solution:

This is nearly identical to Example 4, except that we want to find a value of z such that a random z has probability of .20 of being greater than that value.

The graph is as shown in Figure 9.27. Notice that the difference between this graph and the one for Example 4 is that the given area of .20 is on the *right* since the random z is *greater* than the z we are to find.

FIGURE 9.27

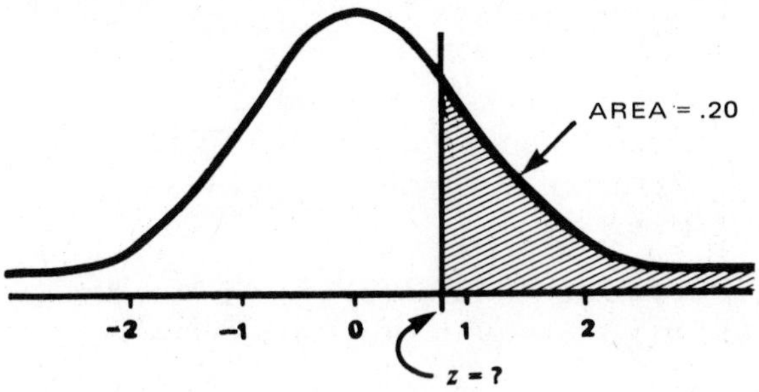

We cannot go directly to the table with our given area, since the table gives only areas *to the left* of a given value of z. So, we must find from the given information what is the area to the left of the required z.

Since the total area under the standard normal curve is 1.0, the area we take to the table is

$$1 - .20 = .80$$

We can now go to the table with an area (.80) and look for the corresponding z. We find that the closest value to .80, when rounded, is .7995. The z corresponding to an area of .7995 is .84—that is,

$$p(z > .84) = .20 ■$$

Compare this with the value obtained in Example 4.

The fact that the answers to Examples 4 and 5 differ only by a sign (plus or minus) can be very useful in using normal-curve tables. The normal distribution is symmetrical [it has the same shape on both sides of the mean $(z = 0)$].

So, for example, if we want to find

$$p(z > 2.5)$$

we can look at the table for

$$p(z < -2.5)$$

and find that it is .0062. So, we have

$$p(z > 2.5) = .0062$$

Another way in which we could find

$$p(z > 2.5)$$

is to find directly from the table that

$$p(z < 2.5) = .9938$$

But this is the area to the *left* of $z = 2.5$. So, we subtract .9938 from 1.0000 and have

$$p(z > 2.5) = 1.000 - .9938 = .0062$$

EXERCISES

Assume z is produced by a normal probability model. In every case, first draw a simple sketch of the given information and the area or z value you are to find.

1   Use the table of normal curves to find each value.

(a) $p(z < 1.96)$ (b) $p(z < -1.96)$
(c) $p(z < 1.0)$ (d) $p(z < -1.0)$
(e) $p(z < .5)$ (f) $p(z < -.5)$
(g) $p(z < 0)$

2   Use the table of normal-curve areas to find each value.

(a) $p(z > 1.96)$ (b) $p(z > -1.96)$
(c) $p(z > 1.0)$ (d) $p(z > -1.0)$
(e) $p(z > .5)$ (f) $p(z > 0)$

3   Use the table of normal-curve areas to find each value.

(a) $p(z < .68)$ (b) $p(z > .68)$
(c) $p(z < -.68)$ (d) $p(z > -.68)$

4   Use the table of normal-curve areas to find the missing value.

(a) $p(z < ?) = .95$ (b) $p(z < ?) = .90$
(c) $p(z < ?) = .98$ (d) $p(z < ?) = .66$
(e) $p(z < ?) = .50$

5   Use the table of normal-curve areas to find the missing value.

(a) $p(z < ?) = .05$ (b) $p(z < ?) = .25$
(c) $p(z < ?) = .01$ (d) $p(z < ?) = .10$
(e) $p(z < ?) = .5$

6   Use the table of normal-curve areas to find the missing value.

(a) $p(z > ?) = .95$ (b) $p(z > ?) = .90$
(c) $p(z > ?) = .99$ (d) $p(z > ?) = .05$
(e) $p(z > ?) = .01$ (f) $p(z > ?) = .10$

7   Find the missing value. But first decide whether the missing value is positive, negative, or zero.

(a) $p(z < ?) = .68$ (b) $p(z < ?) = .16$
(c) $p(z < ?) = .80$ (d) $p(z < ?) = .20$
(e) $p(z > ?) = .20$ (f) $p(z > ?) = .15$
(g) $p(z > ?) = .68$ (h) $p(z > ?) = .98$
(i) $p(z > ?) = .5$

8   Find the missing value. But first decide whether the missing value is positive or negative.

(a) $p(z > ?) = .40$ (b) $p(z < ?) = .88$
(c) $p(z > ?) = .93$ (d) $p(z < ?) = .25$

9 Find the desired value. First decide whether you are going to the table of normal-curve areas with a z value or a probability (that is, an area).

(a) $p(z < 2.38)$ (b) $p(z > -1.65)$
(c) $p(z < ?) = .005$ (d) $p(z > ?) = .975$

10 The data on chest measurements of soldiers (Exercise 1 of Section 9.5) have a mean of 35 inches and a standard deviation of 2 inches. Assuming these data to be normally distributed, how many of the 1516 soldiers would you expect to have chests with measurements of between 33 and 37 inches, inclusive? How many of the soldiers actually were in this interval?

9.7 APPLICATIONS OF THE NORMAL CURVE

If real-world data, such as test scores or weights of persons or things, can be assumed to be generated by a normal probability model, the table of normal-curve areas can be applied in a number of different ways.

■ **Example 1:**   The life of a certain type of battery has been found to be normally distributed with a mean of 200 hours and a standard deviation of 15 hours (this sort of information could be gathered, for example, by keeping records at the factory on the lifetime of samples of batteries over a long period of time). What proportion of these batteries can be expected to last less than 220 hours? (Remember: Probabilities can be thought of as proportions, so we can think of this as a probability problem.)

Solution:

The assumed distribution of the lives of batteries is shown in Figure 9.28. The shaded region shows the batteries that have a life less than 220 hours. We can calculate this area using the table of normal-curve areas if we know the z value corresponding to 220 hours.

$$z = \frac{220 - 200}{15} = \frac{20}{15} = 1.33$$

FIGURE 9.28

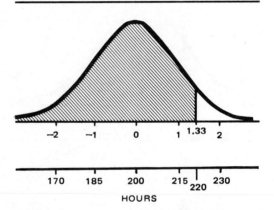

We now know a value for z. Using this value, we can get the area of the shaded region in Figure 9.28 from the table of normal-curve areas (again, assuming that the battery lives are generated by a normal probability model). That is, we can find

$$p(z < 1.33)$$

From the table, we find the area to be .9082, so the proportion of batteries with lives less than 220 hours (which corresponds to a z of 1.33) is .91 (rounded). That is, about 91% of the batteries can be expected to last less than 220 hours. ■

■ **Example 2:**   What proportion of the batteries from Example 1 can be expected to last more than 220 hours?

Solution:

Again, assuming the battery lives to be distributed normally, we have the graph shown in Figure 9.29.

FIGURE 9.29

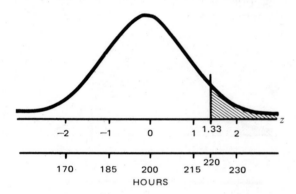

The shaded region represents the proportion of batteries with lives greater than 220 hours.

Since the z corresponding to 220 hours is 1.33, we want to find

$$p(z > 1.33)$$

We know from the table that

$$p(z < 1.33) = .91$$

so

$$p(z > 1.33) = 1.0 - .91 = .09$$

We expect about .09 (or 9%) of the batteries to last more than 220 hours. ■

■ **Example 3:**   The SPWEHIQS Club (Society of People with Extremely High IQs) requires people to take an intelligence test as a condition of joining the club and restricts membership to the top 5%, as measured by this test. Suppose that it has been found that the scores on the intelligence test have a mean of 100 and a standard deviation of 12 and are normally distributed for very large groups of people. What is the lowest score on this test that would be acceptable for admission to SPWEHIQS?

Solution:

The graph in Figure 9.30 shows the distribution of the scores of the intelligence test.

FIGURE 9.30

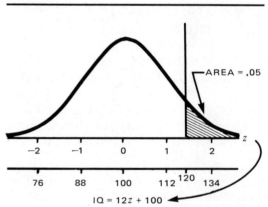

The shaded region on the right of the graph indicates 5%, or .05, of the total area is under the curve. According to the normal-curve table, this area corresponds to a z score of 1.65. This z score can be changed to an intelligence test score by a stretch factor of 12 (the standard deviation of the test score) and a slide factor of 100 (the mean of the test scores).

So, the test score corresponding to a z score of 1.65 is

$$T = 12(1.65) + 100 = 19.8 + 100 = 119.8$$

Rounding this score to 120, we find that a score of 120 would be the lowest acceptable for admission to the club (since the club accepts only the top 5% for membership). ■

■ **Example 4:**   A certain insect has a mean length of 1.2 centimeters and a standard deviation of .12 centimeters. If there are estimated to be 1000 of these insects in a terrarium, how many would be expected to be less than 1 centimeter in length?

Solution:

Assume the lengths are normally distributed. Figure 9.31 is a sketch of the distribution of the insect lengths.

FIGURE 9.31

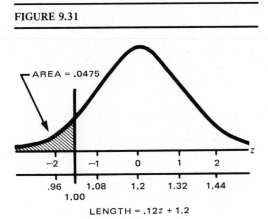

LENGTH = .12z + 1.2

The shaded part of the graph begins at the 1-centimeter mark and includes the region to the *left* of this point (since the problem specifies insects having lengths *less* than 1 centimeter.

The z score corresponding to 1 centimeter is found by the transformation

$$z = \frac{1 - 1.2}{.12} = \frac{-.2}{.12} = -1.67$$

According to our tables, the area to the left of $z = -1.67$ is .0475. So, .0475 (4.75%) of the 1000 insects are expected to be less than 1 centimeter long. That is,

$$.0475 \times 1000 \doteq 48 \text{ insects} \blacksquare$$

EXERCISES

1   The ages of a certain group of teachers are normally distributed with a mean of 40 years and a standard deviation of 6 years. What percentage of the group is expected to be 45 years of age or older?

2   The group of 100 high school girls (Table 9.25) had a mean weight of 118.67 pounds with a standard deviation of 16.9 pounds. What fraction of the girls would be expected to weigh 130 pounds or more if the weights are normally distributed? Refer back to Table 9.25 and see how many of the girls actually weighed 130 pounds or more.

3   A set of final examination scores has a mean of 52 and a standard deviation of 6. The scores are normally distributed. If a teacher wants to assign a grade of A to the top 15% of the scores, what score should be the lowest A? If the bottom 15% are to be F, what test score should be the highest F?

4   Table 9.31 shows the frequency distribution of the weights of 100 newly minted pennies.[9] These were measured accurately to the nearest tenth of a milligram and put into class intervals of width .02 grams. That is, the table shows that 7 pennies weighed 3.07 grams. Actually, this means that 7 of the pennies had a weight between 3.0600 and 3.0799.

TABLE 9.31
WEIGHTS OF NEWLY MINTED PENNIES WEIGHED
ON AN ACCURATE BALANCE[10]

WEIGHT (GRAMS)	NUMBER OF CENTS	WEIGHT (GRAMS)	NUMBER OF CENTS
2.99	1	3.11	24
3.01	4	3.13	17
3.03	4	3.15	13
3.05	4	3.17	6
3.07	7	3.19	2
3.09	17	3.21	1
		Total	100

Draw a frequency distribution for these weights. Do they appear to follow the normal distribution? Using a mean of 3.11 grams and a standard deviation of .043 grams, calculate the probability of a weight of 3.09 grams— that is, the probability that a penny falls between 3.08 and 3.10. Compare this probability to the frequency of such pennies. Similarly, compare the pennies below 3.05 grams and those above 3.15 grams with the normal distribution.

5   Key Problem—*How Much Does a 40-Pound Box of Bananas Weigh?*   In order to help ensure that the boxes of bananas weigh at least 40 pounds upon arrival at their destination, the packing plant might adopt this rule: Pack boxes to have a weight of 41.5 pounds of bananas with a maximum permissible range of 3 ounces above or below 41 pounds, 8 ounces (that is, pack boxes to have at least 41 pounds, 5 ounces, but no more than 41 pounds, 11 ounces, of bananas). With this rule and the shrinkage in travel, the distribution of box weights upon arrival may be assumed to be approximately normal with a mean of 41 pounds and a standard deviation of 4 ounces.[11]

Suppose 30 million boxes of bananas are packed a year. Given this packing plant rule and assumption about the distribution of box weights upon arrival, how many boxes would be expected to weigh less than 40 pounds upon arrival? [*Note:* Tables tell us that for a standard normal z,

$$p(z < 4.00) = .000032]$$

9.8 NON-NORMAL DATA

Data may produce other curves than normal when smoothed.

■ **Example 1–Testing Light Bulbs:** 50 light bulbs are tested by keeping them lighted until they burn out.

Solution:

The data produced are as in Table 9.32.

TABLE 9.32

HOURS LIGHTED	FREQUENCY	RELATIVE FREQUENCY (P)
0–100	18	.36
100–500	14	.28
500–1000	10	.20
1000–1500	5	.10
1500–2000	3	.06
Totals	50	1.00

The smoothed relative frequency polygon looks like the one in Figure 9.32. This curve is not bell-shaped. (You might even call it half a bell!)

FIGURE 9.32

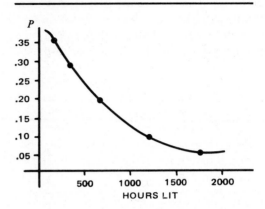

This graph can be used to make probability statements about the light bulbs. But since we do not have a table of areas for this curve, we will estimate these areas by the relative frequencies. For example,

$$p(H < 100)$$

is estimated by

$$P(H < 100) = .36$$

which tells us that a bulb picked at random has an estimated probability of .36 of lasting less than 100 hours.

Similarly, we estimate

$$p(H > 1000)$$

by

$$P(H > 1000) = .10 + .06 = .16$$

which states that a bulb has a probability of .16 of lasting more than 1000 hours. ■

■ **Example 2—Heights of College Basketball Players:**   If we found the heights of a sample of college basketball players, we would get a nonnormal curve when smoothing the relative frequency polygon. It would look something like Figure 9.33

Notice that this graph is not symmetric. That is, unlike the normal curve, the left side of this graph is shaped differently from the right side. The graph is *skewed,* or twisted. This skew can be either to the left or to the right, as shown in Figure 9.34. ■

FIGURE 9.33

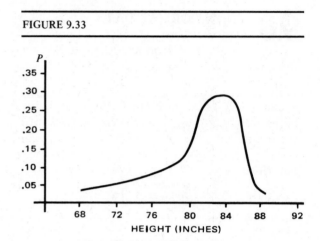

HEIGHT (INCHES)

FIGURE 9.34

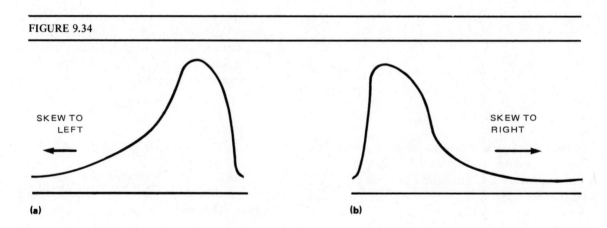

SKEW TO LEFT

SKEW TO RIGHT

(a)

(b)

■ **Example 3—When's the Next Bus?** Suppose that arrival times of a bus are kept for 100 days and the data in Table 9.33 result.

TABLE 9.33

TIME	f
12:00–12:01	21
12:01–12:02	19
12:02–12:03	23
12:03–12:04	18
12:04–12:05	19
	100

We draw a frequency polygon and a relative frequency polygon of the data, as shown in Figure 9.35.

It appears that the bus arrives anytime from 12 noon to 12:05 P.M. with about equal likelihood. So we draw in the smooth curve, which in this case is a straight horizontal line.

Here, again, probabilities are estimated by finding areas under the curve. For example,

suppose we wish to find the probability that the bus arrives between 12:02 and 12:04. We need the area of the shaded region of the graph in Figure 9.36.

FIGURE 9.36

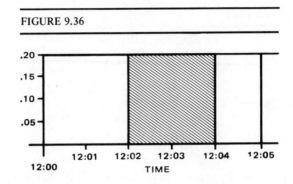

From the smooth curve, we find that

$$p(\text{bus arrives between } 12:02 \text{ and } 12:04) = \frac{2}{5} = .4 ■$$

FIGURE 9.35

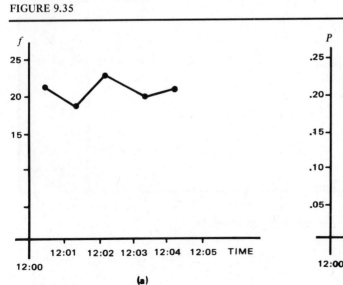

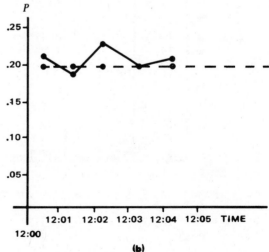

(a)   (b)

■ **Example 4—Have You Read This Month's Magazine?**   Reading habits concerning magazine A (published monthly) were surveyed among 200 readers (Table 9.34).

TABLE 9.34

NUMBER OF ISSUES READ	NUMBERS OF PERSONS
0	50
1	24
2	8
3	2
4	1
5	1
6	0
7	0
8	1
9	3
10	15
11	40
12	55
	200

The results suggest that the magazine is either read very little or it is read faithfully almost every month. The distribution of these results is bimodal. (See Chapter Two.) The two modes are 0 and 12. Such a distribution is said to be U-shaped, for obvious reasons. (See Figure 9.37.) ■

FIGURE 9.37

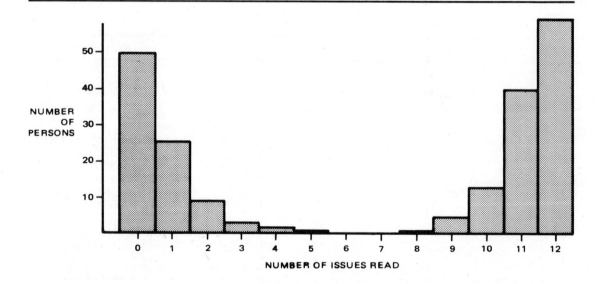

NUMBER OF ISSUES READ

EXERCISES

1   Use the data reported on testing light bulbs (Table 9.32) to estimate these probabilities.

(a) $p(H < 500)$ (b) $p(H > 1500)$

2   The number of deaths per year per Army Corps of the Prussian Army due to being kicked by a horse are reported in the following table. The smooth curve is called a Poisson Distribution and has been studied extensively by statisticians. Draw a relative frequency polygon of these data. From these data, estimate these probabilities of K, the number of persons kicked to death by a Prussian Army horse, per year, per Army Corps.

(a) $p(K = 0)$ (b) $p(K > 1)$

Deaths

K	f
0	109
1	65
2	22
3	3
4	1

3   The weights, W, of 30 football players are as follows. Draw a relative frequency polygon of the data. Do the data appear to be normally distributed? Why or why not?

190 232 233 235 257 237
230 215 230 238 254 232
231 218 242 240 245 242
225 200 253 237 240 201
228 199 228 240 221 243

From the data, estimate these probabilities.

(a) $p(W < 200)$ (b) $p(200 < W < 220)$

4   Mr. Commuter's arrival times, T, at work for a period of several weeks are recorded below.

7:53 7:54 7:58 7:51 7:56 7:50 7:54 7:57
7:51 7:51 7:57 7:50 7:52 7:51 7:54 7:55
7:59 7:55 7:55 7:59 7:52 7:53 7:57 7:59
7:52 7:59 7:53

(*Note:* 7:53 means seven fifty-three, or 7 minutes before eight.)

Draw a relative frequency polygon of the data and then draw in a smooth curve. What shape does this distribution appear to have? Estimate the following, using first the relative frequency polygon and then, as a better estimate, the smoothed curve.

(a) $p(T > 7:55)$ (b) $p(T < 7:53)$

Now, assuming that the smooth curve which you drew was a straight (level) line, estimate these theoretical probabilities:

(c) $p(T > 7:55)$ (d) $p(T < 7:53)$

5   Use the data from Example 4 (Table 9.34) to find these probabilities. First, write each statement as an equation using P.

(a)   The probability that no issues of the magazine are read.

(b)   The probability that one or no issues are read.

(c)   The probability that all twelve issues are read.

Measure-ment

KEY PROBLEM

Measuring g

Objects fall to earth because of gravity. How to measure the effect of the pull of gravity (*g*) has been studied for hundreds of years. Some very clever experiments have been designed to measure *g*, such as timing the fall of a steel bar using a photoelectric cell.

One well-known way to measure gravity is to use information about the relationship between the length of a pendulum and how long it takes it to swing back and forth. You may have noticed that short pendulums swing back and forth quickly and long pendulums swing back and forth slowly. A famous example of a long pendulum is the Focault pendulum, sometimes on display in a museum of science. This pendulum also demonstrates Earth's rotation.[1]

In this chapter, we will see how ideas from measurement can be used with other ideas we have learned in statistics to find the value of the acceleration of gravity, *g*.

4   The compressive strength of cement is the amount of squeezing pressure it can withstand before crumbling. The following data give results (in pounds) of tests done on the same batch of cement by 50 laboratories. What percentage of the labs reported tests showing the cement having strength 300 pounds or more greater than the average reported value?

4466	3914	4084	4135	4084
4154	4123	4030	4120	3626
3771	3889	4180	4141	4524
3810	3777	4072	3871	4181
3948	4081	4334	4046	4130
3676	3888	4126	4128	4058
3589	4135	4018	3675	4251
3842	4409	4134	4017	3888
3922	4300	4447	4228	3907
3825	4172	4005	4241	4339

10.2   MEASURING CAREFULLY

Now let's look at another set of data, which was obtained as follows. The statistics class of Section 10.1 was given some training on how to measure. They practiced reading the scale on the ruler to the nearest millimeter. They placed the stack of books on a sturdy table and put a heavy weight on it. They made sure that the stack of books was straight and that the ruler was held snugly against the books.

Then they measured the stack of books and obtained the following results (in centimeters):

23.6	23.7	24.2	23.9	23.9
24.1	24.2	23.9	24.1	23.4
23.7	24.0	23.8	23.5	24.1

The data are summarized in Table 10.2.

TABLE 10.2

STEM	LEAF
22	
23	4, 5, 6, 7, 7, 8, 9, 9, 9
24	0, 1, 1, 1, 2, 2
25	

Notice how this set of data compares with the previous set.

	First Set	Second Set
Median	23.8	23.9
Mean	23.7	23.9
Range	2.4	.8
Mean deviation	.41	.21
Variance	.35	.064
Standard deviation	.59	.25

We see that the second set of data has considerably less variation in it than does the first set. Notice that the standard deviation of the second set (.25) is less than one-half of the standard deviation of the first set (.59). The extra care taken in measuring is reflected in the smaller variance and standard deviation for the second set of data.

Such careful measurements can be obtained, for example, using high-precision instruments such as calipers or micrometers.

EXERCISES

1   In the key problem for this chapter (estimating the acceleration of gravity), ten students measured the length of a pendulum and obtained the results in Table 10.3.[3]

TABLE 10.3

STUDENT	MEASUREMENT OBTAINED (CENTIMETERS)
1	175.2
2	179.0
3	170.0
4	165.1
5	171.5
6	175.8
7	172.0
8	175.3
9	165.5
10	175.8

Find the mean, range, and mean deviation of these measurements. If a good estimate of the length of the pendulum were required, what value would you use?

2   In the statistics class at New Trier High School, the following pairs of students measured the length (in inches) of a chalkboard. Their results are shown in Table 10.4.

TABLE 10.4

TEAM	MEASUREMENT
Steve–Scott	223
Heather–John	241
Mark–Art B.	239
Janine–Wendy	249
O. J.–Art R.	223
Karen–Kristin	224
Mike A.–Mike B.	238

What is the mean of these measurements? How much variation is in these measurements?

3   Table 10.5 shows two sets of ten measurements made when finding the diameter of a human hair (being examined in a crime laboratory). The first set was made by a usual method, involving calipers. The second set (on the same piece of hair) was made with a new method involving laser beams. Find the mean and standard deviation of each set of measurements. Which set indicates less error in measuring.

TABLE 10.5

METHOD A					METHOD B				
85	79	68	62	70	70	77	75	74	76
76	74	70	75	74	76	72	76	75	73

4   A laboratory is trying to decide which of two scales to purchase. Both are claimed to be high-precision instruments. Ten people weigh a mineral sample with Scale A, and then with Scale B, using great care with each scale. The results are shown in Table 10.6.

TABLE 10.6

SCALE A					SCALE B				
74	77	79	77	75	80	77	75	78	68
75	74	77	77	79	77	74	69	74	78

Which scale appears to have less error? Explain.

5   The value of gravity has been calculated by a laboratory in Ottawa, Canada, using two methods.[4] See Figure 10.2.

You read the graph by adding the mid-value of the intervals as third- and fourth-place decimals to 980.6100. So, for example, in the upper graph, one measurement of g was 980.6105 cm per second per second (see left-most bar).

Which of the methods, A or B, do you think gives measurements with less error? Why?

FIGURE 10.2

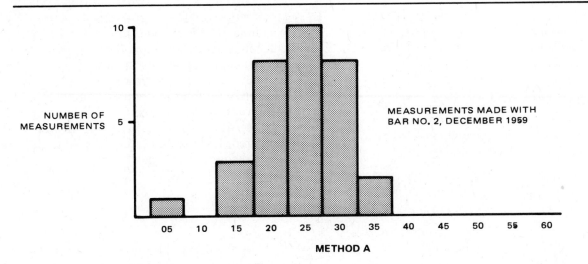

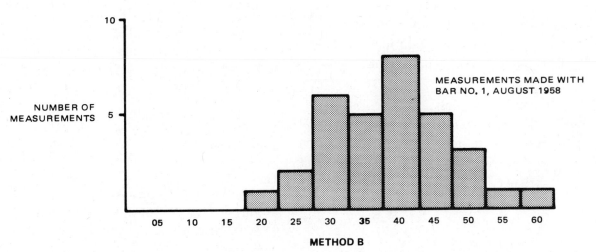

MID-VALUES OF INTERVALS ARE 3RD AND 4TH PLACE DECIMALS TO BE ADDED TO 980.6100 CM/SEC.

┌───┐

PROJECT CORNER

This activity involves taking two sets of measurements of some object, such as the length of the front chalkboard in your classroom or the perimeter of a desk or worktable.

First set of measurements: Each student is to get one measurement of the objects, without paying much attention to rules of good measurement.

Second set of measurements: Each student measures the object again, with much care, as outlined in this section.

Make a box-and-whisker plot of each set of data. (It will be helpful to first make a stem-and-leaf plot of each set.)

Compare the median of each set of measurements, the range, and the length of the boxes (the middle 50%) of the data.

What do you conclude about the effects of observing rules of good measurement?

└───┘

10.3   STANDARD ERROR

Let's go back to our problem of measuring the stack of textbooks. We have talked about error of measurement. We have recognized that all measurements contain error.

What can we do to reduce the error of measurement, after we have been as careful as we can be? The idea we use is like the one we first used in Chapter Two. That is, rather than using any one result, we use the average (mean) of a set of results.

An application of this idea comes from the key problem for this chapter, measuring the acceleration of gravity, g. As stated at the beginning of this chapter, finding a value for g depends upon having values for L, the length of the pendulum, and T, the time the pendulum takes to make one complete swing back and forth. In order to obtain a more precise measure for T, it is recommended that the time for the pendulum to make 50 swings be measured. Let's say that this is found to be 132.2 seconds. Then the average, or mean, of these 50 observations is

$$T = \frac{132.2}{50} = 2.64 \text{ seconds}$$

(Strictly speaking, this value of T should be rounded to 2.6 seconds, since we have the original measurement only to the nearest tenth of a second.)

It is likely that this measure of 2.6 seconds has less error than the individual 50 observations that were made to find T.

So, we are better off to find the mean value of a set of measurements and to use it for the desired value than to use any individual measurement.

Thus, if it is important to obtain measurements as close as possible to the actual weight, or length, or time, we make several measurements with very precise instruments.

Think of it in this way. We could have the class measure a stack of books on many different occasions (30, for instance) and on each occasion find the mean of the set of measurements.

These *means* of the class's observations could then be recorded in a table such as Table 10.7. *Remember: This table shows 25 means of sets of 30 measurements each.*

TABLE 10.7

23.9	24.0	24.0	23.9	24.0
24.0	23.9	24.1	23.8	24.0
23.9	23.8	23.9	24.0	23.7
24.0	23.7	23.9	23.9	23.8
23.9	23.8	23.9	24.0	24.1

These means also vary, since they, too, have error. But they vary less than the original measurements.

The standard deviation of the means of sets of data is a very important idea in statistics and is given the name *standard error of the mean* or, briefly, *standard error*.

The standard error of the mean based on these 25 sets of measurements (of 30 measurements each) is .10. The mean (average) of these 25 means (sometimes called the *grand mean*) is 23.9. This is a very good estimate of the "true" height of the stack of books (assuming, of course, no bias in the measurements or measuring device—see Section 10.5).

The idea of the standard error of a statistic, such as the mean, is a very powerful one because it enables us to make some statement about how precise we believe our statistic to be. For example, if we want to take an opinion poll, we will want to know the precision of our estimate of what proportion of persons in the region, or the country, have certain opinions.

In our examples in this section, we have computed the standard error of the mean by taking repeated measurements (say 25 sets of 30 measurements each, which is 750 measurements) and finding the standard deviation of these 25 means. It implies, for example, that for our opinion poll we would have to repeat the poll many times, and see how the results of the polls varied from one another.

In Chapter Eleven it will be shown that through statistical principles that we need only take *one* sample (for example, a set of measurements or an opinion poll). From one carefully selected sample, we can calculate not only the mean, for instance, but the estimated standard error of that mean. That is, we can use information from the sample to tell us how precise is our measurement.

For now, however, we will calculate standard errors using many samples so that we can become clear about the meaning of this very important concept of standard error.

EXERCISES

1   Here are the means of ten sets of measurements of the weight of a newly minted penny (in decigrams). Each set contained 16 individual measurements using the same scale. Draw a stem-and-leaf table of these data. What is the standard error of these data? What would you believe to be the true weight of the penny?

29.43	28.43	29.25	29.81	29.43
29.37	29.62	28.56	29.43	29.43

2   Exercise 1, above, was repeated, but a different scale was used. Find the standard error for this set of measurements.

29.43	29.68	28.37	29.50	30.18
30.62	29.12	29.06	30.12	29.62

Which scale do you believe to be more precise?

3   A laboratory is asked to find out how much radiation is produced by a sample of waste water from a nuclear power plant. A team of scientists takes 16 measurements and reports the average (mean) of the measurements as shown below. This is repeated 25 times to give these 25 means:

50.93	49.50	49.25	48.43	51.68
48.18	47.25	48.87	50.37	49.93
50.00	50.81	48.06	50.25	48.62
49.81	49.68	51.18	51.81	49.93
50.06	49.00	48.62	49.25	47.87

Put these data into a stem-and-leaf table and draw a graph of the results. What is the mean and estimated standard error of these data? What would you estimate the true amount of radiation to be?

4   Suppose the laboratory decided to take 64 measurements at a time and report the average of those for the sample of waste water in Exercise 3, above. Then this was repeated 25 times to give these results:

50.08	49.16	49.56	49.76	49.72
49.88	50.08	49.60	49.96	49.84
49.08	48.84	51.00	49.44	49.84
50.20	49.00	49.56	48.20	49.24
49.84	48.32	49.45	49.92	49.32

Put these data into a stem-and-leaf table and draw a graph of the results. What are the mean and estimated standard error for these data? How do you interpret the differences in these two sets of data (in Exercises 3 and 4)?

5   Ten students in The Netherlands were involved in an experiment to calculate the value of g, the acceleration of gravity, using a pendulum. One of the values needed to calculate g is T, the time for a pendulum to make a complete swing back and forth. In order to obtain precise measurements for T, the students measured the time required for the pendulum to swing back and forth 50 times. Table 10.8 shows their results for one of the pendulums.[5]

TABLE 10.8	
STUDENT	50T (IN SECONDS)
1	132.5
2	133.7
3	130.8
4	129.0
5	131.1
6	132.8
7	131.6
8	132.1
9	128.9
10	132.8

What is a precise estimate for T?

SPECIAL INTEREST
FEATURE

*Statistics and
Chemistry*[6]

About 100 years ago, a supposed "error" of measurement led to the discovery of the rare gas, argon. Chemists have long been able to remove the oxygen, carbon dioxide, and moisture from the earth's atmosphere. The remaining gas, it was believed, was solely nitrogen.

In 1890, the British scientist, Lord Rayleigh, did a study in which he compared nitrogen obtained from the atmosphere with nitrogen produced by heating the chemical ammonium nitrite. He filled a bulb of carefully determined capacity with each gas in turn under standard conditions— sea level pressure and a temperature of 0° Celsius. The weight of the filled bulb minus its weight with the gas removed gave the weight of the gas itself.

The study produced these results:

Weight of nitrogen from atmosphere	2.31001 grams
Weight of nitrogen from ammonium nitrite	2.29849 grams
Difference	0.01152 grams

Lord Rayleigh was faced with a problem: Was the difference, which was very small, due to measurement error or to a real difference in the weights of the different kinds of nitrogen?

He repeated the study and finally concluded that the difference was greater than any measurement error and, therefore, due to differences in the gases he was measuring. Further investigation led Lord Rayleigh to conclude that the nitrogen from the air contained other gases heavier than nitrogen. Under this assumption, he soon succeeded in isolating the rare gas, argon. This gas proved to be one of a family of gases in the earth's atmosphere whose existence at the time was not even suspected. For his discovery Rayleigh was awarded a Nobel Prize in 1904.

10.4   SYSTEMATIC ERROR (BIAS)

Sometimes, when measuring, a systematic error will creep into our results. For example, we may be using a scale that over-weighs. That is, the scale shows a weight that is more (maybe several ounces) than the actual weight of the object we are weighing. In this case, our results may be carefully obtained (their variance and standard deviation are small), but they are *biased* (the true weight is less than what is given by the scale).

This idea relates to one we discussed in Chapter Nine where we had transformation rules for data. One rule stated that if you add the same slide factor to each individual score, the mean is increased by that amount but the *variance is unchanged.*

Suppose, for example, that a newly minted penny is weighed by an unbiased scale on 10 occasions with these results (in decigrams):

Stem	Leaf
30	5, 6, 6
31	0, 0, 0, 2, 3, 4, 5

Mean = 31.0 decigrams
Standard deviation =   .4 decigrams

If we then use another scale with a constant error of .2 gram (2 decigrams) to weigh the penny, we would expect to get these results:

Mean = 33.0 decigrams
Standard deviation =   .4 decigrams

The mean is increased by the constant error of 2 decigrams. But, as expected, the standard deviation would remain about the same.

If there is a constant error, or bias, added to a set of measurements, as with a defective scale, the *precision* of the instrument will not be affected. The mean of the weights given by the scale will be increased by the constant error, but the standard deviation of the measurement will remain essentially the same.

How, then, can we check a measuring device for bias? One way is to compare the measurement it gives with that given by a device known to be free of bias. For example, the time tone given by the Canadian Broadcasting Company from the Dominion Observatory in Ottawa provides a reference point for correct time for the entire country.

Another procedure for determining bias is to use more than one measuring device. If a laboratory were analyzing the amount of impurity in a drug, they might use two scales. If the two scales agree, this is a good indication that both are free of bias. Both could, of course, have the same systematic error, but this is a very unlikely situation and can usually be ignored.

EXERCISES

1   A thermometer with a known constant error of $+5°$ F is used to give ten measurements of the temperature of a greenhouse.

80.9	79.7	79	80.6	78.9
78.7	79.5	81.1	79.2	80.4

Find the mean and standard deviation of these temperatures. Then correct the mean for the constant error of the thermometer.

2   A second thermometer is used to measure the temperature of a greenhouse and gives these results on ten occasions (assume the greenhouse temperature remains constant):

| 81.7 | 81.9 | 81 | 81.2 | 80.2 |
| 80.7 | 81.3 | 79.7 | 81.1 | 80.7 |

Which of the two thermometers appears to be more precise (have less error)? Does it appear that the second thermometer is biased?

3   A chemist is not sure whether a certain scale is biased, so she uses two scales when weighing the amount of precipitate found in a flask (Table 10.9). Which of the scales appears to have less error? What do you believe to be the true weight of the precipitate? Why?

TABLE 10.9

SCALE A				
49.2	48.0	51.5	52.5	51.9
50.6	46.5	52.0	49.0	49.5
50.4	50.5	49.5	55.4	48.4
SCALE B				
54.1	50.5	49.6	48.0	50.2
52.3	48.6	51.2	49.8	48.5
49.6	52.0	51.1	51.2	48.0

4   Twenty measurements are made of the length of an insect (in millimeters).

41	28	37	38	40
39	39	40	42	39
38	37	42	40	38
42	35	40	39	38

What is the estimated true length of the insect? How could the precision of these measurements be improved upon?

10.5   ACCURACY

After all this discussion about measuring an object, it is reasonable to ask: What is the *real*, or *actual*, length of the chalkboard? What is the *actual* weight of the penny? What is the *true* value of gravity in Ottawa?

These are hard questions. You see, the actual length of an object can only be estimated. We can estimate more and more closely, but how do we know when we have reached that actual, or true, value (or are close enough for all practical purposes)?

The closer a measurement is to the *true* value, the more *accurate* is that measurement.

Since we can never know *exactly* the value we are seeking when we measure, the best we can do is *estimate* how close our measurements are to the actual value (that is, how accurate they are).

Let's go back to the problem of measuring the stack of textbooks. When the class first measured the stack, there was a lot of variation in their measurements (see Section 10.1). The range of lengths was 2.4 centimeters and the standard deviation was .59 centimeters.

We had the feeling that the class was not very accurate in measuring the height of the stack of books. So, in Section 10.2, they measured again, being much more careful. This time, the range of measurements was only .8 centimeters and the standard deviation was .25 centimeters.

As we learned in Section 10.3, this first set of measurements contained a considerable amount of error. We also decided that if we wanted to estimate the *true* height of the stack of books, we should use the mean, or average, of the second set of measurements. We were more confident about using the mean as an estimate of the true height than we were in using any of the individual measurements.

When the class used more care in measuring, the standard deviation of the measurements became smaller. That is, the individual measurements had less error and we had more confidence in their accuracy.

Being careful when measuring, and being accurate, are two somewhat different ideas. Consider the example of the lab technician who is carefully measuring the length of a pendulum. The required length should be in centimeters and should be obtained from using a meter stick. But, inadvertantly, the technician uses a yardstick (marked in inches) instead. The results may have rather low variance (indicating high precision), but they will indeed be biased!

So we see that there are two components of good measurements. First, they have small variance, or standard deviation. Second, measurements should be free from bias. Steps should be taken to try to see if there is bias in the measurements and, if there is, to remove it.

EXERCISES

1   A statistics class took 20 *sets* of 25 measurements of the length of a chalkboard to give a total of 20 means (averages). The means of those sets (in inches) are as follows:

240.4	238.5	237.8	241.0	238.6
236.6	240.0	238.5	241.3	241.1
240.4	241.6	241.1	240.7	240.0
241.6	238.8	238.6	240.5	240.2

Construct a stem-and-leaf table of these measurements. Estimate the standard error of these measurements by finding the standard deviation of the 20 average (mean) measurements. Estimate the true length of the chalkboard from these data.

2   A safety inspector is examining the strength of sewing thread. She instructs her laboratory assistants to take 25 measurements of the strength of the thread and find the mean of those measurements. This is repeated 20 times, giving the following 20 averages (means). What is the estimated true strength of the thread? What is the standard error of the measurements?

49.48	49.28	49.72	50.12	49.48
49.80	49.68	49.68	49.80	49.96
49.52	50.28	48.84	49.84	50.48
49.16	49.48	48.84	48.52	49.08

3   A cartographer (map maker) is measuring the area of a lake, which he has drawn on a map. (This is done by using a device called a *planimeter*.) Since he is interested in the accuracy of his drawing, he is willing to take many measurements and to be very careful. So he takes 25 measurements and finds the mean. Then he repeats this for a total of 20 means as given below. What is the estimated true area of his drawing using the planimeter? What is the standard error of his measurements?

39.08	38.52	39.64	39.80	39.00
39.28	38.40	39.40	39.36	39.76
39.16	39.00	39.68	39.00	38.60
40.04	38.60	38.60	38.92	39.44

4   A team of scientists is estimating the value of gravity on another planet. They take 36 measurements and average the results. Then they repeat this, taking additional groups of 36 measurements, until they have 16 averages (means), as given below.

63.9444	64.4444	63.8333	63.7778
65.1667	64.7778	64.3889	65.1389
64.5833	63.2500	64.8056	64.3056
64.500	64.6111	64.9444	64.5278

What is the estimated value of gravity on that planet? What is the standard error?

5   Two classes measure the length of a field known to be exactly one kilometer long. They obtain these results:

Class I		Class II	
1.11	1.11	0.97	0.99
1.12	1.09	1.02	1.02
1.10	1.10	1.01	1.02
1.10	1.11	1.04	0.96

Which class has produced results with the smallest standard error? Which class seems to have more bias in their results? Which set of data is more accurate?

10.6 APPLICATIONS OF MEASUREMENT
(Finding g—the Acceleration of Gravity[7])

A pendulum can be used to measure the value of g, the acceleration of gravity. You may have noticed that short pendulums swing back and forth quickly and long pendulums swing back and forth slowly. The relationship between the length, L, of a pendulum and the time, T, it takes to go back and forth is given by the formula

$$T = 2\pi \sqrt{\frac{L}{g}}$$

Since the relationship between L and T involves g, we can solve this equation for g.

$$g = 4\pi^2 \cdot \frac{L}{T^2}$$

We know the value of π, 3.14159 (approximately). So, if we have accurate values for L and T, we can obtain a good estimate for g.

The data reported in Table 10.10 were obtained by students at a university in the Netherlands. Ten students obtained measures of L and T for each of 5 pendulums (labeled A, B, C, D, E).

FINDING VALUES FOR L, THE LENGTH OF THE PENDULUM

Each of the ten students measured each pendulum and obtained the results shown in Table 10.10.

TABLE 10.10

	PENDULUM (LENGTH IN CENTIMETERS)				
STUDENT	A	B	C	D	E
1	175.2	151.5	126.4	101.7	77.0
2	179.0	150.0	125.0	100.0	75.0
3	170.0	149.3	124.8	100.4	76.4
4	165.1	149.8	125.0	100.0	75.0
5	171.5	150.0	124.9	100.0	75.0
6	175.8	—	125.0	99.7	75.0
7	172.0	150.0	125.0	100.0	75.0
8	175.3	149.9	125.0	100.0	75.0
9	165.5	150.0	125.0	100.0	75.0
10	175.8	150.7	125.8	100.8	74.6

TABLE 10.11

	VALUES OF 50T (IN SECONDS)				
STUDENT	A	B	C	D	E
1	132.5	123.4	112.8	101.2	88.2
2	133.7	122.3	111.3	99.8	85.8
3	130.8	133.5	113.1	100.5	87.6
4	129.0	133.8	112.2	100.0	86.8
5	131.1	122.6	111.9	100.1	86.8
6	132.8	—	112.0	100.1	86.8
7	131.6	122.8	112.0	100.2	86.8
8	132.1	122.2	111.8	100.0	86.5
9	128.9	122.7	113.0	100.3	86.9
10	132.8	123.0	112.3	100.7	86.7

Some remarks can be made about these measurements. For example, students 2 and 7 did not report any tenths in their results. Are these accurate results, or did these students simply prefer to avoid decimal fractions?

We will consider the student values in more detail a little later.

FINDING VALUES FOR T, THE TIME TAKEN FOR THE PENDULUM TO COMPLETE ONE SWING

It is difficult to measure the time of just one swing of a pendulum, even if it moves slowly. So, the students determined the time required for 50 complete swings back and forth, in seconds. Later, each value was divided by 50 to get T. For now, the time required for 50 swings is called $50T$.

The values of $50T$ for each student and for each pendulum are given in Table 10.11.

Now we can use our formula to calculate the value of g, based on each student's measurements for L and T. (Remember that the table above is for 50 times T!)

For example, for pendulum A and student 1, we have

$L = 175.2$ centimeters

$T = \dfrac{132.5}{50} = 2.65$ seconds

$g = 4\pi^2 \cdot \dfrac{L}{T^2} = 4(3.14159)^2 \cdot \dfrac{(175.2)}{(2.65)^2}$

$\doteq (39.4783)\left(\dfrac{175.2}{7.0225}\right)$

$\doteq 984.9$ centimeters per second per second

Similarly, a value for g can be calculated for each of the other pendulums and for each of the other students—a total of 49 values of g, as given in Table 10.12. (There is one missing value—student 6 did not report results for pendulum B.)

TABLE 10.12

STUDENT	A	B	C	D	E
			PENDULUM		
1	984.9	981.9	980.4	980.1	976.9
2	988.3	989.8	995.9	990.0	1005.5
3	980.7	981.9	980.2	981.1	982.6
4	979.2	980.4	980.0	987.0	982.5
5	984.8	984.9	984.5	985.0	982.5
6	983.8	–	983.5	982.0	982.5
7	980.2	981.7	983.5	983.0	982.5
8	991.5	990.7	987.0	987.0	989.3
9	983.1	983.3	983.5	983.0	980.2
10	983.8	983.1	984.5	981.1	979.5

We now comment a little more about the values obtained by each student. To make this job easier, we will slide each of the values 980 points to the left (subtract 980 from each value in the table). The result is shown in Table 10.13.

TABLE 10.13

STUDENT	A	B	C	D	E	AVERAGE
			PENDULUM			
1	4.9	1.9	.4	.1	-3.1	.84
2	8.3	9.8	15.9	10.9	15.5	14.08
3	.7	1.9	.2	1.1	2.6	1.30
4	-.8	.4	0	7.0	2.5	1.82
5	4.8	4.9	4.5	5.0	2.5	4.34
6	3.8	–	3.5	2.0	2.5	2.95
7	.2	1.7	3.5	3.0	2.5	2.18
8	11.5	10.7	7.0	7.0	9.3	9.10
9	3.1	3.3	3.5	3.0	.2	2.62
10	3.8	3.1	4.5	1.1	-.5	2.40
Mean	4.03	4.19	4.30	4.02	4.40	4.163

Some observations can be made about the values of g (which in the Netherlands is 981.3 centimeters per second per second). Notice, for example, that student 2's values tend to be considerably greater than those obtained by other students. On the average, they are 14.08 greater than 980 and almost 13 units above the true value for g. Student 2's measurements do indeed appear to be biased! Also, notice that for student 1, the value for g decreases systematically for shorter pendulums. This would seem to be evidence of bias, as well.

MEASURING THE THICKNESS OF A SHEET OF PAPER

The ideas from Chapter Eight on linear regression can be combined with measurement concepts to measure the thickness of a sheet of paper.

Of course, one page by itself is too thin to measure using an ordinary ruler. So, we will find the thickness of many pages (a couple of hundred, for instance) and then divide by the number of pages to find the thickness of one page.

We choose the pages of a book for measuring and use a form like the one in Table 10.14. Each person should make at least four measurements, including different groups of pages.

Use a rule marked off in millimeters, and estimate to 0.1 millimeter. This may be difficult to do at first. If so, estimate to .5 millimeter.

You can tell how many pages you are measuring by reading the page number on the front page in the stack you choose and the page number of the page facing the last page in your stack. These should both be odd numbers. The difference between these two page numbers is an even number and, assuming no special inserted material, is *twice* the number of sheets of paper in the stack. So, divide this number by two.

Now, pinch the stack of papers firmly between your fingers and thumb and put the ruler across the stack of pages. Record the measurement in the form illustrated in Table 10.14.

Find the thickness of an individual sheet by dividing the stack thickness by the number of sheets. Then find the average (mean) of these four measurements. What reasons can you give for the differences in the thickness of a sheet of paper in the book you have chosen?

Table 10.14 summarizes the measurement of thickness of paper for 10 students at Nanaimo District Senior Secondary School. These measurements can now be put into Table 10.14.

TABLE 10.14

NAME: DATE:

BOOK: *Mathematics Teaching*, by Travers, Pikaart, Suydam, and Runion. (Harper and Row, 1976).

OBSERVATION NUMBER	PAGE NUMBER FRONT/BACK	NUMBER OF PAGES	NUMBER OF SHEETS	TOTAL THICKNESS (MILLIMETERS) (X)	THICKNESS PER SHEET (MILLIMETERS) (Y)
1	171/347	176	88	7.8	.089
2	101/489	388	194	15.6	.080
3	71/479	408	204	16.9	.083
4	61/553	492	246	19.8	.080
5	211/555	344	172	14.2	.083
6	97/187	90	45	4.5	.100
7	53/231	178	89	7.4	.083
8	71/295	224	112	8.6	.077
9	267/327	60	30	3.9	.130
10	335/411	76	38	4.4	.116
				Total	.921
				Mean	.092

Finally, we will fit these data to a straight line. We wish to estimate the thickness of a stack of paper (Y) given the number of sheets in a stack (X).

1. Find the mean data point for these data.

2. Find the covariance for X and Y.

3. Find the variance of X.

4. Find the slope of the least squares best-fit line:

$$m = \frac{\text{covariance}}{\text{variance } X}$$

5. Find the Y-intercept:

$$C = \bar{Y} - m\bar{X}$$

The results of the calculations are as follows:

1. Mean of X, the number of sheets: 121.8

   Mean of Y, the total thickness of the sheets: 10.3 millimeters

   Mean data point: (121.8, 10.3)

2. Covariance: 402.5

3. Variance of X: 5357.8

4. Slope of least squares best-fitting line:

$$m = \frac{402.5}{5357.8} = .07$$

5. Y-intercept:

$$C = \bar{Y} - m\bar{X} = 10.3 - .07(121.8) = 1.8$$

So, the equation of the least squares regression line is

$$Y' = .07X + 1.8$$

Let's use this equation to estimate the total thickness (excluding the cover) of a book of 150 sheets (300 pages):

$$Y' = .07(150) + 1.8 = 12.3$$

So, the estimated thickness of the stack of sheets is 12.3 (rounded to 12.0) millimeters.

EXERCISES

1   Use the least squares regression equation found in this section to estimate the thickness (excluding covers) of books having these number of sheets:

200   350   425   500

2   Use the method of this section to measure the thickness of the paper used in this book. You should get at least 15 measurements like those recorded in the form in Table 10.15. Then find the least squares regression equation for estimating the thickness of given numbers of sheets.

3   A publisher believes that students prefer to carry a book that is no more than 1 centimeter (10 millimeters) thick. Use the regression equation

$$Y' = .07X + 1.8$$

to estimate the largest number of sheets that a book 1 centimeter thick should have. Then use the equation you found in Exercise 2 of this section to estimate the largest number of pages that could be in such a book using paper having the thickness of the sheets in this book.

4   The data in Table 10.15 were collected on the thickness of sheets from a history book. Estimate the thickness of a sheet of paper in this book. Then find the least squares regression equation for estimating the thickness of a stack of sheets given the number in the stack.

TABLE 10.15

OBSERVATION NUMBER	PAGE NUMBER FRONT/BACK	NUMBER OF PAGES	NUMBER OF SHEETS	TOTAL THICKNESS (MILLIMETERS) (X)	THICKNESS PER SHEET (MILLIMETERS) (Y)
1	15/415	400		16.9	
2	235/295	60		3.3	
3	101/289	188		11.0	
4	25/145	120		2.2	
5	17/373	356		16.4	
6	21/373	352		14.1	
7	47/157	110		7.1	
8	53/263	210		9.5	
9	5/397	392		15.7	
10	33/295	262		11.4	
11	41/393	352		10.2	
12	107/345	238		6.9	
13	213/339	126		5.3	
14	11/417	406		14.1	
15	1/709	708		26.2	
16	115/255	140		7.9	
17	105/413	308		11.6	
18	11/481	470		18.8	
19	217/389	172		9.9	
20	43/219	176		7.8	

CHAPTER 11

Estimation

KEY PROBLEM

How Many Deer Are in the Park?[1]

How do you find the number of deer in a park, or fish in a pond, when you can't count them directly? This is an important problem to persons who study the environment, or who manage parks or wildlife refuges.

A statistical method called *capture-recapture* may be used. Briefly, the method involves *capturing* a random sample of the animals to be counted. For example, a random sample of deer from a forest or park could be taken and counted. Each animal is then marked in some way, such as with a metal tag on an ear. The animals are then released. Later, another random sample of deer is captured and the number of tagged deer in this second sample is noted. The size of the new sample and the number of recaptured (tagged) deer is used to estimate how many deer are in the park.

In this chapter, we will study ways in which statistical ideas are used in estimation problems.

11.1 SAMPLES AND POPULATIONS REVISITED

In Chapter Five, we discussed *samples* and *populations*. A *sample* is a collection of people or things that is chosen from, and hence believed to represent, a larger collection of people or things. The larger collection is called a *population*.

■ **Example 1:**   1200 people selected from across the country are interviewed in order to determine the popularity of an entertainer. What is the sample and what is the population that the sample is believed to represent?

Solution:

The sample is the 1200 people chosen from across the country. The population is all of the people in the country. ■

■ **Example 2:**   A small bottle of water is taken from a swimming pool and tested for the presence of bacteria. Identify the sample and the population.

Solution:

The sample is the bottle of water taken from the swimming pool. The population is all of the water in the swimming pool. ■

STATISTICS AND PARAMETERS

We now need two technical terms. A numerical piece of information provided by a sample is called a *statistic*. The corresponding quantity in the population from which the sample was taken is a *parameter*. In Example 1, the proportion of the 1200 people in the poll who favor a particular singer, say, is a *statistic*. The proportion of people across the country (the proportion of people in the population) who favor the singer is a *parameter*. In Example 2, the amount of bacteria in the bottle of water is a *statistic*. The amount of bacteria in the swimming pool is a *parameter*.

Note: In order to find a statistic, we go to a sample and perform some kind of calculation. We may count what proportion of people in the sample favor a singer, or, using a microscope or chemical test, find the amount of bacteria in the bottle of water taken from the swimming pool.

But we usually cannot measure a parameter directly. We cannot count the number of people across the country who favor the singer. It would take too long, and before we finished, people would have changed their minds! Also, it would be too expensive. Likewise, it is not practical to try to count the bacteria in the entire swimming pool of water!

Since for one reason or another, we do not measure parameters directly, we must estimate them from samples. (The word *parameter* comes from the Greek word *meter,* "to measure," and *para,* which means "along side of.")

Statistics (from samples) can be calculated directly. *Parameters* (from populations) are estimated from information provided by samples.

SOME SHORTHAND NOTATIONS

Since it is important to distinguish between sample values and population values, different symbols are used for each.

We use our alphabet when referring to sample values. For example:

Sample mean = $\overline{X}$
Sample variance = S^2
Sample standard deviation = S

When we are referring to population values, we use Greek letters. For example:

Population mean = μ (Greek letter mu, or m)
Population variance = σ^2 (Greek letter sigma, or lower case s)
Population standard deviation = σ

Suppose we wish to estimate the following:

1. The temperature of Lake Michigan near Chicago. A thermometer is used to measure the temperature of the lake at Chicago at noon on August 21.

2. A patient's white blood cell count. A drop of the patient's blood is examined under a microscope.

In each case, we are using information from a sample (a statistic) to estimate the characteristic (a parameter) of the population from which the sample was taken.

Let's think of each example in these terms.

1. The temperature of the water (the parameter) in Lake Michigan (the population) is estimated as 50 degrees (the statistic). This estimate was obtained, we suppose, by measuring the temperature of the water in several locations in the Chicago area. The water surrounding the thermometer makes up the sample or samples involved.

2. The number of white blood cells (parameter) in the patient's body (population) is estimated as 12,000 per cubic millimeter (statistic) of blood. The blood count is done by a laboratory technician, who takes a drop (sample) of blood from a finger or ear lobe and examines it under a microscope.

Let's see one way in which sampling ideas are used.

■ **Example 3:**　Telephone bills may be based on samples of long-distance calls. The bill in Figure 11.1 shows that for the month a random sample of 67 long-distance calls cost $166.92. How can this information be used for calculating the total bill, based on the charges for the 67 calls?

Solution:

Using the information from the sample of calls, we find that the mean (average) cost of a call was

$$\frac{\$166.92}{67} = \$2.49$$

Now, since the company has recorded that 342 long-distance calls were made, the total bill is

$$342 \times \$2.49 = \$851.58 \blacksquare$$

Notice the savings of time and effort which this sampling procedure has made possible. The costs of each call are not listed. A much shorter bill is needed for the customer. The whole billing procedure has been streamlined!

FIGURE 11.1

DEPARTMENT OF ADMINISTRATIVE SERVICES

TELECOMMUNICATION DIVISION
19.01 PERCENT SAMPLE

NETWORK USAGE ITEMIZED AS FOLLOWS:　　　　　　　　**BILLING DATE　06/01/80**

USAGE FOR 241-4196

DATE	TIME	NUMBER CALLED	DURATION	LONG DISTANCE CHARGE
04/29	1623	201-2 . . . 3300	2	1.90
04/26	1116	202-5 45	13	6.50
04/23	0959	212-	1	.44
04/14	1434		3	. . .
			2	. . .
			. . .	. . .
			. . .	1.62
			8	4.03
			9	4.63

SAMPLE NETWORK USAGE TOTAL　166.92

TOTAL NUMBER OF CALLS MADE = 342

EXERCISES

1   For each sample, identify the sample, statistic, population, and parameter involved. (More than one answer may be correct.)

(a)   The high temperature at Los Angeles International Airport on August 27 was 73 degrees F.

(b)   In a poll of 64 college freshmen at a large college, it was found that 65% were opposed to raising the legal drinking age from 18 to 21 years.

(c)   The pollen count for Urbana, Illinois, was 228 grams per cubic yard of air on August 25.

(d)   A container full of corn taken from a truck unloading at a grain elevator was found to have 35% moisture.

2   A poll of 20 people at a shopping center was taken to rate the taste of a new cola drink. The results (1=terrible to 5=terrific) were:

$$4, 3, 3, 4, 4, 3, 2, 3, 5, 3$$
$$4, 2, 4, 3, 3, 2, 3, 4, 4, 3$$

What is the mean preference score of the 20 people in the poll? What is the range of preference scores in the poll? What parameter could the mean preference score of the poll estimate?

3   A random sample of 25 college students is surveyed about their political views. Five of them have liberal views. What parameter could this statistic be used to estimate?

4   In the newspaper account in Figure 11.2, what samples and populations are involved? What factors could effect how well the samples represent the populations?

FIGURE 11.2

ANCIENT TREE RINGS TELL A TALE OF PAST AND FUTURE DROUGHTS[2]

by WALTER SULLIVAN   TUCSON, ARIZ.

By measuring annual tree rings in groves scattered across the Western states, scientists here are seeking to define the likelihood of prolonged and extreme drought for each of nine river basins in that region.

At least 10 trees are sampled at each site. A hollow drill extracts a pencil-sized core of wood without seriously harming the tree. The core is then studied under magnification to record climate-induced variations in tree-ring width back to the time when the tree began growing. Trees on well-drained slopes are preferred since they respond quickly to a drought.

An objective of the project, funded in part by $286,000 from the National Science Foundation, is to determine whether there is evidence for what some hydrologists call a "Noah effect." This would be a weather extreme beyond known precedent, like the 40 days of rain that, according to Genesis, flooded the world in the days of Noah.

A variant, known as the "Joseph effect," would be a condition—such as a drought—far more prolonged than any on record. The reference, again from Genesis, is to the seven years of famine predicted by Joseph.

It is assumed that, if radical departures from normal behavior have taken place in the past, they may occur again and the frequency of occurence may be estimated.

It is known from other studies that major changes can occur. For example, during the "climatic optimum" 6,000 years ago, when regional temperatures rose 2 to 4 degrees Fahrenheit, the western grasslands moved east through Iowa and Illinois into Indiana and Ohio.

5   Refer to Figure 11.3. Identify the sample and population. Give a statistic from the report. Tell the parameter it is estimating.

FIGURE 11.3

> FACT-FILE

FRESHMAN CHARACTERISTICS AND ATTITUDES

A U.S. NATIONAL PROFILE BASED ON RESPONSES OF
188,000 STUDENTS WHO ENTERED COLLEGE IN THE FALL OF 1982[3]

──────── CHARACTERISTICS ────────

Age by December 31, 1982:

16 or younger	0.1%
17	2.5%
18	74.2%
19	18.9%
20	1.8%
21	1.3%
22	0.6%
23-25	0.4%
26-29	0.1%
30 or older	0.0%

Residence preferred during fall term:

With parents or relatives	19.4%
Other private home or apartment	25.8%
College dormitory	43.8%
Fraternity/sorority house	5.0%
Other campus housing	3.8%
Other	2.2%

Reasons noted as very important in selecting college attended:

Relatives' wishes	6.6%
Teacher's advice	4.0%
Good academic reputation . . .	53.5%
Offered financial assistance . . .	16.7%
Not accepted anywhere else . .	2.7%
Advice of former student	14.9%
Offers special programs	25.5%
Low tuition	20.6%
Advice of guidance counselor . .	7.7%
Wanted to live at home	11.1%
Friend's suggestion	7.2%
Recruited by college	4.3%

Current religious preference:

Jewish	3.0%
Protestant	33.7%
Roman Catholic	38.9%
Other	17.2%

──────── ATTITUDES ────────

	Men	Women	Total
Political views:			
Far left	2.1%	1.5%	1.8%
Liberal	18.5%	19.4%	18.9%
Middle of the road	55.9%	63.7%	59.8%
Conservative	22.0%	14.9%	18.4%
Far right	1.5%	0.6%	1.0%
Agree strongly or somewhat that:			
Government isn't protecting consumer	64.4%	73.1%	68.8%
Government isn't controlling pollution	74.9%	82.2%	78.6%
Government should discourage the use of energy	74.5%	80.8%	77.6%
Military spending should be increased	47.9%	29.9%	38.8%

6   Find a newspaper article in which sample data (statistics) are used (or assumed to be used) to estimate population data (parameters). Give examples from the newspaper article.

7   A telephone company takes a 15% random sample of a customer's long-distance telephone calls. The total charge for this sample of 86 calls is $112.82. Based on this sample, what is the estimated telephone bill for all of the long-distance calls?

8   A telephone company takes a 20% random sample of a customer's long-distance calls. There were 53 calls in the sample, costing a total of $92.50. Based on this sample, what is the estimated total bill for long-distance calls?

11.2   ESTIMATING POPULATION MEANS

Suppose you want to find the average (mean) height of the 16-year-old males in a school. You probably would not set out to find the height of each and every 16-year-old male in the school. Instead, you would take a random sample of, say, 50 or 60 16-year-old males and find their mean height. You would take this statistic as a good estimate of the parameter, which is the mean height of all 16-year-old males in the school.

We used a similar approach when estimating the total long-distance telephone charges in Section 11.1. The method used is based on the assumption that the mean telephone charge for the sample of telephone calls is a good estimate of the mean charge for all calls.

How can we help ensure that our statistics, which we get from samples, are good estimates of the population parameters?

Two important ideas are involved, as was discussed in Chapter Ten.

1.   Our estimated value should have a small standard error.

2.   Our estimated value should contain little or no bias.

STANDARD ERROR

Throughout this book, we have used the idea that in order to get better estimates of quantities, such as probabilities, we should use large numbers of trials. For example, when tossing a fair coin, the more tosses we had the closer the proportion of heads got to .5.

We now pursue this idea a little more by conducting an experiment. We wish to examine the effect that sample size has on how well a sample statistic ($\bar{X}$) estimates a population parameter (μ).

We will draw samples of various sizes from a population with a mean known to be 50 (that is, $\mu = 50$) and a variance known to be 100 (that is, $\sigma^2 = 100$).

Note: You should remember that this is an artificial example since we do not usually know population values (parameters) in the real world.

The procedure is simple. Think of the population as a huge box containing a very large number of identical tickets (or, perhaps, marbles). Each ticket is numbered in such a way that the mean of all of the numbers on the tickets is exactly 50.0 and the variance of all of these numbers is exactly 100.0. That is,

$$\mu = 50.0$$
$$\sigma^2 = 100.0$$
$$\sigma = 10.0$$

Now we draw a random sample of four from the box. That is, we shake the box, pull out a ticket, and write down the number on the ticket. We then replace the ticket in the box, shake the box, and take out another ticket. We repeat until we have our random sample of four observations (or numbers). We then find the mean of these four numbers and record it in a stem-and-leaf table. This activity is repeated until we have produced 100 means from samples of size 4. Table 11.1 gives the results of such an activity.

We notice that the sample means cluster somewhere around 50, the population (true) mean. We notice also that the average (mean) of all 100 sample means is close to 50 (in fact, it is 50.5).

TABLE 11.1

Population values: This population has a mean of 50.0 ($\mu = 50.0$) and a standard deviation of 10.0 ($\sigma = 10.0$).

STEM	LEAF	f
35		0
36		0
37	6	1
38		0
39	8, 9	2
40	0	1
41		0
42	1, 1, 3, 4, 5, 6, 8	7
43	0, 0, 7	3
44	1	1
45	3, 7	2
46	3, 3, 3, 4, 4, 5, 7	7
47	0, 1, 3, 3, 6, 6, 7	7
48	0, 1, 2, 2, 5, 5, 5, 8, 8	9
49	1, 2, 5, 7, 7, 7, 7, 8, 9	9
50	0, 1, 3, 6, 6	5
51	1, 3, 9	3
52	3, 3, 5, 6, 6, 6, 6, 7, 8, 9	10
53	3, 5, 5, 8, 8, 8	6
54	0, 4, 8, 8	4
55	2, 2, 4, 4, 7, 8, 8	7
56	6, 6, 7, 9	4
57	0, 2, 6	3
58	0, 6, 8, 8	4
59	1, 7	2
60	0, 9	2
61		0
62	1	1
63		0
64		0

Total 100

Sample values: Sample of size 4 ($N=4$) were drawn

Mean of 100 sample means = 50.5

Standard deviation of the 100 sample means (standard error) = 5.3

We should notice something else about this experiment, though. Suppose we had only one sample of size 4. How confident could we be that this sample would give a good (reliable) estimate of the population mean of 50?

We see that there is a lot of variation in the sample means for this small sample. One sample had a mean as small as 37.6. Another sample had a mean as large as 62.1. We also note that the sample means tend to cluster about their mean (which is 50.5).

You may recall from Chapter Two that a measure of how much a set of data is "spread out" around its mean is given by the *standard deviation*.

As we learned in Chapter Ten, the standard deviation of sample means is called *standard error*. We see from Table 11.1 that the standard error for the 100 sample means is 5.3.

Let's see the effect of taking a larger, random sample from the same population. Table 11.2 gives the results of taking samples of size 16. Table 11.3 gives the results when the sample size is increased to 36.

TABLE 11.2

Population values: $\mu = 50.0$; $\sigma = 10.0$

STEM	LEAF	f
41		0
42		0
43	4	1
44	4, 5, 7	3
45	0, 1, 6	3
46	1, 2, 6, 7	4
47	0, 0, 1, 3, 8, 9, 9	7
48	0, 2, 3, 4, 5, 5, 6, 6, 7, 7, 8, 9, 9, 9	14
49	3, 3, 3, 3, 4, 4, 5, 5, 6, 6, 6, 7, 7, 9	14
50	0, 0, 0, 1, 1, 2, 2, 3, 3, 3, 3, 6, 6, 7, 8, 9	16
51	0, 0, 1, 1, 1, 3, 4, 5, 6, 6, 7, 7, 9	13
52	0, 0, 1, 3, 3, 4, 4, 5, 6, 9, 9	11
53	0, 1, 1, 2, 5, 5, 5, 6, 9	9
54	3	1
55	7, 9	2
56	8, 8	2
57		0
58		0

Total   100

Sample values: $N = 16$; Mean = 50.2; Standard error = 2.5

TABLE 11.3

	Population values: $\mu = 50.0$; $\sigma = 10.0$	
STEM	**LEAF**	*f*
44		0
45		0
46	4	1
47	1, 2, 2, 3, 5, 6, 7, 9, 9, 9	10
48	0, 2, 3, 3, 3, 3, 3, 4, 4, 5, 5, 5, 6, 6, 6, 7, 7, 8, 9, 9	20
49	0, 1, 2, 2, 2, 2, 2, 2, 3, 4, 4, 4, 5, 5, 5, 5, 6, 6, 6, 7, 7, 7, 7, 8, 9	25
50	1, 1, 2, 2, 3, 4, 5, 5, 5, 5, 5, 5, 6, 6, 7, 7, 7, 8, 9	19
51	0, 2, 3, 4, 4, 4, 4, 5, 6, 9	10
52	0, 1, 1, 1, 3, 4, 5, 6, 9	9
53	2, 3, 5, 9, 9, 9	6
54		0
55		0

Sample values: $N = 36$; Mean = 49.9; Standard error = 1.7   Total   100

The pattern of outcomes is similar to what happened before. The means of the samples tend to cluster around the population mean (which, remember, we happen to know is 50).

We notice something else. As the sample gets larger, the means cluster more closely together.

Let's look at what happened for samples of size 16 (Table 11.2). We see that 59 of the 100 random samples had a mean between 48 and 52. Now look at the results for samples of size 36 (Table 11.3). Here we find that 75 of the 100 samples had a mean between 48 and 52.

In other words, as the sample size gets larger and larger, the standard error of the sample means gets smaller and smaller.

Table 11.4 summarizes the standard error of the sample means, based on the results of the sampling activities given in Tables 11.1 through 11.3. Remember that the standard error is the standard deviation of the 100 sample means in each activity. The smaller standard error (standard deviation of the sample means) shows that the sample means are closer and closer together.

TABLE 11.4

	SAMPLE SIZE		
	$N = 4$	$N = 16$	$N = 36$
Mean of $\bar{X}$	50.5	50.2	49.9
Standard deviation of $\bar{X}$ (Standard error)	5.3	2.5	1.7

Remember: In each case, 100 random samples have been drawn from a population with a mean of 50.0 and a standard deviation of 10.0.

BIAS

A good sample estimate should have little or no bias. We learned in Chapter Ten that bias in a measure is a *systematic* error that has crept in, to make that measure too big or too small.

Table 11.4 tells us that the mean (average) of the sample means for the various sample sizes is very close to the population mean ($\mu = 50.0$). Apparently, the sample means did not have much bias in them. On the average, they approached the population mean μ. The sample mean is an *unbiased* statistic.

EXERCISES

For Exercises 1 through 3, refer to the stem-and-leaf tables (Tables 11.1 through 11.3).

1   We are drawing a random sample from a population with a mean of 50 and a standard deviation of 10.

(a)   What proportion of the sample means is within two units of the population mean (that is, between 48 and 52, inclusive) for a sample size of 4?

(b)   Answer (a) if the sample size is 16.

(c)   Answer (a) if the sample size is 36.

2   XYZ Manufacturers produce a flashlight battery that is believed to have a life of 45 hours. Past records of tests of the batteries show the standard deviation of the battery lives is 12 hours. Suppose a quality control inspector takes a sample of 16 batteries and finds the life of each battery. Suppose also that 100 such samples are drawn. What proportion of these samples would be expected to have a mean life greater than 47 hours? Less than 40 hours?

You may use Table 11.5 of sample values, produced in a manner similar to the way in which the data of Tables 11.1 through 11.3 were produced. From a population with a mean of 45 and a standard deviation of 12, random samples of size 16 were drawn and the means recorded. This was repeated for a total of 100 samples.

TABLE 11.5

Population values: $\mu = 45.0$; $\sigma = 12.0$

STEM	LEAF	f
35		0
36		0
37	1, 8	2
38		0
39	2, 2, 5, 7	4
40	2, 3, 5, 5, 6	5
41	3, 4, 4, 5, 6, 7, 7, 8	8
42	0, 4, 4, 5, 7, 8, 9	7
43	1, 1, 3, 3, 4, 4, 5, 6, 8, 9	10
44	2, 2, 3, 3, 3, 4, 4, 6, 7, 7, 7, 8	12
45	0, 0, 1, 2, 2, 2, 3, 4, 4, 5, 5, 6, 7, 7, 8, 9	16
46	0, 1, 2, 4, 6, 9	6
47	1, 2, 2, 2, 2, 3, 4, 5, 5, 6, 7, 7, 8, 9	14
48	1, 4, 5, 5	4
49	0, 1, 3, 3, 5, 8	6
50	4, 8, 9, 9	4
51	8	1
52	0	1
53		0
54		0

Total   100

Sample values: $N = 16$; Mean = 44.9; Standard error = 3.2

3   Suppose in Exercise 2, above, a sample of 25 batteries was used rather than 16. Use Table 11.6 to estimate the proportion of samples whose mean will be within two hours of the assumed value of 45 hours for XYZ batteries. This table was produced by drawing 100 random samples from a population with a mean of 45 and a standard deviation of 12. This table records the *means* of the samples only. (Computer program SAMPLE, from Appendix G was used.)

TABLE 11.6

Population values: $\mu = 45.0$; $\sigma = 12.0$

STEM	LEAF	f
38		0
39		0
40	2, 2, 7, 9	4
41	2, 2, 3, 6, 7, 7, 9	7
42	1, 3, 6, 7, 9, 9	6
43	0, 0, 0, 1, 1, 1, 2, 2, 2, 3, 4, 4, 5, 6, 7, 7, 8, 8, 9, 9, 9	21
44	0, 1, 1, 1, 3, 3, 4, 6, 7, 7, 8, 9, 9, 9	14
45	0, 0, 0, 1, 2, 4, 5, 6, 6, 7, 7, 7, 8, 9, 9	15
46	0, 0, 1, 1, 1, 3, 3, 4, 5, 6, 8	11
47	0, 1, 1, 2, 3, 3, 4, 4, 5, 6, 9	11
48	0, 0, 1, 3, 3, 3, 8, 9	8
49		0
50		0
51	5	1
52	0, 5	2
53		0
54		0

Sample size: $N = 25$ Total   100

4 Suppose that in the case of XYZ batteries, a sample of 100 batteries is tested. What is the estimated proportion of samples whose mean will be within two hours of the assumed value of 45 hours? Use the data in Table 11.7.

TABLE 11.7

Population values: $\mu = 45.0$; $\sigma = 12.0$

STEM	LEAF	f
39		0
40		0
41	8, 8, 9	3
42	4, 4, 8, 9, 9	5
43	1, 2, 2, 3, 5, 5, 5, 5, 6, 7, 7, 7, 8, 8, 8, 8, 9	17
44	0, 0, 1, 1, 3, 3, 3, 3, 4, 4, 4, 4, 4, 5, 5, 5, 6, 6, 6, 7, 7, 8, 8, 8, 8, 9, 9, 9	28
45	0, 0, 0, 0, 0, 0, 0, 1, 1, 2, 2, 3, 4, 4, 5, 5, 5, 6, 6, 6, 7, 7, 9, 9, 9	25
46	0, 0, 2, 2, 2, 3, 4, 4, 5, 5, 6, 7, 8, 8, 9	15
47	1, 2, 4, 4, 5, 6, 8	7
48		0
49		0

Total 100

SPECIAL INTEREST FEATURE

Statistics and Pollsters [4]

The New York Times poll is based on telephone interviews conducted from August 23 through August 28 with 2,063 adults in New York State. Of those, 638 were enrolled Democrats and 408 were enrolled Republicans. The remainder of those interviewed were enrolled in other parties, were registered without party enrollment, or were not registered at all.

For the purposes of the look at the New York Democratic primary race for the Senate nomination, only the responses from the 638 Democrats were used.

The sample of telephone exchanges called was selected by a computer from a complete list of exchanges in the state. The exchanges were chosen to insure that each part of the state was represented in proportion to its population. For each exchange, the telephone numbers were formed by random digits, permitting access to both listed and unlisted residential numbers.

The results have been weighted to account for household size and to adjust for variations in the sample relating to race, sex, age, education, and the proportions of the state population living in New York City and upstate.

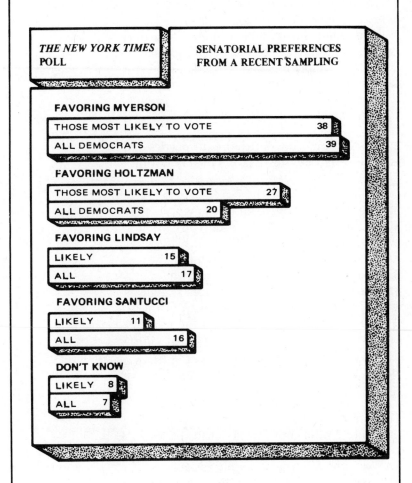

THE NEW YORK TIMES POLL

SENATORIAL PREFERENCES FROM A RECENT SAMPLING

FAVORING MYERSON

| THOSE MOST LIKELY TO VOTE | 38 |
| ALL DEMOCRATS | 39 |

FAVORING HOLTZMAN

| THOSE MOST LIKELY TO VOTE | 27 |
| ALL DEMOCRATS | 20 |

FAVORING LINDSAY

| LIKELY | 15 |
| ALL | 17 |

FAVORING SANTUCCI

| LIKELY | 11 |
| ALL | 16 |

DON'T KNOW

| LIKELY | 8 |
| ALL | 7 |

In theory, for questions answered by all 638 enrolled Democrats, one can say with 95 percent certainty that the results differ by no more than 4 percentage points in either direction from what would have been obtained by interviewing all enrolled Democrats in the state. The theoretical-error margin does not take into account an additional, unmeasurable, margin of error resulting from various practical difficulties in taking any survey of public opinion.

11.3   ESTIMATING POPULATION VARIANCES

So far in this chapter, we have considered only the problem of estimating population means from sample means. But information from a sample can also be used to estimate the variance of the population from which the sample was taken. If a sample has a large variance, for example, then we would expect the variance of the population to be large.

In this section, we will explore properties of sample variances as estimators of population variance.

We have already done a lot of the work for this activity, but we did not tell you! When we got the 100 samples in Section 11.2, we calculated not only the mean of all of the sample means, but the variance and the standard deviation for each random sample as well. Table 11.8 shows the variances of the 100 samples of size 4 taken from the population with a mean of 50 and variance of 100.

Recall how we calculate variance (see Chapter Two).

The sample variances are given in the table to the nearest whole number. To read the table, multiply the stem by 10 and add a leaf. So, for example, the largest sample variance obtained was 222. Likewise, the smallest sample variances obtained were 1, 5, and 6. That is, out of 100 samples, 3 samples had very small variances, indeed! (Keep in mind that in all cases, sampling was from a population with a known variance of 100.)

TABLE 11.8

Population values: $\mu = 50.0$; $\sigma^2 = 100.0$

STEM	LEAF	f
0	1, 5, 6	3
1	0, 0, 1, 1, 2, 2, 3, 3, 3, 9, 9	11
2	0, 4, 4, 6, 7, 7, 7, 9	8
3	1, 4, 4, 4, 8, 9	6
4	0, 2, 3, 4, 6, 7, 7, 8, 8	9
5	1, 1, 1, 2, 5, 6	6
6	1, 3, 4, 7, 8, 9	6
7	0, 1, 3, 6, 8, 9	6
8	0, 1, 3, 4, 4, 7, 8	7
9	1, 2, 3, 7, 9	5
10	0, 4, 5, 5, 7, 7, 7, 9	8
11	5, 6, 7	3
12	1, 2, 3	3
13	0, 0, 0, 4, 4, 5, 5, 6, 6	9
14	1, 3	2
15	8	1
16	0, 0, 4	3
17	0, 7, 7	3
18		0
19		0
20		0
21		0
22	2	1
23		0
24		0
		Total 100

Sample values:
Size, $N = 4$; Mean of sample variances = 76.2

Notice what has happened here. The sample variances have a fairly large range (the range is $222 - 1 = 221$) and the average (mean) value of these 100 values is 76.2.

Let's look at the results as the sample size is increased, first to 16 (Table 11.9) and then to 36 (Table 11.10).

TABLE 11.9

Population values: $\mu = 50.0$; $\sigma^2 = 100.0$

STEM	LEAF	f
2		0
3		0
4	2	1
5	1, 4, 4, 5, 6, 8, 8, 9	8
6	0, 4, 7, 8, 8, 9, 9, 9	8
7	0, 1, 1, 2, 3, 4, 7, 8, 8, 9, 9	11
8	1, 1, 2, 2, 3, 3, 4, 5, 5, 5, 7, 7, 8, 9	14
9	0, 1, 1, 3, 3, 4, 5, 6, 6, 6, 9, 9, 9	13
10	1, 1, 1, 1, 1, 2, 2, 2, 3, 3, 3, 4, 5, 5, 7, 7, 8, 9, 9	19
11	0, 0, 0, 1, 2, 3, 4, 4, 4, 5, 5, 6, 6, 6, 9	15
12	1, 1, 4, 6	4
13	0, 1, 3	3
14	1, 1, 3, 8	4
15		0
16		0

Sample values: $N = 16$; Mean of sample variances = 93.9   Total   100

Our estimates of the population variance have improved somewhat. The mean of the 100 sample estimates is 93.9. (Remember, the population variance is 100.) The range of estimates is now $184 - 42 = 106$.

Let's try again, with random samples of size 36 (Table 11.10).

TABLE 11.10

Population values: $\mu = 50.0$; $\sigma^2 = 100.0$

STEM	LEAF	f
4		0
5		0
6	0, 3, 6, 7, 7, 7	6
7	3, 4, 5, 7, 7, 9, 9, 9	8
8	0, 0, 1, 2, 3, 4, 4, 5, 5, 6, 6, 7, 7, 8, 8, 8, 9, 9, 9	19
9	0, 0, 0, 0, 1, 2, 2, 2, 2, 3, 3, 3, 5, 5, 5, 5, 7, 8, 9, 9	20
10	0, 1, 1, 1, 1, 2, 2, 2, 2, 3, 3, 3, 4, 4, 4, 5, 5, 5, 6, 6, 8, 8, 8, 8, 9	25
11	1, 1, 1, 2, 2, 4, 4, 4, 5, 6, 8, 8, 8, 9	14
12	0, 2, 3, 3, 5, 5, 7	7
13	5	1
14		0
15		0

Sample values: $N = 36$; Mean of sample variances = 97.0   Total   100

Note now that the mean of the sample estimates of the population variance is 97.0 and the range in these estimates has decreased to $135 - 60 = 75$. We summarize these results in Table 11.11.

TABLE 11.11

	SAMPLE SIZE		
	$N = 4$	$N = 16$	$N = 36$
Mean of sample variances	76.2	93.9	97.0
Range of sample variances	221.0	106.0	75.0

Now let's discuss the results of our sampling activity according to our two ideas about good estimators.

Standard Error

We can see from Tables 11.8 through 11.10 that as the sample size increased from 4 to 36, there was less variation in the sample estimates of the population variances. The range in the estimates went down from 221 to 75. This indicates that the estimates were closer together. We could have calculated the *variance* (and standard deviation) of these variance estimates, but that would be too complicated for our purposes here. All we need to know is that an increase in sample size has reduced the amount of variation in the estimates of the population variance. Therefore, the standard errors have been reduced, even though we have not given the actual numbers for the standard errors.

Bias

We expect that the mean value of the statistic, for repeated samples, will get closer and closer to the population value, if there is little or no bias in that statistic. But we have seen that, particularly for the small samples ($N = 4$ or $N = 16$), the mean value of the sample variances *underestimates* considerably the population variance of 100.

Since small samples underestimate the variance of the population from which they were taken, we say that sample variances are *biased* estimates of population variance.

There is a way, however, to help remove the bias from a sample variance. Remember from Chapter Two how we find variance:

$$\text{Variance} = \frac{\text{sum of } (X - \bar{X})^2}{N}$$

This is the formula we used to find the sample variances in Tables 11.8 through 11.10. Now, to give an unbiased estimate of the population variance, we use $N - 1$ in the denominator of the formula instead of N. That is,

$$S^2 = \frac{\text{sum of } (X - \bar{X})^2}{N - 1}$$

Using this formula, we recalculated the sample variances for each of the three sets of 100 random samples and obtained the values in Table 11.12. Notice now that for each set, the mean of the 100 sample variances is close to 100, the population variance.

Note also that the effect has been more dramatic for the small samples. In particular, for samples of size 4, the sample variances have increased, on average, from 76.2 (see Table 11.8) to 101.6.

TABLE 11.12(a)

Population values: $\mu = 50.0$; $\sigma^2 = 100.0$

STEM	LEAF	f
0	1, 7, 8	3
1	3, 3, 4, 5, 6, 6, 7, 7, 8	9
2	5, 6, 7	3
3	3, 3, 5, 6, 6, 7, 8	7
4	1, 5, 5, 6	4
5	1, 2, 4, 6, 7, 9	6
6	2, 2, 2, 4, 4, 8, 8, 8, 9	9
7	3, 4	2
8	2, 4, 5, 9	4
9	0, 2, 3, 5, 8	5
10	2, 4, 5, 7, 8	5
11	1, 2, 3, 6, 7	5
12	1, 2, 4, 9	4
13	2, 3, 8	3
14	0, 1, 3, 3, 3, 5	6
15	4, 4, 6	3
16	2, 3, 4	3
17	3, 3, 3, 8, 9	5
18	0, 0, 1, 1, 7	5
19	1	1
20		0
21	1, 3, 3, 8	4
22	6	1
23	5, 6	2
24		0
25		0
26		0
27		0
28		0
29	3	1
30		0
31		0

Sample values: $N = 4$; Mean of $S^2 = 101.6$ Total   100

TABLE 11.12(b)

Population values: $\mu = 50.0$; $\sigma^2 = 100.0$

STEM	LEAF	f
2		0
3		0
4	4	1
5	4, 8, 8, 9, 9	5
6	2, 2, 3, 4, 9	5
7	1, 2, 2, 3, 4, 4, 5, 6, 6, 7, 8, 9	12
8	2, 3, 4, 4, 4, 6, 6, 7, 7, 8, 9, 9	12
9	1, 1, 1, 3, 3, 4, 5, 6, 7, 7, 9, 9	12
10	0, 1, 2, 2, 3, 5, 6, 6, 7, 8, 8, 8, 8, 9, 9, 9	16
11	0, 0, 0, 1, 1, 2, 4, 4, 5, 6, 7, 7, 7, 8, 8	15
12	0, 1, 2, 2, 2, 2, 3, 4, 4, 4, 5, 9, 9	13
13	2, 4, 8, 9	4
14	2	1
15	0, 1, 2, 8	4
16		0
17		0

Sample values: $N = 16$; Mean of $S^2 = 100.3$ Total   100

TABLE 11.12(c)

Population values: $\mu = 50.0$; $\sigma^2 = 100.0$

STEM	LEAF	f
4		0
5		0
6	2, 5, 7, 9, 9, 9	6
7	5, 6, 7, 9	4
8	0, 1, 1, 2, 3, 3, 4, 4, 5, 6, 7, 7, 8, 8, 9	15
9	0, 0, 0, 0, 1, 1, 1, 2, 2, 2, 2, 3, 3, 4, 5, 5, 5, 5, 5, 5, 7, 8, 8, 9	24
10	0, 1, 1, 2, 3, 4, 4, 4, 4, 5, 5, 5, 5, 6, 6, 6, 7, 7, 7, 8, 8, 8, 9, 9	24
11	1, 1, 1, 2, 2, 4, 4, 4, 5, 5, 7, 7, 7, 8	14
12	0, 1, 1, 2, 2, 3, 6, 6, 7, 9, 9	11
13	1, 9	2
14		0
15		0

Sample values: $N = 36$; Mean of $S^2 = 99.8$ Total   100

Note also that the effect has been more dramatic for the small samples. In particular, for samples of size 4, the sample variances have increased, on average, from 76.2 (see Table 11.8) to 101.6.

From now on in this book, whenever the variance of a sample is required, the formula giving the *unbiased* estimate of the population variance should be used. That is,

$$S^2 = \frac{\text{sum of } (X - \bar{X})^2}{N - 1}$$

When a sample standard deviation is required, the formula to use is

$$S = \sqrt{\frac{\text{sum of } (X - \bar{X})^2}{N - 1}}$$

In Chapter Two, we gave this example: Seven students obtained the following scores on a quiz:

$$8, \ 6, \ 6, \ 7, \ 4, \ 7, \ 4$$

and we found these statistics:

$$\text{Mean score} = \frac{\text{sum of scores}}{N} = \frac{42}{7} = 6.0$$

$$\text{Variance} = \frac{\text{sum of } (X - X)^2}{N} = \frac{14.0}{7} = 2.0$$

$$\text{Standard deviation} = \sqrt{\text{variance}} = \sqrt{2.0} \doteq 1.41$$

Now we can view this example from the point of view of sampling.

■ **Example:**   A random sample of seven students was taken from a large mathematics class. The students got these scores on a quiz:

$$8, \ 6, \ 6, \ 7, \ 4, \ 7, \ 4$$

Use this information to estimate the mean and variance of the entire mathematics class on the same quiz.

Solution:

Population mean—The sample mean is an unbiased estimate of the population mean. The mean quiz score of the sample of seven students was 6.0. This is an unbiased estimate of the mean score of the entire class on the quiz.

Note: We know that for small samples, the means have a relatively large variance and standard error. If we took a larger sample, we could reduce this standard error.

Population variance—We learned that the formula

$$\frac{\text{sum of } (X - \bar{X})^2}{N}$$

gives a biased estimate of the population variance, especially for small samples. An *unbiased* estimate of the population variance is

$$S^2 = \frac{\text{sum of } (X - \bar{X})^2}{N - 1} = \frac{14.0}{6} \doteq 2.7$$

Notice that the value of 2.7 is quite a bit larger than the value of 2.0, that we got for the variance in Chapter Two. This is because we are now using the sample information to *estimate* the variance of the population. The sample variance $S^2 = 2.7$, obtained from the formula that has $N - 1$ in the denominator, is a better estimate of the population variance than is 2.0.

The corresponding sample standard deviation is

$$S = \sqrt{\frac{\text{sum of } (X - \bar{X})^2}{N - 1}} = \sqrt{2.7} \doteq 1.6 \ ■$$

A rule of thumb is that a *small sample* is one whose size is less than 10.

EXERCISES

1   Ten male army recruits take a fitness test and get these scores:

15, 10, 12, 14, 14, 16, 12, 11, 11, 13

Assume these recruits to be a random sample of all male army recruits.

(a)   What is an unbiased estimate of the mean fitness score of army recruits on this test?
(b)   Give biased and unbiased estimates of the variance of scores of all recruits on this test.
(c)   What is the sample standard deviation of these scores?

2   A gasoline pump inspector finds these errors (in cubic inches) when 20 pumps deliver 5 gallons of a gasoline.

3, 4, 3, 0, 0, −2, −1, 2, 1, 0
−2, 1, 0, −2, 5, −6, 3, −2, −1, −2

Assume these are a random sample of all pumps in the city.

(a)   Find the unbiased estimate of the mean error of the city's pumps.
(b)   Find biased and unbiased estimates of the variance of errors in the pumps.
(c)   What is the sample standard deviation of the errors in the pumps?

3   For the preference poll data in Exercise 2 of Section 11.2, do the following (assume the poll to be a random sample of consumers in a midwestern city):

(a)   Find the unbiased estimate of the mean preference of the population for the new cola.
(b)   Find a biased and unbiased estimate of the variance of preferences for the cola in the population of consumers.
(c)   What is the sample standard deviation of the preferences?

11.4   THE CENTRAL LIMIT THEOREM

A theorem is a statement of some mathematical property or principle that can be proven by logical arguments. The *Central Limit Theorem* is a very important theorem in statistics. We will not prove it in this book, but we will look at examples that suggest the theorem is indeed true.

Let's do some more sampling. We will take 100 random samples of various sizes from a population whose mean (μ) is 63.5 and whose standard deviation (σ) is 12.0. Again, we need to remember that when doing statistics in the real world, we usually do not know the mean or standard deviation of a population; we use statistics to help us find out what such population values are! But we do this sampling to demonstrate important statistical ideas.

We will use sample sizes of 4, 16, and 36. The results of drawing these 300 samples are given in Table 11.13.

Let's make some observations about the results of our sampling as we look at the stem-and leaf tables. (*Remember:* What we see recorded in each table are means of 100 samples.)

1.   The mean of the 100 sample means is close to the population mean (which we happen to know is 63.5). For samples of size 4, the mean of the 100 samples is 63.3, and so on. This should not surprise us, since we have learned that the sample mean is an unbiased statistic. That is, as we take more and more random samples from a population, the mean of all of the sample means gets closer and closer to the mean of the population.

TABLE 11.13(a)

Population values: $\mu = 63.5$; $\sigma = 12.0$

STEM	LEAF	f
49		0
50		0
51	8	1
52	7	1
53	1, 5, 6	3
54		0
55	2, 3, 3, 5, 6, 7	6
56	4, 6, 6, 6, 9	5
57	1, 1, 1, 3, 8, 9	6
58	1, 3, 5, 8, 8, 9	6
59	1, 1, 3, 4, 5, 5, 6, 8	8
60	3	1
61	3	1
62	2, 2, 5, 5, 8	5
63	2, 2, 4, 5, 5, 5, 7, 9	8
64	0, 0, 0, 2, 4, 7, 9, 9	8
65	1, 1, 3, 3, 4, 7	6
66	1, 3, 5, 5, 6, 6, 9, 9	8
67	0, 2, 7, 9	4
68	0, 2, 3, 5	4
69	3, 5, 5, 6, 6, 8	6
70	8	1
71	1, 2, 4, 5, 6, 6, 8, 9	8
72		0
73	2, 9	2
74	1	1
75	3	1
76		0
77		0

Sample values: $N = 4$; Mean = 63.3; Standard error = 5.7   Total 100

TABLE 11.13(b)

Population values: $\mu = 63.5$; $\sigma = 12.0$

STEM	LEAF	f
54		0
55		0
56	9	1
57	7	1
58		0
59	0, 2, 2, 5, 6, 6, 7, 9	8
60	0, 2, 5, 7, 7, 7	6
61	1, 2, 5, 8, 9, 9, 9	7
62	0, 1, 3, 3, 4, 5, 5, 6, 7, 8, 8	11
63	2, 2, 2, 2, 3, 3, 3, 3, 3, 4, 4, 4, 4, 4, 5, 6, 6, 7, 7, 8, 8, 9, 9, 9, 9	25
64	2, 2, 3, 4, 4, 5, 6, 6, 7, 8	10
65	0, 0, 0, 3, 3, 3, 5, 7	8
66	0, 1, 2, 4, 4, 5	6
67	0, 4, 4, 5, 6, 7	6
68	2, 3, 3, 4, 5, 6, 6, 9	8
69	5	1
70	2	1
71		0
72	1	1
73		0
74		0

Sample values: $N = 16$; Mean = 63.9; Standard error = 2.9   Total   100

TABLE 11.13(c)

Population values: $\mu = 63.5$; $\sigma = 12.0$

STEM	LEAF	f
56		0
57		0
58	1, 6, 9	3
59	7, 8, 9	3
60	0, 0, 3, 6, 7, 7, 7, 8, 8, 9	10
61	0, 1, 2, 4, 5, 8, 9	7
62	0, 0, 2, 2, 2, 2, 4, 5, 6, 6, 7, 7, 7, 8, 8, 8	16
63	1, 1, 1, 1, 2, 3, 3, 4, 4, 5, 5, 7, 8, 9	14
64	0, 0, 1, 1, 2, 2, 2, 3, 3, 4, 4, 4, 4, 4, 4, 5, 5, 6, 6, 6, 7, 8, 8	23
65	0, 0, 0, 2, 3, 3, 4, 4, 4, 4, 5, 5, 8, 9	14
66	0, 3, 4, 5, 5, 8, 8	7
67	1, 2, 4	3
68		0
69		0

Sample values: $N = 36$; Mean = 63.4; Standard error = 2.0   Total   100

2. As the sample size gets larger, the distribution of the sample means looks more and more like a normal distribution. That is, the shape of the curve suggested by the leaves of the stem-and-leaf tables looks more and more like a bell as the sample size gets larger.

3. As the sample size gets larger, the variation in the sample means gets smaller. More specifically, as the sample size gets larger, the standard error (the standard deviation of the sample means) gets smaller.

We have already observed a relationship between sample size and standard error in Table 11.4. We will look at this relationship again, but this time using information that will show another important relationship.

Table 11.14 summarizes the sampling activities shown in Table 11.13. It also shows the population standard deviation σ (in the top row of the table) and suggests a relationship between

$$\frac{\sigma}{\sqrt{N}}$$

and the standard deviation of the 100 sample means (that is, the standard error). (*Remember:* N is the sample size.)

Notice what we have found in Table 11.14. It is that the standard error (bottom line of the table) in each case is close to the value of the quantity:

$$\frac{\sigma}{\sqrt{N}}$$

It is important to note that

$$\frac{\sigma}{\sqrt{N}}$$

is a theoretical value. It comes from the parameter σ and the sample size N.

We are now ready to state the Central Limit Theorem.

 Central Limit Theorem: If we take more and more random samples of size N from a population with a mean μ and standard deviation σ then, as the sample size gets larger, the shape of the distribution (that is, of the smooth relative frequency polygon) of the sample means becomes more and more like a normal distribution whose mean is μ and whose standard deviation is

$$\frac{\sigma}{\sqrt{N}}$$

The Central Limit Theorem enables us to solve a variety of statistical problems.

TABLE 11.14			
Population standard deviation (σ)	12.0	12.0	12.0
Sample size (N)	4	16	36
$\sqrt{N}$	2	4	6
$\dfrac{\sigma}{\sqrt{N}}$	6.0	3.0	2.0
Standard error (from Table 11.13)	5.7	2.9	2.0

■ **Example 1**:   Suppose many random samples of size 36 are drawn from a population with a mean of 50 and standard deviation of 10. Within what distance of the mean (μ = 50) would we expect two-thirds of the sample means to be?

Solution:

In order to solve this problem, you may need to review Chapter Nine. We will use information we know about area under the normal curve.

The Central Limit Theorem tells us that the sample means are normally distributed around the population mean. It also tells us that the standard deviation of these sample means has the (theoretical) value of

$$\frac{\sigma}{\sqrt{N}}$$

With this information, we can draw a picture of what we would expect the distribution of many, many sample means to be like (see Figure 11.4).

In the problem we are to solve, we are asked to find the distance from the population mean of 50 within which we would expect to find two-thirds of the sample means. Recall from Chapter Nine that one standard deviation either side of the mean of a normal distribution includes about two-thirds (actually 68%) of the area under the curve.

What is the standard deviation of the distribution of the sample means in our example? From the Central Limit Theorem, we know that it is

$$\frac{\sigma}{\sqrt{N}}$$

And since $\sigma = 10.0$ and $N = 36$, we have

$$\frac{\sigma}{\sqrt{N}} = \frac{10.0}{\sqrt{36}} = \frac{10.0}{6} \doteq 1.7$$

See Figure 11.5.

FIGURE 11.5

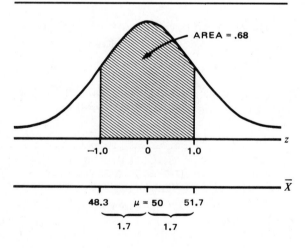

FIGURE 11.4

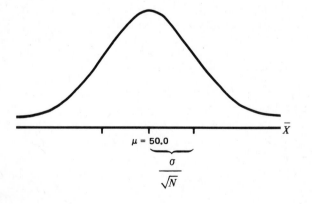

If we go 1.7 units either side of the mean of 50, we have the interval from

$$50 - 1.7 = 48.3$$

to

$$50 + 1.7 = 51.7$$

So, we expect about two-thirds (68%) of the sample means to be between 48.3 and 51.7 for random samples of size 36. ■

■ **Example 2:**   Suppose many random samples of size 25 are drawn from the same population, as in Example 1. Now, within what distance of the mean of 50 would we expect two-thirds of the sample means to be?

Solution:

Before we find the answer, let's think about what we have learned about sampling. We now know that as the sample size gets larger, the sample means get closer and closer to the population mean. Since here we are taking

smaller samples (of size 25 instead of 36) we would expect the sample means to be further apart than in Example 1. So, we expect two-thirds of the sample means to be in a larger interval around the population mean of 50 than in Example 1. (See Figure 11.6.)

We need to know the standard deviation of the distribution of means of samples whose size is 25. By the Central Limit Theorem, it is

$$\frac{\sigma}{\sqrt{N}} = \frac{10}{\sqrt{25}} = \frac{10}{5} = 2.0$$

Refer to Figure 11.7.

If we go 2.0 units either side of 50, we have the interval from

$$50 - 2 = 48.0$$

to

$$50 + 2 = 52.0$$

and we expect two-thirds of the sample means to be in the interval 48.0 to 52.0. ■

FIGURE 11.6

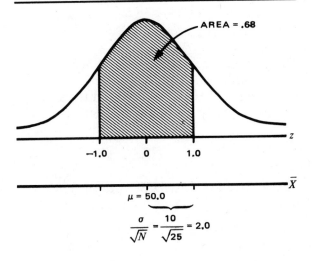

FIGURE 11.7

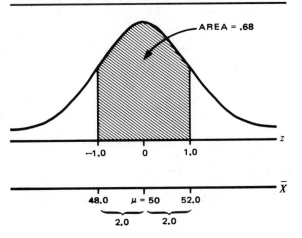

■ **Example 3:**   Suppose many random samples of size 36 are drawn from a population with a mean of 50 and a standard deviation of 10. Within what distance of the mean ($\mu = 50$) would we expect 95% of the sample means to be?

Solution:

The graph of the sample means is similar to those for Examples 1 and 2. But now we need to find an interval within which .95 of the sample means are expected to be located. By going to the table of normal curve areas, we find that .95 of the area under the curve is between $z = -1.96$ and $z = 1.96$ (Figure 11.8).

FIGURE 11.8

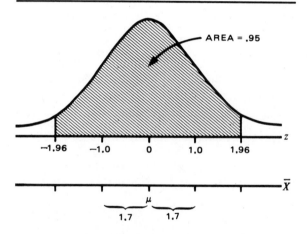

We now need to find the score that corresponds to a z score of 1.96. In Chapter Nine, we learned how to change z scores to those having any specified mean and standard deviation (Section 9.3). The rule was

Score = z score $\times$ standard deviation + mean

From Figure 11.9, we can see that the required mean is 50.0, the population mean. But what is the standard deviation? It is the standard error of the sample means. That is,

$$\text{Standard error} = \frac{\sigma}{\sqrt{N}} = \frac{10}{6} \doteq 1.7$$

FIGURE 11.9

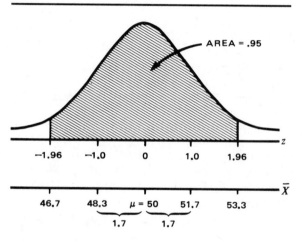

So, the score corresponding to a z score of 1.96 is

$$1.96 \times 1.7 + 50$$
$$\doteq 3.27 + 50 = 53.3$$

This is the score at the upper end of the required interval. To find the score at the lower end of the interval, we need the score corresponding to a z of -1.96. That is,

$$-1.96 \times 1.7 + 50$$
$$\doteq -3.27 + 50 \doteq 46.7$$

Hence, we expect that 95% of the sample means will be between 46.7 and 53.3. ■

EXERCISES

1   Table 11.15 summarizes the sampling activities reported in Tables 11.1 through 11.3. Recall that in each case, 100 random samples of size N were taken from a population with mean $(\mu)=50$ and standard deviation $(\sigma)=10$. Does it seem (from Table 11.15) that the sample mean is an unbiased estimate of the population mean? Explain.

2   Referring to Exercise 1, above, what is the theoretical standard error for the sampling activities summarized in Table 11.15 for random samples of size

(a)   4   (b)   16   (c)   36

3   Compare each theoretical standard error obtained in Exercise 2, above, with its corresponding standard error obtained from finding the standard deviation of the 100 sample means. Do this for random samples of size

(a)   4   (b)   16   (c)   36

4   Many random samples of size 36 are taken from a population with a mean of 63.5 and a standard deviation of 12. Within what interval around the mean would we expect two-thirds of the sample means to be?

5   Answer Exercise 4, above, if the random samples are of size 25 instead of 36.

6   Answer Exercise 4 of this section if the random samples are of size 4.

7   Answer Exercise 4 of this section if the random samples are of size 100.

8   Many random samples are drawn from a population with a mean of 63.5 and a standard deviation of 12. Within what interval around the mean would we expect to find 95% of the sample means if the size of the random samples is

(a)   36   (b)   25
(c)   4   (d)   100

TABLE 11.15

Population standard deviation (σ)	10.0	10.0	10.0
Sample size (N)	4	16	36
Square root of sample size $(\sqrt{N})$	2	4	6
Standard deviation of 100 sample means	5.3	2.5	1.7
Mean of 100 sample means	50.5	50.2	49.9

11.5   CONFIDENCE INTERVALS FOR μ (LARGE SAMPLES)

The problem of estimating μ, the population mean, using information from a random sample taken from that population, is very common in statistics.

EXAMPLES

■ **TV Show Popularity:**   The popularity of a TV show is determined from 100 carefully selected households around the country. The mean rating was 31 and the standard deviation of the ratings was 4.3.

■ **Amount of Moisture in Harvested Corn:** The amount of moisture in the harvested corn was found by measuring the amount in 36 samples of corn taken from trucks unloading at the elevators. The mean amount of moisture was 38.2% with a standard deviation of 2.3%.

■ **Cost of a Long-Distance Call:**   The cost of a long-distance call was estimated by taking a sample of 67 such calls. Their mean cost was $.89 with a standard deviation of $.15.

There are two approaches to estimating the mean of a population using information from a random sample.

POINT ESTIMATES

Our knowledge of sampling tells us that the mean of a random sample is an *unbiased* estimate of the mean of the population from which it was taken. If the sample is large, we can use the sample mean as a good estimate of the mean of the population from which it was taken. This sample value is called a *point estimate* of the population mean, since it is a single number.

Examples of Point Estimates:

■ The national popularity of the TV show is estimated to be 31.

■ The amount of moisture in the corn being hauled is estimated to be 38.2%.

■ The average cost of a long-distance telephone call for that month was estimated as $.89.

INTERVAL ESTIMATES

Another way of estimating population means is to find an interval in which μ is likely to be. With the help of the Central Limit Theorem, we can state the probability of success in correctly locating the population mean, μ, in an interval. This method applies to large samples (greater than 20). How to deal with small samples is discussed in Chapter Twelve.

■ **Example 1:**   The popularity of a TV show, based on a random sample of 100 households from around the country, has a mean rating of 31.0 and standard deviation of 4.3. Use this sample information to find an interval within which we could be 95% *confident* of locating the population mean—the mean popularity of the show for the whole country.

Solution:

This is a powerful and informative way of estimating μ. For example, it allows us to provide different intervals depending upon how much confidence in the estimate is desired. Also, it emphasizes the fact that a sample estimate is being used and, like all sampling situations, is subject to error.

Let's take the example of the TV show and recall what the Central Limit Theorem tells us about this sampling situation. Suppose we draw many random samples, of size N, from the population of viewers across the country. We know that the distribution of all of these means approaches the normal curve and has a mean of μ with a standard deviation of

$$\frac{\sigma}{\sqrt{N}}$$

What is the value of

$$\frac{\sigma}{\sqrt{N}}?$$

We know N, of course. We were told that a national sample of 100 viewers was used. And even though we are not told the population standard deviation, σ, we know that the sample standard deviation is

$$S = 4.3$$

We use this as a point estimate of σ.

So, we can estimate the standard error (standard deviation of the sample mean) as

$$\frac{4.3}{\sqrt{100}} = .43$$

See Figure 11.10.

FIGURE 11.10

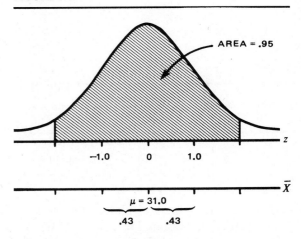

In Section 11.4, we saw how we could use the Central Limit Theorem to help us find an interval about the population mean within which we would expect to find, say, 95% of a large number of random samples taken from that population.

We now wish to take a different view of the sampling situation. We wish to have a procedure that will enable us to be, say, 95% *confident* in locating the mean of the population from which the sample was taken.

We can give 7 steps to follow in finding a confidence interval for estimating a population mean from a sample mean.

Step 1: Draw a graph of the sampling situation, according to the information given (see Figure 11.11).

FIGURE 11.11

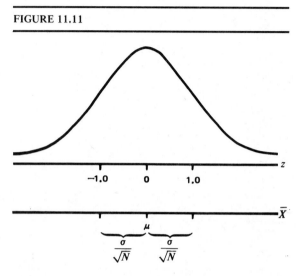

Step 2: Shade in the area under the curve to correspond to the required level of confidence for the desired interval. In Example 1, we are to find a 95% confidence interval. Refer to Figure 11.12.

FIGURE 11.12

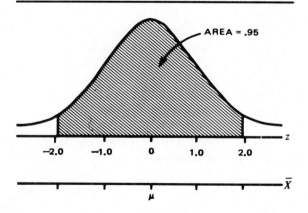

Step 3: Find, from a table of normal curve areas, the z scores (positive and negative) that enclose the given area (in this case, .95) under the curve. See Figure 11.13.

FIGURE 11.13

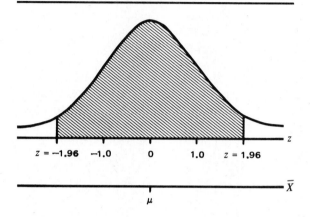

Step 4: Find, from the given information, the estimated theoretical standard deviation of the sample means. Since the sample standard deviation is 4.3 and the sample size is 100, we have, as already noted,

$$\frac{\sigma}{\sqrt{N}} \doteq \frac{4.3}{\sqrt{100}} = .43$$

See Figure 11.14.

FIGURE 11.14

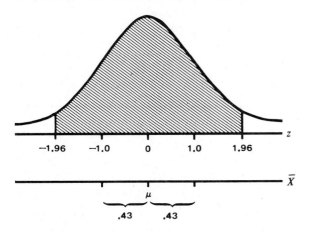

.43   .43

FIGURE 11.15

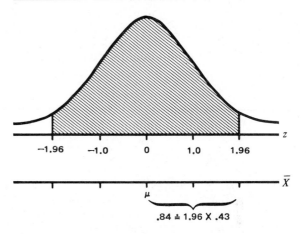

.84 ± 1.96 X .43

Step 5: Use the information from Steps 3 and 4 to find the *upper end* of the desired interval about μ.

In Step 3, we found that a z of 1.96 was required to give an area of .95 under the curve.

As shown in Figure 11.15, the score corresponding to a z score of 1.96 is

$$\text{Upper score, } U = 1.96 \times .43 + \mu$$
$$\doteq .84 + \mu$$

Step 6: Now find the *lower end* of the desired interval about μ (Figure 11.16). The score corresponding to a z score of -1.96 is

$$\text{Lower score, } L = -1.96 \times .43 + \mu$$
$$\doteq -.84 + \mu$$

FIGURE 11.16

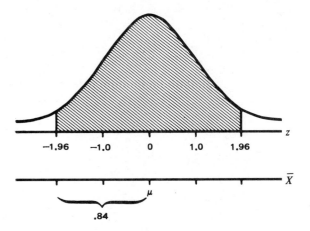

.84

Step 7:   Substitute $\overline{X}$, the sample mean, for μ, the population mean. We do not know μ. We are trying to estimate it from what we know about one sample which we have taken from the population. We do know, however, that the mean of this sample (which we were told is 31) is an unbiased estimate of the population mean μ. So we use 31 as an estimate of μ and obtain the interval

$$U = .84 + 31 = 31.84$$

and

$$L = -.84 + 31 = 30.16$$

This interval, from 30.16 to 31.84, is called a *confidence interval* for μ (Figure 11.17). We are 95% confident that by following this procedure we have succeeded in locating the population mean μ. ∎

FIGURE 11.17

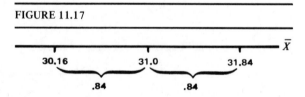

■ **Example 2:**   Suppose that in Example 1, we wished to be 90% confident of locating μ, instead of 95% confident. What interval would we use?

Solution:

Step 1:   The graph of the sampling situation is the same as for Example 1.

Step 2:   Shade in the area corresponding to the specified level of confidence for the interval. In Example 2, we are to use a 90% level of confidence. The area is shown in Figure 11.18.

FIGURE 11.18

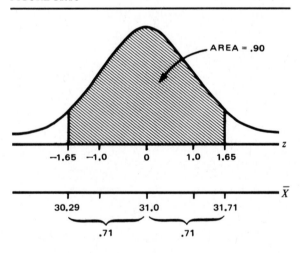

Step 3:   Find the z scores corresponding to the given level of confidence. From a normal curve table, the required values are $z = 1.65$ and $z = -1.65$.

Step 4:   Find the estimated theoretical standard deviation of the sample means. This is unchanged from Example 1.

$$\frac{\sigma}{\sqrt{N}} \doteq \frac{4.3}{\sqrt{100}} = .43$$

Step 5:   Find the upper end of the interval about μ.

$$U = 1.65 \times .43 + \mu$$
$$\doteq .71 + \mu$$

Step 6:   Find the lower end of the interval about μ.

$$L = -1.65 \times .43 + \mu$$
$$\doteq -.71 + \mu$$

Step 7:   Substitute $\overline{X}$, the sample mean, for μ, the population mean. We estimate μ by the sample mean 31.0. So, we have the 90% confidence interval.

$$U = .71 + 31.0 = 31.71$$
$$L = -.71 + 31.0 = 30.29$$

So, we are 90% confident that the population mean (the national popularity of the TV show) is between 30.29 and 31.71, based on the information from our sample of size 100. ∎

EXERCISES

1   For the example of the TV show popularity ratings, find an interval in which you are 99% confident of locating μ. That is, for a sample of 100 observations, a sample mean of 31, and a sample standard deviation of 4.3, find a 99% confidence interval for μ.

2   Find a 90% confidence interval for estimating the true (population) mean amount of moisture in the corn being sampled in the example on page 332. The sample information is:

Size = 36

Mean amount of moisture = 38.2%

Sample standard deviation =   2.3%

Then find a 95% confidence interval for μ.

3   Find a 99% confidence interval for the population mean charge of the long-distance calls in the example on page 332. The sample information is:

Size = 67

Mean = $.89 per call

Standard deviation = $.15

4   A set of 64 independent measurements of the length of a micro-organism is made. The mean value of the measurements is 27.5 millimeters and the sample standard deviation is 3.2 millimeters. Find these confidence intervals for the true length of the micro-organism.

(a)   90%   (b)   95%   (c)   99%

5   From the examples given here, what do you conclude about the effect on a confidence interval of increasing the size of sample being used? For example, of increasing the sample size from 25 to 100?

6   From the newspaper account in Figure 11.19, give point estimates and confidence intervals for the population mean that is being estimated. If this cannot be done, tell what information is lacking. (Note date of poll from footnote.)

FIGURE 11.19

SEPARATISTS LEAD POLL[5]

MONTREAL (AP)–The separatist side took a surprise lead Sunday in the final opinion poll before Quebec votes on whether to start down the road to independence.

Tuesday's referendum in the largely French-speaking Canadian province will probably shake the North American status quo no matter who wins, since even the "no" side is demanding fundamental changes in Canada's political system.

The poll, published by the Montreal newspaper Dimanche-Matin, showed 40 percent in favor of the proposal by Quebec's separatist government and 37 percent opposed, a reversal of the results of a survey by the same polling organization a week earlier.

The remaining 23 percent either were undecided or refused to answer.

The balance between the "oui" and "non" sides in the polls seesawed for months, but in recent weeks the "no's" had maintained a slight lead, making the pro-Canada federalists increasingly confident of victory.

In view of the latest poll, however, analysts were rating the race a toss-up.

---

7   Key Problem—*How Many Deer Are in the Park?*[6]   Suppose that in our example, 12 deer are caught at points at random throughout the park. These deer are marked with ear tags and released. Now the park contains

marked and unmarked deer. At a later time,
30 deer are captured at random points
throughout the park. Suppose that in this
sample, 3 deer are marked. That is, our sample
from the population of deer in the park has

$$\frac{3}{30} \text{ or } 10\%$$

that are marked. If we now assume that this
sample is representative of the entire population of deer, then we can say that 10% of the
entire deer population is marked.

However, we know that in all there are 12
marked deer in the park. So, of what number
is 12 deer 10%?

$$.10N = 12$$

or

$$N = 12 \times 10 = 120$$

We estimate there are 120 deer in the park.
What factors will influence the goodness of
our estimate?

Hypothesis Testing

KEY PROBLEM

The Quest for the Perfect Figure[1]

For thousands of years, the *golden rectangle* has been believed by many to be the perfect figure. Its width-to-length ratio has been thought to be the most pleasing to the eye. The Parthenon of ancient Greece (Figure 12.1) was designed in the proportion of the golden rectangle. Patterns similar to the golden rectangle have been found in other ancient cultures. Blocks of stone used in the pyramids of Egypt have faces proportioned after the golden rectangle. Documents we use, like driver's licenses and credit cards, have proportions that approximate those of the golden rectangle.

The Shoshoni Indians from the southwestern United States decorate their leather goods with beaded rectangles. We can use the ideas from this chapter to check whether, from a statistical point of view, the patterns of the Shoshonis are similar to those of ancient civilizations.

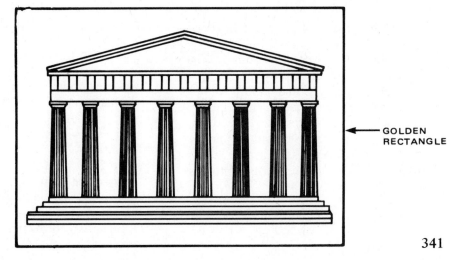

GOLDEN RECTANGLE

FIGURE 12.1

THE PARTHENON IN ATHENS, GREECE, COMES VERY CLOSE TO FITTING INTO A GOLDEN RECTANGLE.

12.1   MAKING STATISTICAL DECISIONS–REVISITED

We will use the ideas from Chapter Eleven to help us solve the following problem.

■ **Example 1:**   The manufacturer of XYZ batteries claims that their flashlight batteries last 15 hours. A consumers' group doubts the claim, for they have received many complaints about XYZ batteries burning out quickly. So the group goes to the factory and picks a random sample of 36 batteries. They then keep a record of how long each lasts and find that the 36 batteries last a mean time of 13.5 hours and that the standard deviation is 5.0 hours.

How conclusive is this evidence that XYZ batteries do not last as long as the company says they do?

Solution:

First, this is a statistical problem. We have to use information from a sample (in this case, 36 flashlight batteries) to make a conclusive statement about a population (all flashlight batteries manufactured by XYZ).

The consumers' group will not test *all* the thousands of XYZ batteries at the factory. They would not be allowed to, nor would it be sensible to do so. Instead, we are using information obtained from a *sample* to make decisions about the *population* that the sample is believed to represent. (Indeed, assuming that XYZ batteries will be manufactured the same way in the future, the population actually includes future batches of XYZ batteries, too!)

Clearly, the XYZ company could have been unlucky, in that the consumers' group could have obtained a poor sample (that is, batteries that, on the average, just *happen* to have shorter lives than typical for the population). But again, the consumers' group may have obtained a particularly *good* sample (that is, XYZ batteries may in general last much *less* than 15.0 hours–close to 13.5 hours, for example).

We will follow steps similar to those in Chapter Five to check the claim of XYZ. But we can now give a slightly different interpretation to these steps, on the basis of what we have learned about statistical methods.

Step 1–*Choose a model (define a population):* We begin by assuming that the claim of XYZ is true. That is, we *assume* that the sample of 36 batteries was taken from the population of all XYZ batteries with an average (mean) life of 15.0 hours. So, we are specifying a population with a mean of 15 hours.

In order to specify the population more fully, we need to know its standard deviation as well. We are not given this information about the population, but we do have the sample standard deviation, $S = 5.0$. We also know from Chapter Eleven that since the sample size is rather large ($N = 36$), 5.0 is a good estimate of the population standard deviation.

So, our model is assumed to be a population with a mean of 15 hours and a standard deviation of 5.0 hours (that is, $\mu = 15$ hours and $\sigma = 5.0$ hours).

Step 2–*Trial (define a sample):*   We are to carry out a sampling exercise that consists of drawing a random sample of size 36 from our assumed population. (The size of the sample is 36, since 36 batteries were tested in our problem.)

This sampling exercise could be done by filling a large bin with thousands of tickets of identical size, representing thousands of XYZ batteries. Each ticket would be marked with a number representing the life of one of the batteries. The mean of all values marked on the tickets is 15.0 and the standard deviation is 5.0.

Step 3—*Outcome of trial (calculate a statistic):* We are interested in the mean life of our sample of 36 batteries. So, we find the mean obtained in Step 2.

Step 4—*Repeat the trials (obtain many samples):* We repeat Steps 2 and 3 many times. As before, we use a computer to do this.

The program SAMPLE (Appendix F) first creates a population of numbers with a mean (μ) and standard deviation (σ), which are specified by the program user. It then draws a random sample (whose size is also specified by the user) from this population and calculates the mean and standard deviation of this sample.

Step 5—*Estimate probability of obtained statistic (or less):* We wish to estimate how likely it is that, for a population with a mean of 15 hours and a standard deviation of 5.0 hours, a sample mean of 13.5 (or less) would be obtained by chance.

Step 6—*Make a decision about the model (about the population mean, as defined in Step 1):* If the probability of getting a sample, by chance, that has a mean of 13.5 hours or less *is small* (less than .10), we decide the claim is not acceptable. We doubt that XYZ batteries have a mean life of 15 hours.

If, however, it is *somewhat likely* that we get a sample with a mean life of 13.5 or less, we have no basis for rejecting the claim. We are unable to reject the claim that the batteries have a mean life of 15 hours (even though it may, indeed, be false!).

The stem-and-leaf table for 100 samples is given in Table 12.1.

We see from this table that 6 of the 100 sample means were 13.5 (or less). So, we find that

$$P(\text{sample mean of 13.5 or less}) = \frac{6}{100} = .06$$

That is, we estimate that samples with means of 13.5 hours or less happen by chance .06 of the time. This is an unusual outcome (probability is less than .10), so we *doubt* the manufacturer's claim that their batteries last 15.0 hours. (They probably last *less* than 15 hours). ■

TABLE 12.1

Population values: $\mu = 15.0$; $\sigma = 5.0$

STEM	LEAF	f
13	0,2,3,4,5,5,6,6,6,7,7,7,8,8,8	15
14	0,0,1,1,1,1,1,2,3,3,3,4,4,4,5,5,5,5,5,5,6,6,7,7,7,7,7,7,7,7,7,7,7,8,8,8,8,8,9,9,9,9,9	43
15	0,0,0,1,1,2,2,3,3,3,3,3,3,4,4,4,4,4,4,4,4,5,5,6,6,6,6,6,7,8,8,8,8,9	32
16	1,2,4,5,7,7,7,7,7,8	10

Sample size: $N = 36$ Total 100

Suppose now that we are asked to make a decision when the problem is slightly different.

■ **Example 2:**   The Silver Bell Mining Company is trying to find out the quality of iron ore in a certain deposit. The company is willing to mine the site only if the ore contains more than 12% iron. A random sample of 49 pieces of ore is taken and the amount of iron in the sample is found. The results are:

Mean amount of iron	13.4%
Standard deviation	6.1%

Do we have statistical evidence that the true amount of iron ore in the mine exceeds 12%?

Solution:

We want to estimate how likely it is that, by chance, we would find a random sample of ore with an average (mean) of 13.4% iron if the actual amount of iron in the mine is only 12%. Since amounts *greater* than 13.4% would also cast doubt on the claim of only 12%, we will estimate the probability of obtaining a random sample containing 13.4% *or more* of iron ore.

Step 1—*Choose a model*:   We assume that there is in fact only 12% iron ore in the mine. This is our assumed value of μ, the population

mean. The population standard deviation is assumed to be 6.1%, the estimate provided by the sample.

Step 2—*Trial (define a sample)*:   We draw a random sample of size 49 from the population of iron ore with an assumed mean of 12% and standard deviation of 6.1%. Again, the computer program SAMPLE has been used.

Step 3—*Outcome of the trial*:   When the sample of 49 pieces of ore is drawn, find its mean and record it in a stem-and-leaf table.

Step 4—*Repeat the trials*:   Repeat Steps 2 and 3 for a total of 100 times. The results are in Table 12.2.

Step 5—*Estimate the probability of the obtained statistic (or more)*:   We wish to estimate the probability of getting, from a population with a mean of 12% (and standard deviation of 6.1%), a random sample of size 49 whose mean is 13.4% *or more.*

From the stem-and-leaf table of the sample means (Table 12.2), we see that samples with a mean of 13.4% *or more* were obtained 16 times. That is,

$$P(\text{sample mean of 13.4\% or more}) = \frac{16}{100} = .16$$

TABLE 12.2

Population values: $\mu = 12.0$; $\sigma = 6.1$

STEM	LEAF	f
9	7,9	2
10	0,0,2,2,4,4,5,5,7,7,7,8,9	13
11	0,0,0,0,1,1,1,2,2,3,3,3,4,5,5,6,6,7,7,7,8,8,9,9,9,9,9	28
12	0,0,0,0,0,0,0,1,1,1,2,2,2,2,2,2,3,4,5,6,6,6,6,6,7,8,9,9	28
13	0,1,1,1,1,2,2,2,3,3,3,3,3,4,4,4,5,5,5,5,6	21 ⎫
14	1,3,5,5,6,7,9	7 ⎬ 16 sample
15	1	1 ⎭ means ⩾ 13.4

Sample size: $N = 49$ | | Total   100

Step 6–*Make a decision:* This is not an unusual event (.16 is greater than .10). So, we decide that we do *not* have enough evidence to conclude that the actual mean quantity of iron ore in the mine is different from 12%. *Note:* The amount of iron ore *could* be less than 12%. ∎

As a final example, we will do a problem in which we are provided the individual sample data. Our first task is to find the sample mean and standard deviation as we did in Chapter Eleven.

■ **Example 3:** The Nite-Nite Bulb Company produces a flashlight bulb, which they claim lasts for 50 hours. A consumer group takes a random sample of 16 bulbs from 1 week's stock and finds they last the following lengths of time, to the nearest hour. Should Nite-Nite's claim be accepted or rejected?

$$53 \quad 53 \quad 49 \quad 53 \quad 44 \quad 43 \quad 44 \quad 52$$
$$52 \quad 43 \quad 44 \quad 47 \quad 54 \quad 46 \quad 54 \quad 45$$

Solution:

We first need to calculate the sample mean and standard deviation.

Mean

$$\overline{X} = \frac{\text{sum of } X}{N} = \frac{776.0}{16} = 48.5 \text{ hours}$$

Standard deviation

$$S = \sqrt{\frac{\text{sum of } (X - \overline{X})^2}{N - 1}} = \sqrt{\frac{288.0}{15}} = \sqrt{19.2} \doteq 4.4 \text{ hrs.}$$

Now that we have the basic statistics that we need, we follow the familiar six steps.

Step 1–*Choose a model:* We assume that the light bulbs at Nite-Nite actually do last 50 hours. And we use our sample information to estimate the standard deviation of these light bulbs to be 4.4 hours.

Step 2–*Trial:* A trial consists of drawing a random sample of 16 (light bulbs) from our population, which has a mean of 50 hours and a standard deviation of 4.4 hours. Use the program SAMPLE.

Step 3–*Outcome of the trial:* We find the mean of the sample of 16 observations. That is, on the average, how long did our sample of 16 light bulbs last? We record this mean in a stem-and-leaf table.

Step 4–*Repeat the trials:* Do 100 trials with the help of SAMPLE. Table 12.3 shows the results for our 100 trials.

TABLE 12.3

Population values: $\mu = 50.0$; $\sigma = 4.4$

STEM	LEAF	f
47	1,3,5,7	4 ⎫ 13
48	0,0,1,1,4,5,5,5,5,6,7,7,9,9,9,9	16 ⎭
49	0,0,1,1,1,2,2,2,2,3,3,3,4,4,4,5,5,5,5,7,7,8,8,8,9	26
50	0,0,0,0,0,1,1,1,1,1,1,2,2,3,4,4,4,6,6,6,6,6,6,6,6,6,7,7,7,8,9,9	32
51	0,0,1,1,1,2,2,2,3,3,4,4,5,5,6,6,6,9	18
52	0,6	2
53	3,5	2

Sample size: $N = 16$ Total   100

Step 5—*Estimate the probability of the obtained statistic (or less)*:   We want to make a decision about the company's claim that their light bulbs last 50 hours. Our sample of 16 light bulbs has a mean life of 48.5 hours. Samples with mean lives *less than* 48.5 hours would also cast doubt upon their claim. So we estimate the probability of obtaining, by chance, samples of light bulbs with a mean of *48.5 or less.*

From the stem-and-leaf table, we see

$$P(\text{sample mean of 48.5 or less}) = \frac{13}{100} = .13$$

Step 6—*Make a decision*:   We found that it is not unusual (the probability is greater than .10) to get a random sample of size 16 with a mean of 48.5 hours or less from a population with a mean of 50 hours and a standard deviation of 4.4 hours. We decide that we do *not* have enough evidence to reject the company's claim. ■

EXERCISES

1   Suppose instead, the sample of batteries in Example 1 had a mean life of 14.5 hours. (Assume the claimed population mean remains as 15 hours and the standard deviation remains as 5.0 hours.) What decision would you make about XYZ's claim?

2   Suppose that in Example 3, above, the sample of 16 light bulbs had had a mean life of 47.5 hours. (Assume the claimed population mean remains as 50 hours and the standard deviation remains as 4.4 hours.) What decision would you make about the company's claim?

3   A mining company takes 40 pieces of ore from a vein and measures the percentage of iron in each. It is found that the mean percentage of iron in the 40 pieces is 14.0 with a standard deviation of 6.1. Does this sample cast doubt upon the belief of the company that the amount of iron in that vein was only 12%? Use Table 12.4 of 100 sample values of size 40 (produced by the computer program SAMPLE) drawn from a population with a mean of 12 and a standard deviation of 6.1.

TABLE 12.5

STEM	LEAF	f
228	2	1
229	0,0,2,2,6,7,7,8,9,9	10
230	0,1,1,1,2,3,3,4,4,4,5,5,5,6,6,6,6,6,6,7,8,8,9,9	24
231	0,0,0,0,0,1,1,1,1,1,1,1,2,2,2,2,2,2,2,3,3,3,3,4,4,4,4,4,4,5,6,6,6,7,7,7,7,8,8,8,8,9,9,9,9,9	46
232	0,0,1,1,1,2,2,2,2,2,2,3,4,5,6,7	16
233	0,1,4	3
		100

4   The Bureau of Standards in a large city runs a test of the accuracy of gasoline pumps. One gallon of gasoline (as measured by the pump) is put directly into a container marked in cubic inches (one gallon = 231 cubic inches). The amount of gasoline obtained, in cubic inches, is recorded. A random sample of 36 pumps around the city is used.

Mean amount of gasoline = 229.0 cubic inches
Standard deviation =   2.1 cubic inches

Does this sample information cast doubt on the honesty of the city's pumps? Use Table 12.5, which gives means of 100 random samples of size 36 drawn from a population with a mean of 231 and standard deviation of 2.1.

5   Suppose in Exercise 4, above, the mean amount of gasoline for the sample of 36 pumps was 230.3 cubic inches. What would you conclude about the honesty of the city's pumps?

6   A school district is interested in whether student achievement in mathematics in their district is significantly lower than the nation-wide mean of 85. A random sample of 50 students takes the test and gets a mean of 83 and a standard deviation of 7.3. Use these data to help answer the question. They were produced by taking 100 random samples of size 50 from a population with a mean of 85 and a standard deviation of 7.3 (Table 12.6).

TABLE 12.6

STEM	LEAF	f
82	7,7,7	3
83	2,2,3,4,5,5,7,8,9,9	10
84	0,0,2,2,2,2,2,3,4,4,5,6,6,6,7,7,7,7,7,8,8,8,8,8,8,9,9	27
85	0,0,0,0,0,1,2,2,2,3,3,3,3,3,4,4,4,5,5,5,5,5,5,6,6,7,7,7,8,8,8,8,8,9,9,9,9	37
86	0,0,1,1,1,1,2,2,2,3,3,3,3,4,4,5,6,7,7	19
87	0,1,8	3
88	6	1
		100

TABLE 12.4

STEM	LEAF	f
9	5	1
10	3,3,4,5,5,8,8,8,8,9	10
11	0,0,0,2,2,2,2,3,3,3,3,4,4,4,4,5,5,5,5,5,5,5,5,6,6,6,6,6,6,7,7,7,7,8,8,9,9,9,9,9,9	38
12	0,0,0,1,1,1,1,2,2,2,2,2,3,3,3,3,4,4,5,5,5,5,5,5,5,6,6,7,8,8,9,9,9,9,9	34
13	0,0,1,2,2,2,3,3,3,4,4,4,7,7	14
14	0,2,3	3
		100

Traditionally, accountants have required the utmost accuracy in the records of business and other organizations. So, they insist on doing much of their work on a complete set of records. Recently, however, sampling methods have shown to be highly satisfactory for accounting purposes while at the same time costing considerably less. A knowledge of statistics is needed in order to do sampling properly.

Studies have been done to compare the results of working with a complete set of records and with a sample of the records. For example, airlines use statistical sampling to estimate their share of the revenue of tickets for passengers traveling on two or more airlines. During a four month period, three companies tested sampling. At one point, about 12% of the interline tickets was used and the sampling error was less than .1%. The savings in terms of clerical work were estimated to be nearly $75,000 annually for some of the larger carriers and more than $500,000 for the airline industry.

Sampling may, in fact, give *more accurate* results than using all records. For example, taking inventory in a large company involves training many persons for the task. It may be better to do inventory on a sample of items, which requires fewer persons, who could be better trained.

12.2 TESTING HYPOTHESES ABOUT μ (LARGE SAMPLES)

A hypothesis is a belief or claim about something. We can form a hypothesis that last winter was the worst (for example, the coldest) Chicago has ever had. Or we can form the hypothesis that a new drug will be more effective than the old ones in curing headaches. Or we can hypothesize that a president's popularity is very low.

In statistics, we make statements about populations based on samples that represent those populations. This is done by making a claim (a hypothesis) about the population (for example, about its mean, μ), and then deciding about the truth of that claim by using information from the sample.

EXAMPLES

■ **New Drug:** We want to see whether a new drug is "more effective" than an old drug. By more effective, it would seem that we mean "cures more people." We take a sample of headache sufferers and give them the new drug. We compare the number of cured persons with those cured using the old drug.

■ **President's Popularity:**   The opposition party claims that the president's popularity is at an all time low. So, we take a sample of, for instance, 1000 voters from across the country and get their opinion about the president. Assuming the 1000 people to be representative of the country, we then see how high a rating they give the president.

In each of these examples, there is a sample, which provides data, and a population about which we make a statement, using the information from the sample.

So far, all this should be familiar to you, since it is really just a repetition of what we did in Section 12.1. However, in Section 12.1, we actually drew random samples of a certain size from a population with a certain mean and standard deviation with the help of a computer. The result was an estimated probability of obtaining a sample mean of a certain size.

Now we are ready to find theoretical probabilities to help us solve statistical problems. These methods are much more like those most commonly used by statisticians in their daily work.

Let's go back to the problem from Section 12.1.

■ **Example:**   The manufacturer of XYZ batteries claims that their flashlight batteries last 15 hours. A consumers' group doubts the claim, for they have received many complaints about XYZ batteries burning out quickly. So the group goes to the factory and picks a random sample of 36 batteries. They then keep a record of how long each lasts and find that the 36 batteries last a mean time of 13.5 hours and that the standard deviation is 5.0 hours.

How conclusive is this evidence that XYZ batteries do not last as long as the company says they do?

Solution:

To answer this problem, we use our familiar six steps, but we use some ideas from the Central Limit Theorem, too.

Step 1—*Choose a model:* We begin by assuming that XYZ's claim is true. That is, we assume that our sample of 36 batteries is a random sample from a population with a certain mean and standard deviation. The *population* mean is the mean life of all of XYZ's batteries—which they claim is 15 hours. ("Our batteries last for 15 hours.") We estimate the population standard deviation by the standard deviation of the lives of the 36 batteries in our sample, which, we are told, was 5.0 hours.

Step 2—*Trial:* A trial consists of drawing a random sample of size 36 from a population with a mean (μ) of 15 hours and a standard deviation of 5.0 hours.

We will not actually draw samples, but the idea of a trial is still important.

Step 3—*Outcome of the trial:* In this problem we are asked to make a decision about XYZ's claim concerning how long their batteries will last (that is, about the mean of a population) based on how long a sample of 36 batteries lasted (that is, the value of the mean of a sample).

So, we are interested in the means of the samples we talked about in Step 2.

Step 4—*Repeat the trials:* In Section 12.1, we drew 100 samples and recorded the sample means in a stem-and-leaf table. We will not now talk about the means of 100 samples, but instead we will consider what we would expect to happen, based on our knowledge of sampling, if we were to draw hundreds and hundreds of samples of size 36 from our assumed population.

We learned from Chapter Eleven that we would expect the sample means to cluster around the population mean (which we assumed to be 15 hours) and that this clustering will be bell-shaped or normal. We also learned that the standard deviation of this distribution of sample means (the standard error) is

$$\frac{\sigma}{\sqrt{N}}$$

[Since we have a large sample ($N=36$), S gives a good estimate of σ.]

With this information, we estimate the theoretical standard error as

$$\frac{\sigma}{\sqrt{N}} = \frac{5.0}{\sqrt{36}} = \frac{5.0}{6} = .83$$

We now draw a graph of the expected shape of the distribution of the hundreds and hundreds of sample means (Figure 12.2).

Step 5—*Find p(obtained statistic or less):* We are to find the probability of getting (in a random sample) a sample of batteries with a mean life of 13.5 hours or less from the model population.

FIGURE 12.2

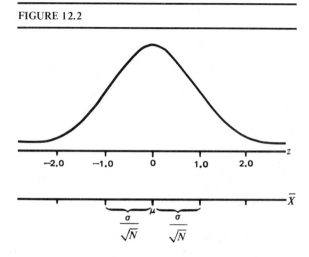

There is a new feature in Step 5. In Step 4 we did not actually produce the hundreds of random samples, but instead we drew a smooth (theoretical) curve to show what the distribution of means would look like. So, the result is a theoretical probability, not an estimated probability.

Figure 12.3 shows the smooth curve we got in Step 4. We then mark the value of the mean of the sample bulbs and shade in the area *to the left* of that point (because we want the probability of getting a sample mean *less than* or equal to 13.5 hours).

FIGURE 12.3

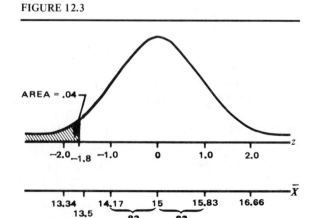

Since we know from the Central Limit Theorem that the theoretical (smooth) curve is normal, we can use the table of normal curve areas to find p (13.5 or less).

In order to find this area, we need to know what z score corresponds to a sample mean ($\overline{X}$) of 13.5. To change from $\overline{X}$ to z, we use the transformation rule

$$z = \frac{\text{sample mean } (\overline{X}) - \text{population mean } (\mu)}{\text{standard deviation of sample means}}$$

$$= \frac{13.5 - 15.0}{.83} = \frac{-1.5}{.83}$$

$$= -1.81$$

From the table of normal curve areas (Appendix C), we find that the area to the *left* of -1.81 is .0351, which rounds to .04.

Thus,

p(sample mean less than or equal to 13.5) = .04

Compare this with the estimated probability which we found in Section 12.1:

P(sample mean less than or equal to 13.5) = .06

Step 6—Decision: We found in Step 5 that the theoretical probability of getting a random sample of 36 with a mean of 13.5 or less from population with mean of 15 was .04.

Since this is a rare event (it happens less than .10 of the time), we reject the claim that the population mean is 15. We are *doubtful* that XYZ batteries last 15 hours. ∎

EXERCISES

1   Suppose for the XYZ example, we define *unusual* to be an outcome with a probability of .01 instead of .10. What is our decision now about our hypothesized population mean?

2   Suppose XYZ had claimed their batteries last 14 hours. What is our decision now about their claim?

3   Use the method of this section, involving the Central Limit Theorem, to solve Exercise 3 of Section 12.1. Compare the decision you made there with the decision made using the Central Limit Theorem.

4   Solve Exercise 4 of Section 12.1 using the Central Limit Theorem. Compare your decision with the decision you made in Exercise 3 of Section 12.1.

5   Solve Exercise 6 of Section 12.1 using the Central Limit Theorem.

6   A supermarket chain claims that a family of four spends an average (mean) of only $80.00 per week on groceries. A consumer group surveys 64 families and finds that their mean weekly grocery bill is $86.00, with a standard deviation of $9.00. Do we have evidence that the claim of the supermarket is doubtful? Use the Central Limit Theorem.

7   Suppose that in Exercise 6, above, the results found by the consumer group had been based on a sample of size 36 instead of size 64. Would this change the decision made concerning the claim of the supermarket?

8   Key Problem Recap—*The Quest for the Perfect Figure:* Figure 12.1 is an example of the famous golden rectangle. The ratio of its width to its length is approximately .618 or 61.8%.

In this exercise, we will examine data from anthropology in order to see if the Shoshoni Indians appeared to be using the golden rectangle as a pattern for their artwork. The data are width-to-length ratios of beaded rectangles used to decorate leather goods.[3]

69.3	66.2	69.0	60.6	57.0
74.9	67.2	62.8	60.9	84.4
65.4	61.5	66.8	60.1	57.6
67.0	60.6	61.1	55.3	93.3

Based on this sample data, do we have evidence that the Shoshoni ratios differ, on the average, from the golden ratio of only 61.8? You may assume that the mean and standard deviation are reliable estimates of the population mean and standard deviation.

9   The Nite-Nite Bulb Company produces a flashlight bulb which they claim lasts for 50 hours. A consumer group takes a random sample of 36 bulbs from 1 week's stock and finds they last the following lengths of time, to the nearest hour. Should Nite-Nite's claim be accepted or rejected?

53	53	49	53	45	43	42	52	52	43	44	47
53	43	53	46	44	52	48	54	44	49	46	53
45	52	53	41	47	48	50	44	54	46	54	45

12.3   WRONG DECISIONS

Nobody's perfect. No decision-making method is perfect, either. Statistics will not keep you from making wrong decisions. But it gives you a way of knowing your chances of making right or wrong decisions.

Let's take the example of XYZ batteries. We wanted to decide whether their batteries are as good as the company claimed them to be. So, we assumed their claim to be true and estimated our chances of getting a sample of 36 batteries with a mean (average) life of 13.5 hours *or less* under the assumption that XYZ batteries last 15 hours, on the average.

Figure 12.4 shows a graph that is useful in explaining our statistical decision rule.

FIGURE 12.4

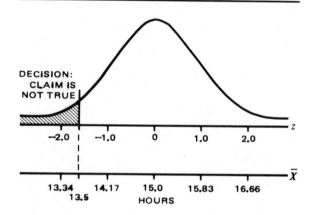

We reject XYZ's claim if the probability of getting a sample of batteries with a mean life of 13.5 hours or less is small, when the claim by XYZ is true. And in Chapter Ten, we said what we mean by small. We agree that small probabilities are .10, .05, or even .01 or smaller.

We also said in Chapter Ten that in this way we were defining an *unusual* event. An unusual event is one that takes place very seldom—for instance, 10% of the time, 5% of the time, or 1% of the time.

We produced a table of 100 sample means (Table 12.1) and estimated that we would get a mean of 13.5 *or less* 6% of the time if XYZ's claim were true. So, we were doubtful about the truth of their claim and *decided* it was not true.

Could we have been wrong? Yes. We realize that 6% of the time, we will get samples with average battery lives as small as 13.5 or smaller even if the population mean is 15.0 hours. But we are willing to take this chance of being wrong. It is, as we agreed, a very small chance.

How could we reduce our chances of making a wrong decision even more? Well, let's keep in mind what kind of decision it is. The decision involves deciding a claim (or hypothesis) is *not* true and rejecting the claim. So, we could choose an even smaller value as our definition of an unusual event. Instead of .10, we could take .05 or .01, or .001, or . . . . What would be the end result of choosing these smaller values? It would be simply that we would never reject the claim. Regardless of the evidence *against* the truth of the claim, we would fail to reject it!

As an extreme example, suppose the consumer group tested the sample of 36 batteries and found that they had an average (mean) life of only 5.0 hours! This would certainly cast doubt upon XYZ's claim that their batteries last 15 hours! But, since we have decided beforehand to avoid making a wrong decision (by wrongly rejecting the claim), we do not reject the claim.

What is the effect of not rejecting the claim? We have opened the door to a new kind of wrong decision: that of *failing to reject a claim that should be rejected!* In the case of XYZ batteries, we surely should reject the company's claim of 15 hours if we obtain a sample with a mean life of only 5.0 hours!

We have, then, two kinds of wrong decisions, or errors.

TYPE I ERROR—Rejecting a claim that is true: This will happen with whatever probability we define as an unusual event. In this book, it is .10 unless stated differently. A value more commonly used by statisticians is .05.

TYPE II ERROR—Accepting a claim that is not true: That is, the claim should have been rejected, but it wasn't. We would hope, of course, that our procedure would enable us to detect such false claims. If we define *unusual* as very small (such as less than .01), we increase our chances of failing to reject a false claim.

Let's discuss these errors in the light of the newspaper article shown in Figure 12.5 on the next page. It is reported in the article that soil, air, and water samples near an accident site (at which a missile exploded) showed evidence of "no significant environmental contamination." We gather from the article that the amount of chemicals (nitrates) in well water was tested and the quantity was not different from amounts expected due to natural causes. That is, it was concluded that the explosion had not effect on the environmental quality.

FIGURE 12.5

TESTS FIND NO HARM TO ENVIRONMENT FROM TITAN BLAST[4]

LITTLE ROCK, Ark., Oct. 1 (UPI)—Tests taken since the explosion of a Titan 2 missile near Damascus on Sept. 19 show that the accident caused no significant environmental contamination, the Air Force said yesterday.

Maj. Gary Fishburn, the chief water quality officer at Brooks Air Force Base in San Antonio, Tex., said nitrates that were found in abandoned wells about a quarter of a mile from the site of the explosion apparently were natural concentrations, not the result of the accident.

The explosion, which occurred when a tool was dropped in maintenance of the missile, piercing the surface and letting fuel vapors escape, resulted in one death. A nuclear warhead was thrown clear of the silo in which the missile was held, but the authorities have indicated that the missile was recovered intact and that no radioactivity escaped as a result of the mishap.

AIR, SOIL AND WATER SAMPLED

Major Fishburn told a news conference that 270 environmentally related samples had been taken from the air, soil, groundwater and surface water and that they showed "no significant environmental contamination has occurred."

"It is highly improbable nitrate concentration in any current well is any result of Air Force activity," he added.

The officer said that since groundwater moves slowly, at a rate of one to 10 feet a day, it would be impossible for contaminants to turn up in wells a quarter-mile away.

He said the Air Force tested one well that is 1,500 feet from the site and found no significant amounts of nitrates there. Asked if the Air Force had tested the wells where the high levels of nitrates had been found, Major Fishburn said he did not know if they were included in the samples.

"We have taken samples from as many wells as possible," he said. "We may have missed a couple."

EVIDENCE 'PRETTY CONCLUSIVE'

Major Fishburn said he believed his evidence was "pretty conclusive," adding, "We do not believe we are going to observe any nitrate rise in the aquifer," the underground layer containing water and into which wells are sunk.

He said some nitrate concentration of 3 to 4 parts per million was found in surface water in the area, mainly in stock ponds, but "we have good reasons for concluding nitrate concentration at that level can accumulate naturally." Naturally occurring nitrates in groundwater have three major sources: fertilizer, septic tanks and manure.

Major Fishburn said the Air Force would drill eight monitoring wells around the silo site, which he said would show quickly if there were any contamination entering through the water table.

SILO RUPTURE UNCONFIRMED

He said the Air Force did not know yet what the condition was at the bottom of the silo and whether it had ruptured, perhaps permitting the hydrazine mixture of fuel to enter the groundwater.

However, the major stated, "We do not anticipate we will allow any significant amounts of hydrazine to leave the area."

Residents of Guy, Ark., about six miles from the silo site, which is 20 miles from Little Rock, have complained of sickness since the explosion, but the Air Force has said its preliminary tests showed no abnormal amounts of toxic chemicals reached the tiny town.

A claim of no harm to the environment has been accepted. So, a Type I error has not occurred. However, if there *is* in fact harm to the environment (for example, maybe their samples just happened to show less contamination than actually exists), then a Type II error has occurred. We do not know whether such an error has taken place, but if it has, its implications are dangerous. For example, persons may drink the well water or hike in the area, believing there is no problem with contamination.

We should note from the article that one way in which the researchers tried to prevent such an error was to take many samples (270 of them) from many possible areas of contamination—air, soil, ground water, and surface water—in order to increase the likelihood of detecting contaminants if they were present.

EXERCISES

1   Suppose in the example of Nite-Nite light bulbs (Section 12.2), we define the probability of an unusual event as .01. Make a decision now about the claim of the company concerning the life of their light bulbs. What is the probability of a Type I error?

2   What effect does increasing the sample size have on the standard error? What effect does increasing the sample size have on Type I and Type II errors?

3   Refer to Exercise 6 of Section 12.2. Suppose a Type I error is made. What would this error be, in words? What would a Type II error be?

4   In the newspaper account[5] in Figure 12.6, what would a Type I error be? What would a Type II error be?

FIGURE 12.6

SPACE MAY YIELD NAUSEA CURE

BY *JON VAN*

A TEST of biofeedback techniques to prevent motion sickness may be the first, and most practical, biological experiment to occur on America's space shuttle.

Because about half of America's astronauts have experienced the debilitating dizziness and nausea of "space sickness," an effective way to control it would be a major step forward for the nation's space effort.

In tests at the Ames Research Center in California, researchers for the National Aeronautics and Space Administration report significant success in teaching people to avoid motion sickness simply by altering the way they think.

---

5   Find a newspaper account of some statistical finding, such as reported in Figure 12.5 or Figure 12.6. Explain, in words, what a Type I and Type II error would be. Interpret the consequences of each error in terms of the findings that are reported.

12.4   STUDENT'S t

We have pointed out several times in this book that if we have large samples (of size 20 or greater) we can make reliable estimates of the mean and variance of the population from which the sample was drawn. We have also noted that if the sample is small, the estimates of the population mean and variance are not reliable.

Special methods to deal with small samples were developed by an Irish chemist by the name of William Sealy Gossett (1876–1937). He published his work under the pen name Student.

Student invented the statistic, which he called t. He defined t as

$$\frac{\overline{X} - \mu}{S/\sqrt{N}}$$

where $\overline{X}$ is the mean of the sample and S is the standard deviation of the sample.

Student's t is defined very much like the z statistic that we have met several times in this book. Roughly speaking, as the sample size gets larger and larger, the distribution of Student's t becomes more and more like that of the standard, normal z statistic. It is important to know, however, that Student's t depends upon drawing random samples from a population that is normally distributed. In all cases in the rest of this book, we make this assumption.

Table 12.7 gives the values of t that are produced by 100 random samples of size 12 drawn from a normal population with a mean of 0.

We can use Table 12.7 to estimate probabilities of obtaining a random t greater (or less) than a given value, just as we have for z statistics and chi-square statistics.

For example:

$$P(t \text{ greater than or equal to } 1.3) = \frac{13}{100} = .13$$

$$P(t \text{ less than or equal to } -2.5) = \frac{2}{100} = .02$$

Caution: Our stem-and-leaf table has negative stems. For example, the smallest value in Table 12.7 is −4.1. Notice also that for the negative stems, the leaves are listed in reverse order from left to right. For example, −1.1 is to the *right* of −1.8, since −1.1 is greater than −1.8.

TABLE 12.7

Population value: $\mu = 0$

STEM	LEAF	*f*
−6		0
−5		0
−4	1	1
−3		0
−2	5,3,1,0	4
−1	8,7,3,3,3,1,1,1,0,0	10
−0	9,8,8,8,8,7,7,7,7,7,6,6,6,6,6,5,5,5,4,4,4,3,3,3,2,2,2,2,1,1,1	30
0	0,0,0,0,0,0,0,0,0,1,1,1,1,1,1,2,2,3,3,3,3,3,3,3,3,4,4,5,5,5,6,6,6,6,7,8,8,8,8,9,9,9	40
1	2,2,3,3,4,4,5,5,9	9
2	0,1,4,5,6	5
3	0	1
4		0
5		0

Sample size: $N = 12$ Total   100

When the curve for the *t* scores is smoothed for samples of size 12, it looks like the one in Figure 12.7.

FIGURE 12.7

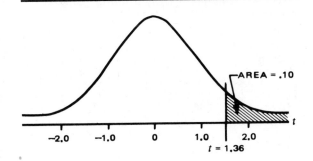

As we saw in Chapter Nine, for the normal curve it is useful to have tables that provide areas under a curve. Tables for areas under a *t* curve have also been developed. For samples of size 12, a portion of the *t* table looks like Table 12.8.

TABLE 12.8
AREAS UNDER THE DISTRIBUTION OF STUDENT'S *t*
FOR SAMPLES OF SIZE 12

p	.10	.05	.025	.01	.005
t	1.36	1.80	2.20	2.72	3.11

This table is read as follows. The upper row gives the probability (area) for the corresponding *t* value in the lower row. So, for example, we have:

$$p(t > 1.36) = .10$$
$$p(t > 1.80) = .05$$
$$p(t > 2.72) = .01$$

From Table 12.7, we can find corresponding estimated probabilities, based on 100 random samples of size 12:

$$P(t \geqslant 1.4) = .11$$
$$P(t \geqslant 1.8) = .07$$
$$P(t \geqslant 2.7) = .01$$

Since the shape of the t distribution (and therefore the area corresponding to that t) depends upon the sample size from which the t was produced, we need to know the sample size before we read a table of areas for t. Before we go to the table that statisticians use, we need to know what is meant by *degrees of freedom* when we are using Student's t. (We already learned about degrees of freedom when we used the chi-square statistic in Chapter Six.)

Recall that to find an unbiased estimate of population variance, we use this formula for calculating the sample variance:

$$S^2 = \frac{\text{sum of } (X - \bar{X})^2}{N - 1}$$

We say that the sample statistic S^2 has $N-1$ degrees of freedom (that is, 1 less than the size of the sample). The t statistic, which uses S, the square root of the sample variance, also has $N-1$ degrees of freedom.

So, for a sample size of 12, the t statistic has $12 - 1 = 11$ degrees of freedom.

We are now ready to read the table of areas for Student's t (Appendix E), a portion of which is shown in Table 12.9.

TABLE 12.9
STUDENT'S t-DISTRIBUTION (VALUES OF t)

df	PROBABILITIES				
	.10	.05	.025	.01	.005
1	3.08	6.31	12.71	31.82	63.66
2	1.89	2.92	4.30	6.96	9.92
3	1.64	2.35	3.18	4.54	5.84
4	1.53	2.13	2.78	3.75	4.60
5	1.48	2.02	2.57	3.36	4.03
6	1.44	1.94	2.45	3.14	3.71
7	1.42	1.89	2.36	3.00	3.50
8	1.40	1.86	2.31	2.90	3.36
9	1.38	1.83	2.26	2.82	3.25
10	1.37	1.81	2.23	2.76	3.17
11	1.36	1.80	2.20	2.72	3.11
12	1.36	1.78	2.18	2.68	3.05

Suppose we wish to find this probability, where t is calculated from a sample of size 9:

$$p(t > 1.86)$$

Since we are dealing with a sample of size 9, we have 8 degrees of freedom (df). We go to the row of the table for $df = 8$ and find the value of 1.86. The probability value at the top of the column in which 1.86 is located is .05. So

$$p(t > 1.86) = .05$$

Similarly, for a *t* with 5 degrees of freedom, we have

$$p(t > 3.36) = .01$$

Now suppose we want to find probabilities for *t*'s which are negative.

■ **Example 1:** $p(t < -1.40)$ where *t* has 8 degrees of freedom.

Solution:

We need to find the shaded area in Figure 12.8. The *t* distribution, like the standard normal *z* distribution, is symmetric. That is, it has the same shape on both sides of $t = 0$. So, we can use the same table for areas for *t*, and simply regard the *t*'s in the table as negative.

FIGURE 12.8

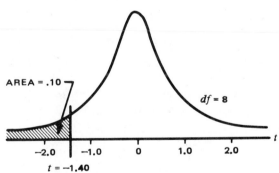

Therefore, from Table 12.9 we have:

$$p(t < -1.40) = .10 \text{ (8 degrees of freedom)} ■$$

For values of *t* that are not in the table of areas, we interpolate using the procedure in Appendix C.

■ **Example 2:** Find $p(t > 3.0)$ where *t* has 5 degrees of freedom.

Solution:

From Appendix E, we have, for 5 degrees of freedom:

$$p(t > 2.57) = .025$$

and

$$p(t > 3.36) = .01$$

Therefore, $p(t > 3.0)$ is a value between .025 and .01.

We note that $t = 3.0$ is the following fraction of the total distance from $t = 2.57$ to $t = 3.36$.

$$\text{Fraction of distance} = \frac{3.00 - 2.57}{3.36 - 2.57} = \frac{.43}{.79} = .54$$

This fraction is about .5. So we can say that $p(t > 3.0)$ is about half way between .025 and .01.

$$p(t > 3.0) = .025 - \frac{.43}{.79}(.025 - .01)$$

$$= .025 - .54(.015) = .025 - .008$$

$$= .017 ■$$

EXERCISES

1   Table 12.10 gives values of t calculated from 100 random samples of size 10. Find these estimated probabilities from Table 12.10.

(a) $P(t > 1.8)$ (b) $P(t > 2.3)$
(c) $P(t > 1.4)$ (d) $P(t < -1.8)$
(e) $P(t < -2.1)$ (f) $P(t < -1.4)$

2   Use Table 12.9 of areas for Student's t to find these theoretical probabilities for a t with 9 degrees of freedom.

(a) $p(t > 1.8)$ (b) $p(t > 2.3)$
(c) $p(t > 1.4)$ (d) $p(t < -1.8)$
(e) $p(t < -2.3)$ (f) $p(t < -1.4)$

3   Use the table of areas for Student's t in Appendix E to find these theoretical probabilities. The degrees of freedom for each t are indicated.

(a) $p(t > 1.80)$ $df = 11$
(b) $p(t < -1.44)$ $df = 6$
(c) $p(t > 1.32)$ $df = 21$
(d) $p(t < -2.06)$ $df = 24$

4   Use the table of areas for Student's t (Appendix E) to find these theoretical probabilities. You will need to use interpolation.

(a) $p(t > 2.0)$ $df = 11$
(b) $p(t < -2.5)$ $df = 16$
(c) $p(t > 2.6)$ $df = 23$
(d) $p(t < -1.9)$ $df = 9$

TABLE 12.10

STEM	LEAF	f
−4		0
−3		0
−2	1	1
−1	9,9,8,6,3,2,2,1,1,1,1,1	12
−0	9,9,9,9,9,8,8,8,8,7,7,6,6,5,5,5,4,4,3,3,3,3,2,2,1,1,1,1,1,1	30
0	0,0,1,1,1,2,2,2,2,2,2,3,3,3,4,4,4,4,4,4,4,5,5,5,5,6,6,6,6,6,7,9,9	34
1	0,0,0,1,1,2,2,3,4,4,5,5,6,6,6,7,7,8,8	19
2	1,2	2
3	5,9	2
4		0
5		0
		100

12.5 CONFIDENCE INTERVALS FOR μ (SMALL SAMPLES)

We can use Student's t to help us find a confidence interval for estimating the population mean μ given the mean and standard deviation of a small sample drawn from that population.

■ **Example:** A savings bank wants to estimate the average amount of money in its savings accounts. A sample of 16 accounts is taken. It is found that the mean amount of savings in the sample is $140.50, with a standard deviation of $23.51. Find a 90% confidence interval for the mean (average) of all savings accounts in the bank.

Solution:

Student's t allows us to follow a procedure for finding the desired confidence interval that is nearly identical to that used for large samples (Section 11.5).

Step 1: Draw a graph of the distribution of t for the given degrees of freedom. Since in this example we are dealing with a sample of size 16, the t has $16 - 1 = 15$ degrees of freedom as shown in Figure 12.9.

FIGURE 12.9

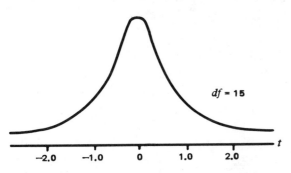

Step 2: Shade in the area under the curve to correspond to the required level of confidence. We are to find a 90% confidence interval (see Figure 12.10).

FIGURE 12.10

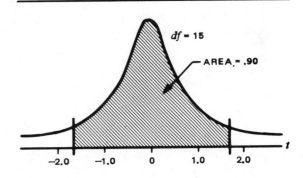

Step 3: Find, from a table of areas for Student's t, the values of t that will enclose .90 of the area under the curve with equal area on each side of $t = 0$ (Figure 12.11).

FIGURE 12.11

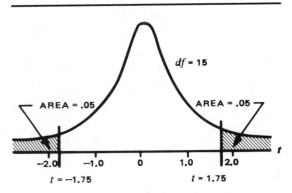

df = DEGREES OF FREEDOM

Since we have an area of .90 under the curve, and since the t distribution is symmetrical (same shape both sides of $t = 0$), we have an area of .45 on each side of $t = 0$. Since, like the standard normal z, the total area under the t distribution is 1.0, we have an area of .05 left in each "tail" of the curve. From Appendix E, with 15 degrees of freedom, we have $p(t > 1.75) = .05$. So our required values of t are $t = 1.75$ and $t = -1.75$.

Step 4:   Find, from the given sample information, the value of

$$\frac{S}{\sqrt{N}}$$

We are told that the sample standard deviation is $23.51 and the sample size is 16. So, we have

$$\frac{S}{\sqrt{N}} = \frac{\$23.51}{\sqrt{16}} = \frac{\$23.51}{4} = \$5.88$$

Step 5:   Find the *upper end* of the interval about μ. Student's t can be transformed just as a z score. So, for the upper end of the interval, we have

$$U = t \times \frac{S}{\sqrt{N}} + \mu$$

$$= 1.75 \times 5.88 + \mu$$

$$\doteq 10.29 + \mu$$

Step 6:   Find the *lower end* of the interval about μ.

$$L = -1.75 \times 5.88 + \mu$$

$$\doteq -10.29 + \mu$$

Step 7:   Substitute $\overline{X}$, the sample mean for μ the population mean. We have the following interval (see Figure 12.12):

$$L = \quad \$10.29 + \$140.50 = \$150.79$$

$$U = -\$10.29 + \$140.50 = \$130.21$$

FIGURE 12.12

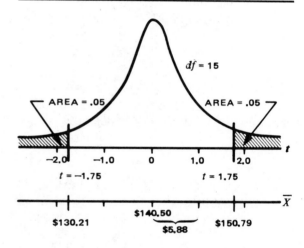

And we are 90% confident that the mean of all savings accounts in the bank is between $130.21 and $150.79. ∎

EXERCISES

Assume data drawn from normal populations.

1   In the example of this section, find the following confidence intervals for the mean of all accounts in the bank, given the same sample information.

(a) 95%   (b) 99%

2   A school counselor wants to estimate the mean IQ of all students in the graduating class. She takes a random sample of 9 students from the class and finds these sample statistics:

Mean = 115
Standard deviation = 11.2

Find a 95% confidence interval for the mean IQ of all the students in the graduating class.

3   A pollster wants to estimate the mean popularity of an entertainer, on a scale from 1 = low to 5 = high. He takes a random sample of 12 people from a shopping center, which is patronized by the entire community. Find a 99% confidence interval for the mean popularity of the entertainer in the community. The sample values he found were:

Mean = 4.1
Standard deviation = 0.2

4   A random sample of 10 measurements of the percentage of gold in a vein has a mean value of 12.3% and a standard deviation of 2.5%. Find a 90% confidence interval for the mean percentage of gold in the vein.

5   **The Age of Discovery:**[6]   Table 12.11 lists 12 famous scientific discoveries and the ages of scientists at the time they made them. Use this sample information to find a 90% confidence interval for estimating the true mean age at which great scientific discoveries are made.

TABLE 12.11

DISCOVERY	DISCOVERER	DATE	AGE
Earth goes around sun	Copernicus	1543	40
Telescope, basic laws of astronomy	Galileo	1699	34
Principles of motion, gravitation, calculus	Newton	1665	23
Nature of electricity	Franklin	1746	40
Burning is uniting with oxygen	Lavoisier	1774	31
Earth evolved by gradual processes	Lyell	1830	33
Evidence for natural selection controlling evolution	Darwin	1858	49
Field equations for light	Maxwell	1864	33
Radioactivity	Curie	1896	34
Quantum theory	Planck	1901	43
Special theory of relativity, $E = mc^2$	Einstein	1905	26
Mathematical foundations for quantum theory	Schroedinger	1926	39

12.6 TESTING HYPOTHESES ABOUT μ (SMALL SAMPLES)

Student's t can also be used to test hypotheses about μ using information from small samples. The procedure is very similar to the one used for large samples (Section 12.2). *Note:* Small samples have size less than 20.

■ **Example:**   The Nite-Nite Bulb Company produces a flashlight bulb that they claim lasts for 50 hours. A consumer group takes a random sample of bulbs from 1 week's stock and finds the following:

Mean life = 48.5 hours
Standard deviation =   4.4 hours

Should Nite-Nite's claim be accepted or rejected?

Solution:

Since the sample of bulbs is small ($N = 16$), we use Student's t. We follow the six steps.

Step 1—*Choose a model:*   We assume that the light bulbs actually do last 50 hours (that is, $\mu = 50.0$).

Step 2—*Trial:*   A trial consists of drawing a sample of size 16 from this population.

Step 3—*Outcome of trial:*   We calculate the value of Student's t, given the sample information.

Step 4—*Repeat the trials:*   We do not actually repeat the trials. But if we did, we know that the distribution of the statistic t becomes closer and closer to the theoretical curve for 15 degrees of freedom.

Step 5—*Find p(obtained statistic, or less):*   Our sample data give us this value for t:

$$t = \frac{X - \mu}{S/\sqrt{16}} = \frac{48.5 - 50.0}{4.4/\sqrt{16}} = \frac{-1.5}{1.1} \doteq -1.36$$

We want to find, from the t distribution:

$$p(t < -1.36) \text{ for 15 degrees of freedom}$$

See Figure 12.13. Appendix E tells us that

$$p(t < -1.36) \doteq .11$$

FIGURE 12.13

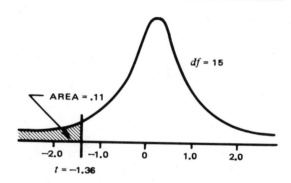

Step 6—*Make a decision:*   Obtaining a t as small as −1.36 or smaller is not a rare event. It happens more than .10 of the time. So we conclude we have insufficient evidence to doubt Nite-Nite's claim that their bulbs last 50 hours (but we are close!). ■

EXERCISES

Assume data drawn from normal populations.

1   A sample of 12 ninth graders has a mean weight of 118 pounds, with a standard deviation of 5 pounds. Use Appendix E to test whether the sample differs significantly from the assumed mean weight of only 114 pounds for the total freshman class.

2   A sample of 16 measurements of the percentage of manganese as a mineral deposit has a mean of 81.3 and a standard deviation of 3.2. Test the claim that the true amount of manganese in the deposit is only 77.0.

3   Here are ten measurements of the designs by Crow Indians (in terms of width-to-length ratio × 100).[7]

        58.7 85.0 52.2 66.6 52.9
        58.8 86.2 44.7 57.1 48.3

Do a statistical test to determine whether the width-to-length ratios of the Crow Indians are significantly less than the golden ratio (recall from the Key Problem that the golden ratio is 61.8).

4   **The Transylvania Effect:**[8]   The possible effect of the moon on human behavior is called the *Transylvania effect*. In order to test possible effects of the moon on mental stability, the data below were recorded on the number of admissions to the emergency room of a Virginia mental health clinic during the 12 full moons from August 1971 to July 1972.

Number of Admissions During Full Moon

5	13
13	16
14	25
12	13
6	14
9	20

The hospital reported that on the average, for the rest of the year when the moon was *not* full, the average number of admissions was 11.2. Do a statistical test to determine whether there was evidence of significantly more admissions when the moon was full.

CHAPTER

13

Advanced Topics

KEY PROBLEM

Birth Order and Delinquency

A person's birth order (whether one is oldest or youngest in the family) has been studied as a possible factor in determining how well one does in college, how successful one is in a career, or even whether it affects one's chances of becoming a delinquent. [A study of successful business women, reported in *Savvy Magazine*, states that 61% of those earning more than $106,000 a year are firstborn or only children. (*Savvy*, January 1983, p. 14.)]

A large sample of girls in a public high school was asked to fill out a questionnaire about their birth order and was also asked questions designed to indicate delinquent behavior.

Table 13.1 shows the results from 1154 high school girls. It shows, for example, that 24 of the girls who were oldest (firstborn) in the family were also delinquent.

TABLE 13.1

| BIRTH ORDER | DELINQUENT | | TOTALS |
	YES	NO	
Oldest	24	450	474
In Between	29	312	341
Youngest	35	211	246
Only Child	23	70	93
Totals	111	1043	1154

Methods in this chapter will be given to determine whether these data indicate a relationship between birth order and delinquency.

367

13.1 CHI-SQUARE: UNEQUAL EXPECTED FREQUENCIES

The chi-square statistic can be used in a remarkable variety of ways. Recall from Chapter Six that a basic pattern for a chi-square table is the following:

Outcome	Expected Frequency	Obtained Frequency	(Difference)^2

This applies, for example, to the rolling of a die. If the die is fair, all outcomes are expected to occur an equal number of times.

So far in this book, all of our chi-square problems have involved equal expected frequencies. Now we will see how chi-square can be used to solve problems where the expected frequencies are unequal.

■ **Example 1:** In a certain city, the racial makeup of the population is:

White	40%
Black	35%
Others	25%

Is the racial composition of the city council representative of the city?

Whites	25
Blacks	10
Others	10

How representative of the racial composition of the city is the makeup of the council?

Solution:

Another way of asking this question is: Does there appear to be *bias* in the makeup of the council which favors, say, whites, for instance, or is the makeup such that it would occur rather often in a random sample?

We will use the chi-square test for this problem and use these headings:

Expected Frequency (E)	Observed Frequency (O)	$E-O$	$(E-O)^2$

We can get the number of observed outcomes from our data about the city council: There are 30 white people, and so on. But how do we get the expected number of outcomes?

Let's go back to our familiar six steps to statistical decision making.

Step 1—*Choose a model*: What model is appropriate for seeing if the makeup of the city council is biased or fair? Models we have used before are fair die, a three-sided spinner, or some other way of getting three equally likely outcomes.

Let's say we used a four-sided fair die and ignored the fourth side. Then, we will get each of the remaining three sides (call them 1, 2, and 3) with equal probability.

So *if* we were to use as our model:

Side 1	White
Side 2	Black
Side 3	Others

Our table of expected frequencies would be

Expected Frequency (E)	
White	15
Black	15
Others	15
	45

That is to say, we would expect, with a fair die, each of the three racial groups to be represented with equal numbers.

However, it is not reasonable to expect *equal* numbers of each of the groups on the council, since there are more white people in the city and fewer blacks and others.

So our model should not be a *fair* die, but a *loaded* die. Our die is loaded so that in the long run, when rolled, we get the following proportions of outcomes for each of the three sides (remember, we ignore side 4):

Side 1 (white)   .40 of the time, since 40% of the city is white

Side 2 (black)   .35 of the time, since 35% of the city is black

Side 3 (other)   .25 of the time, since 25% of the city is others

With this loaded die and 45 people on the city council, our expected outcomes are:

White	40%,		$(.40 \times 45) = 18,$
Black	35%,	or	$(.35 \times 45) = 15.75,$
Other	25%,	or	$(.25 \times 45) = 11.25.$

Step 2—*Trial*:   We will now roll our loaded die 45 times, once for each member of the city council.

Step 3—*Outcome of the trial*:   We will record how many of each of the three outcomes were obtained. Here is a set of sample outcomes for 45 rolls of a die:

Obtained Frequency (*O*)

Side 1	24
Side 2	15
Side 3	6
	45

For this set of outcomes, we calculate the chi-square statistic (Table 13.2). You may want to refer back to Chapter Six to see how this was done.

TABLE 13.2

OUTCOME	EXPECTED	OBTAINED	$E-O$	$(E-O)^2$
1 (white)	18	24	6	36.00
2 (black)	15.75	15	−0.75	.56
3 (other)	11.25	6	−5.25	27.60

From this table, we calculate chi-square using the familiar formula:

$$\chi^2 = \text{sum of } \frac{(\text{expected} - \text{obtained})^2}{\text{expected}}$$

The procedure is exactly the one we used in Chapter Six. The only difference here is that the expected number of outcomes is now, generally, different from row to row.

$$\chi^2 = \frac{(18-24)^2}{18} + \frac{(15.75-15)^2}{15.75} + \frac{(11.25-6)^2}{11.25}$$

$$= \frac{(6)^2}{18} + \frac{(-.75)^2}{15.75} + \frac{(-5.25)^2}{11.25}$$

$$= \frac{36}{18} + \frac{.56}{15.75} + \frac{27.6}{11.25}$$

$$= 2.0 + .04 + 2.45$$

$$= 4.49 \quad \text{(rounded to 4.5)}$$

$$\doteq 4.5$$

This is one value of a chi-square statistic. As you will recall from Chapter Six, we now have to make a decision as to the likelihood of getting a chi-square as large or larger than our obtained chi-square (the one produced by the data we have on the makeup of the city council). So we will generate many chi-squares using our model described in Step 1.

Step 4—*Repeat trials:* We repeat Steps 2 and 3 for a total of 100 trials and obtain the chi-square values in Table 13.3.

Step 5—*Probability of obtained statistic (or greater):* We now use the results of our experiment to estimate how likely it is to obtain a city council with the following makeup, as a random sample from the population of the city:

Whites	25
Blacks	10
Others	10
	45

The chi-square for these data is

Outcome	Expected	Obtained	$E-O$	$(E-O)^2$
White	18	26	8	64.00
Black	15.75	10	−5.75	33.06
Other	11.25	9	−2.25	5.06

$$\chi^2 = \frac{64}{18} + \frac{33.06}{15.75} + \frac{5.06}{11.25}$$

$$= 3.56 + 2.10 + .45 = 6.11 \text{ (rounded to 6.1)}$$

From the stem-and-leaf table, we estimate the probability of a chi-square as large or larger than 6.1 as

$$\frac{6}{100} = .06$$

Step 6—*Make a decision:* Since the probability of getting a chi-square as large as 6.1 or larger is small (less than .10), we decide our model is *not* appropriate. It is unlikely that the racial makeup of the council is a random sample of the city's racial makeup.

We can also use the chi-square table (Appendix A) to find the probability of getting a chi-square as large as 6.1 or larger from a model defined in Step 1. Our table of outcomes has only 2 degrees of freedom. There are three rows in the table (corresponding to three possible outcomes of the die). So, once any two rows are known, the third row is fixed. This is the rule we discussed in Chapter Six (Section 6.6).

We go to the chi-square table with 2 degrees of freedom and find that the probability of a chi-square = 6.1 or more is approximately .05. ■

TABLE 13.3

(Multiply each value by .01.)

STEM	LEAF		f
0	1,1,1,1,1,1,1,1,1,1,4,4,4,4,4,4,4,4,4,5,5,5,5,5,5,5,5,9,9,9,9,9,9,9,9,9,9,9,9,9,9,9,9,9		45
1	2,2,2,2,2,2,2,2,2,6,6,6,6,7,7,7,7		17
2	5,5,5,5,5,5,5,5,5,8,8,8,8,8,8		15
3	3,7,7,7,7		5
4	1,1,1,5,9,9,9		7
5	2,2,2,7,7		5
6	5,9,9	χ^2 greater than	3
7	6,6	or equal to	2
8	5	6.1	1
			100

■ **Example 2:**   Mr. Schnitzel, the psychology instructor, teaches several large classes of introductory psychology and grades on the normal curve. He gave all 200 students the same final examination and assigned the distribution of letter grades in Table 13.4 to the course.

TABLE 13.4

GRADE	NUMBER OF STUDENTS GETTING GRADE
A	25
B	48
C	105
D	18
F	4
	Total 200

According to the course catalog, a normal distribution of grades would result in the following percentages:

Grade	Percent Getting Grade for Normal Distribution
A	5
B	20
C	50
D	20
F	5

Did Mr. Schnitzel's assignment of grades conform to a normal distribution? Or, does the number of students getting the various letter grades differ so much from the normal distribution that we are doubtful that they are normally distributed?

Solution:

We set up the chi-square table in the usual way (Table 13.5).

We now calculate chi-square with the familiar formula:

$$\chi^2 = \text{sum of } \frac{(E-O)^2}{\text{expected number}}$$

$$= \frac{(-15)^2}{10} + \frac{(-8)^2}{40} + \frac{(-5)^2}{100} + \frac{(22)^2}{40} + \frac{(6)^2}{10}$$

$$= \frac{225}{10} + \frac{64}{40} + \frac{25}{100} + \frac{484}{40} + \frac{36}{10}$$

$$= 22.5 + 1.6 + .25 + 12.1 + 3.6$$

$$= 40.05$$

This appears to be a rather large chi-square. But let's check it out in our chi-square table (Appendix A). Since there are five categories in our table (five different letter grades assigned), this chi-square has 4 degrees of freedom. We go to the table and see that a chi-square this large occurs very seldom by chance (less than .001 of the time). So, the difference between the number of grades we expect for a normal curve and what was given by Mr. Schnitzel is too large to expect by chance very often. We conclude, rather, that

TABLE 13.5

OUTCOME (GRADE)	EXPECTED NUMBER OF GRADES	OBTAINED NUMBER OF GRADES	$E - O$
A	5% of 200 = 10	25	−15
B	20% of 200 = 40	48	− 8
C	50% of 200 = 100	105	− 5
D	20% of 200 = 40	18	22
F	5% of 200 = 10	4	6
	200	200	0

the letter grades are *not* normally distributed (it looks like he gives more As and Bs and fewer Ds and Fs than would be expected when grading on the normal curve). ■

We have shown how the chi-square statistic can be used to help make conclusions about outcomes even though the expected number of observations in each cell are not equal. There are restrictions about the use of the chi-square, however, and these are discussed in more advanced texts. One restriction to be aware of is that the expected number of observations in each cell should not be smaller than about 5.

EXERCISES

1   Is is true, as is often assumed, that people prefer the flakiest (lightest) pie crust? An experiment was done at Cornell University to test this assumption.[1] Thirty people were involved. Each person, in turn, was blindfolded and asked to state the order of preference for three pieces of pastry—one light (L), one medium (M), and one heavy (H)—which were presented in random order.

There are six different ways in which order of preference could be given, as shown. Also given are the number of persons choosing each of the six:

Preference Ordering	Number of Persons Choosing (Obtained Frequencies)
LMH	10
LHM	3
MLH	3
MHL	8
HLM	1
HML	5
	30

(a)   Under the assumption that each of the six orderings is preferred equally, what is the expected number of persons choosing each?

(b)   Test the assumption of equal preference using chi-square.

(c)   A model was developed under the assumption of preference for lighter pastry. According to this model, the probabilities of each of the six outcomes being chosen is as follows:

Preference Ordering	Probability of Outcome under Assumption of Preference for Light Pastry
LMH	.23
LHM	.18
MLH	.18
MHL	.15
HLM	.15
HML	.11

Now, what is the expected number of outcomes (that is, expected number of persons preferring each ordering of pie crusts)?

(d)   Test the assumption of preference for light pastry, using the model given in (c).

2   In the study of genetics, scientists are interested in determining how characteristics of living things are inherited from one generation to another. The fundamental principles of heredity were discovered by Gregor Johann Mendel (1822–1884), when he proved by crossbreeding garden peas that there are definite patterns in the way that characteristics such as size, shape, and surface texture are passed on from generation to generation.

One of Mendel's classical experiments involved crossbreeding two kinds of plants, one with round yellow seeds and the other with wrinkled green seeds. Here are the results which should have been obtained, according to his law of heredity, and those results actually obtained.[2] Did the seed-growing experiment support or cast doubt upon his genetic theory?

Type of Seed	Expected Proportion	Expected Number	Obtained Number
Round and yellow	9/16	313	315
Wrinkled and yellow	3/16	104	101
Round and green	3/16	104	108
Wrinkled and green	1/16	35	32
		556	556

3 A biological experiment with flowers yielded the following results. The expected values were based upon a genetic theory of inheritance. Did the results support the theory?

Characteristic of Flower	Expected f	Observed f
AB	180	164
Ab	60	78
aB	60	65
ab	20	13

4 A public opinion poll claims that 60% of the voters favor Mr. Alpha, 30% favor Mr. Beta, and 10% favor Ms. Gamma. The actual election produced these votes:

Candidate	Number of Votes
Alpha	503
Beta	115
Gamma	35

Does it appear that the poll was accurate?

5 A census in a certain city shows the racial composition to be 30% white, 55% black, and 15% others. The city council has the following racial composition (number of persons of each race):

White	20
Black	20
Other	10

How representative of the racial composition of the city is the racial composition of the city council?

6 A typist prepared a manuscript of 167 pages, of which 103 contained no errors. Forty-five pages had one typing error each. Sixteen pages had two errors each. Three pages had three errors. None had more than three errors. The table summarizes this information.[3]

Number of Typing Errors	Number of Such Pages (Observed Outcomes)
0	103
1	45
2	16
3 or more	3
Total	167

There is a probabilistic model, called a *Poisson model,* which has been used to predict the number of typing errors per page. Using this model, the following probabilities are calculated.

Number of Typing Errors	Probability
0	.59
1	.31
2	.08
3 or more	.02

(a) Calculate the expected number of pages having 0, 1, 2, and so on, errors, according to this model.

(b) Use the chi-square to test whether this model is an appropriate one.

7   A probabilistic model has been used to predict occurrences of heavy rainstorms. The following table shows the number of heavy rainstorms reported by 330 weather stations during a 1-year period.[4] (For example, 102 stations reported no heavy rainstorms, 114 reported one heavy rainstorm, and so on.)

The third column shows the probability of numbers of heavy rainstorms according to the model being used. Use chi-square to test how well the model predicts the occurrence of heavy rainstorms.

Number of Heavy Rainstorms	Number of Stations Reporting	Probability
0	102	.301
1	114	.361
2	74	.216
3	28	.086
4 or more	12	.036

8   Geologists sometimes want to know the composition of pebbles in a stream (for example, what fraction of the pebbles are quartzite?). R. Flemal, in 1967, gathered 100 samples of 10 pebbles each from the Gros Ventre River in Wyoming.[5] Table 13.6 shows the results. The right-hand column gives the expected number of samples according to a binomial probability model. Do the chi-square test to see whether the pebbles can reasonably be thought to be distributed according to this model.

TABLE 13.6

NUMBER OF QUARTZITE PEBBLES	OBSERVED FREQUENCY	EXPECTED FREQUENCY
10	6	9.1
9	25	24.7
8	31	30.0
7	28	21.7
6	9	10.3
5	0	3.3
4 or less	1	.9
Totals	100	100.0

13.2   CHI-SQUARE: INDEPENDENCE

Here are headlines from some newspaper stories.

HEADLINES

> **Relationship found between car accidents and achievement in school**
>
> **NEW EVIDENCE STRENGTHENS HEART DISEASE, FAT LINK**
>
> **Is there a connection between birth order and delinquency?**

COMMENTS

The first story suggested that students who do better in school are involved in fewer automobile accidents.

The second story reported on research on relationships between the amount of cholesterol in a person's diet and the chances of that person having a heart attack.

The third story reviewed research on whether there is a relationship between a child's order of birth (oldest, in between, or youngest in the family) and having delinquent behavior.

Each of these stories can be restated using the idea of *independence*. As we saw in Chapter Four, two events are independent if one's taking place has no effect on whether the other takes place. (Or, two events are independent if they are unrelated to each other.)

So, each headline can be restated as a question involving independence, as used in our statistical sense.

Is having a car accident independent of whether a student does well in school?

Are heart disease and amount of fat in the diet independent?

Are birth order and delinquency independent?

■ **Example (Key Problem for this Chapter)—** *Birth Order and Delinquency:* A researcher was interested in finding out whether a child's birth order was related to that child's chances of becoming a juvenile delinquent. In order to study this question, the researcher asked 1154 high school girls to fill out a questionnaire that measured the degree to which they showed delinquent behavior. Each girl was also asked to indicate whether she was

   (1)   the oldest child in her family,
   (2)   in between,
   (3)   the youngest child in her family, or
   (4)   an only child.

The results are recorded in Table 13.7. This table is called a two-way table, since each data value is classified in two ways—birth order and delinquency.

TABLE 13.7

	DELINQUENT		
BIRTH ORDER	YES	NO	TOTALS
Oldest	24	450	474
In Between	29	312	341
Youngest	35	211	246
Only Child	23	70	93
Totals	111	1043	1154

In this table, each of the girls in the study can be classified two ways: (1) whether she is delinquent, and (2) her birth order. So, we see along the bottom total line, for example, that of the 1154 girls in the sample, 111 are classified as delinquent and the remaining 1043 as non-delinquent. In the rightmost column, we see that of the 1154 girls, 474 are the oldest in their family, 341 are somewhere between oldest and youngest (in between) and so on. We can also see that 24 girls are *both* delinquent and oldest child, and so on.

The research question being asked here is, as already stated: Are delinquency and birth order independent? Another way of asking this question is: Is a girl's probability of being delinquent related to (or affected by) her being a first child (or an in between child, and so on)?

Solution:

We will answer this question by assuming *independence* of delinquency and birth order. Now, under this assumption of independence, we want to calculate the number of the 1154 girls we would *expect* to be both delinquent and oldest.

To do this, we use the formula for independent events given in Chapter Four, page 95. According to the formula, for two independent events A and B, we have

$$p(A \text{ and } B) = p(A) \times p(B)$$

In words, the formula says that for independent events A and B, the probability of both A and B occurring is the probability of that A occurs multiplied by the probability that B occurs.

In terms of the birth order and delinquency research, we can let

A = girl is delinquent
B = girl is oldest child

Now, *if* delinquency and birth order are independent, we can state that the probability that a girl is both delinquent and an oldest child is the product of her being delinquent multiplied by the probability of here being the oldest child. We can find from our data that the probability that a girl (picked at random) is delinquent is

$$p(A) = \frac{111}{1154}$$

That is, there are 1154 girls in the sample, and 111 of them are delinquent. So, our chances of picking a delinquent girl at random from the group are 111 out of 1154 or .10 (rounded).

Similarly, we also have the probability that a girl (again, picked at random) is an oldest child is

$$p(B) = \frac{474}{1154}$$

That is, of the 1154 girls, 474 are oldest children.

Now, what is p (girl is *both* delinquent *and* oldest child)? If being delinquent and being the oldest child *are* independent events, then we know that

$$p(\text{delinquent and oldest child})$$
$$= p(\text{delinquent}) \times p(\text{oldest child})$$

But we also know, from our data, that

$$p(\text{delinquent}) = \frac{111}{1154} \doteq .096$$

$$p(\text{oldest child}) = \frac{474}{1154} \doteq .411$$

So, we know by the formula for independent events (again, assuming independence)

$$p(\text{delinquent and oldest child}) = .096 \times .411 \doteq .039$$

Now, how many girls would we *expect* to be both delinquent and an oldest child *if delinquency and birth order* are independent?

As we learned in Chapter Three, the expected number is the product

$$p(\text{delinquent and oldest}) \times \text{number of girls}$$
$$= .039 \times 1154 \doteq 45.4$$

This looks like a rather complicated procedure But there is a simplified method of finding the expected number of observations, under the assumption that the two ways of classifying the data (delinquency and birth order) are independent.

Let's look at the arithmetic we did for finding the expected number of girls who are delinquent and oldest children. It was

$$p(\text{delinquent and oldest}) \times \text{number of girls}$$

$$= \frac{111}{1154} \times \frac{474}{1154} \times 1154$$

which reduces to

$$\frac{111 \times 474}{1154} \doteq 45.6$$

A pattern will help us keep our arithmetic straight. Let's use this pattern to find the expected number of girls who are delinquent and in between in birth order (Table 13.8).

TABLE 13.8

O = Observed; E = Expected

| | DELINQUENT | | |
BIRTH ORDER	YES	NO	TOTALS
Oldest	$O = 24$ $E =$	$O = 450$ $E =$	474
In Between	$O = 29$ $E =$	$O = 312$ $E =$	341
Youngest	$O = 35$ $E =$	$O = 211$ $E =$	246
Only Child	$O = 23$ $E =$	$O = 70$ $E =$	93
Totals	111	1043	1154

Note: Rule for finding expected value is

$$\frac{\text{Column} \times \text{row}}{\text{Total}}$$

So, the expected number of girls who are delinquent and in between is

$$\frac{111 \times 341}{1154} \doteq 32.8$$

The remaining expected values are as follows:

Girls who are delinquent and

$$\text{Youngest} = \frac{111 \times 246}{1154} \doteq 23.7$$

$$\text{Only child} = \frac{111 \times 93}{1154} \doteq 8.9$$

Girls who are not delinquent and

$$\text{Oldest} = \frac{1043 \times 474}{1154} \doteq 428.4$$

$$\text{In between} = \frac{1043 \times 341}{1154} \doteq 308.2$$

$$\text{Youngest} = \frac{1043 \times 246}{1154} \doteq 222.3$$

$$\text{Only child} = \frac{1043 \times 93}{1154} \doteq 84.1$$

We now have for each cell of the table an *observed* number of girls and *expected* number under the assumption of independence between delinquency and birth order (Table 13.9).

TABLE 13.9

O = Observed; E = Expected

| | DELINQUENT | | |
BIRTH ORDER	YES	NO	TOTALS
Oldest	$O = 24$ $E = 45.6$	$O = 450$ $E = 428.4$	474
In Between	$O = 29$ $E = 32.8$	$O = 312$ $E = 308.2$	341
Youngest	$O = 35$ $E = 23.7$	$O = 211$ $E = 222.3$	246
Only Child	$O = 23$ $E = 8.9$	$O = 70$ $E = 84.1$	93
Totals	111	1043	1154

We now need to know how many degrees of freedom our chi-square will have.

DEGREES OF FREEDOM

Table 13.9 has been reproduced as Figure 13.1 to help us think about how many degrees of freedom our chi-square will have. Let's first consider the total number of delinquent girls. This number is fixed, but there are four categories of them. If we know how many there are in any three of them, we can find out how many there are in the fourth category by subtraction. So, there are three degrees of freedom in the rows of the table (one less than r, the number of rows).

Now let's consider the total number of oldest girls. The total number of oldest girls is fixed. So, once we know how many of them are delinquent, we can find out the number of non-delinquents by subtracting. Hence, there is only one degree of freedom in the columns of Table 13.9 (one less than c, the number of columns).

The total number of degrees of freedom (which can be thought of as represented by the unshaded region of Figure 13.1) is

$$(r-1)(c-1) = (4-1)(2-1) = 3 \times 1 = 3$$

See Figure 13.1.

The chi-square statistic can be used in a way very similar to previous methods in order to help us make a decision about whether our assumption (our statistical model) of independence is justified.

Step 1—*Choose a model:* Our model used here is that of two dice being thrown, one for each way of classification. We throw a die that represents the delinquency of a girl chosen at random. (Note that the die is loaded, since the outcomes *delinquent* and *not delinquent* are not equally likely. A girl picked at random from the 1154 girls is much more likely to be non-delinquent than delinquent.) We have:

Die 1: Whether girl is delinquent

$$p(\text{delinquent}) = \frac{111}{1154} \doteq .10$$

$$p(\text{nondelinquent}) = \frac{1043}{1154} \doteq .90$$

Die 2: Birth order of girl (this is a four-sided die whose sides are weighted)

$$p(\text{oldest}) = \frac{474}{1154} \doteq .41$$

$$p(\text{in between}) = \frac{341}{1154} \doteq .30$$

$$p(\text{youngest}) = \frac{246}{1154} \doteq .21$$

$$p(\text{only child}) = \frac{93}{1154} \doteq .08$$

Step 2—*Trial:* A trial consists of rolling the two dice 1154 times, one for each girl in the sample.

FIGURE 13.1

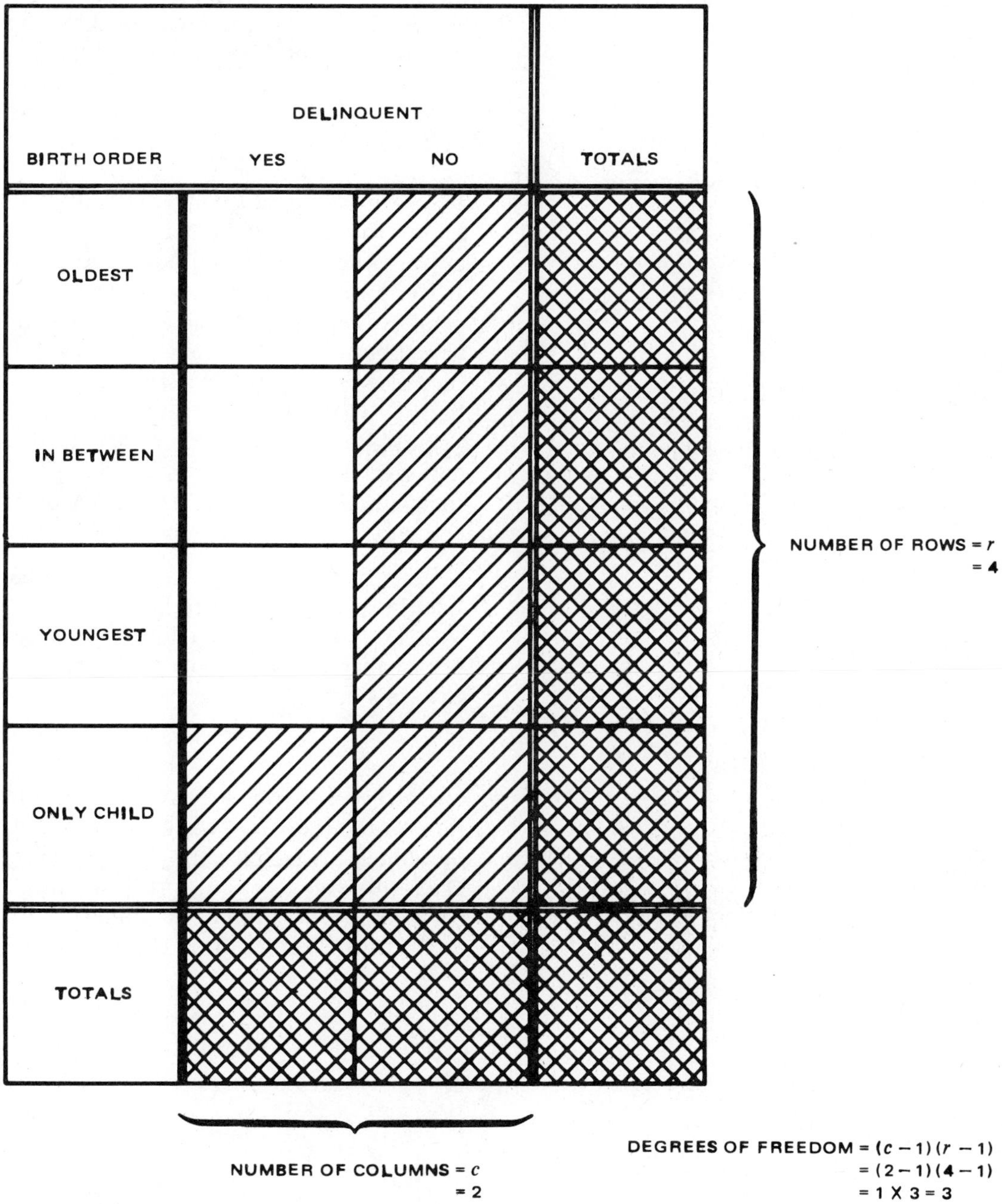
DELINQUENT
BIRTH ORDER
YES
NO
TOTALS
OLDEST
IN BETWEEN
YOUNGEST
ONLY CHILD
TOTALS
NUMBER OF ROWS = r
= 4
NUMBER OF COLUMNS = c
= 2
DEGREES OF FREEDOM = $(c-1)(r-1)$
= $(2-1)(4-1)$
= $1 \times 3 = 3$

Step 3—*Outcome of the trial:* There are eight possible outcomes, corresponding to each of the eight cells of the two way table.

Delinquent {
  oldest
  in between
  youngest
  only child
}

Nondelinquent {
  oldest
  in between
  youngest
  only child
}

Record how each of the 1154 rolls turned out.

Step 4—*Repeat the trials:* Conduct 100 trials and calculate the chi-square statistic for each trial. We will not do this by hand.

Step 5—*Find probability of the obtained chi-square (or greater):* The obtained chi-square is (refer to Table 13.9):

$$\frac{(24-45.6)^2}{45.6} + \frac{(29-32.8)^2}{32.8} + \frac{(35-23.7)^2}{23.7}$$

$$+ \frac{(23-8.9)^2}{8.9} + \frac{(450-428.4)^2}{428.4} + \frac{(312-308.2)^2}{308.2}$$

$$+ \frac{(211-222.3)^2}{222.3} + \frac{(70-84.1)^2}{84.1}$$

$$= \frac{(-21.6)^2}{45.6} + \frac{(-3.8)^2}{32.8} + \frac{(11.3)^2}{23.7} + \frac{(14.1)^2}{8.9}$$

$$+ \frac{(21.6)^2}{428.4} + \frac{(3.8)^2}{308.2} + \frac{(-11.3)^2}{222.3} + \frac{(-14.1)^2}{84.1}$$

$$\doteq 10.23 + .44 + 5.39 + 22.34 + 1.09 + .05$$

$$+ .57 + 2.36$$

$$= 42.47$$

Step 6—*Make a decision:* With 3 degrees of freedom, a chi-square this large (or larger) occurs very seldom (less than .10) by chance. So, we decide there *is* a relationship between delinquency and birth order. ▪

EXERCISES

1   A market research team conducted a survey to see if there was a relationship between personality type and a person's attitude toward small cars. A total of 300 persons were asked to fill out a questionnaire about how they view themselves. On the basis of their answers, they were classified as: (1) cautious conservative, (2) middle-of-the-roader, or (3) confident explorer. The persons were also asked their opinions of small cars: (1) favorable, (2) neutral, or (3) unfavorable.

Their responses are reported in Table 13.10. Use the chi-square test to help you decide if there appears to be a relationship between personality type and a person's attitude toward small cars.[6]

2   Professor Hans Zeisel of the University of Chicago Law School has gathered information on the size of jury lists and the number of women on each list for 46 juries selected in Massachusetts between April 1966 and October 1968.[7] The number of women chosen by each of seven judges from the total number

TABLE 13.10

SELF-PERCEPTION	OPINION OF SMALL CARS		
	FAVORABLE	NEUTRAL	UNFAVORABLE
Cautious conservative	80	10	10
Middle-of-the-roader	58	8	34
Confident explorer	49	9	42

of people on the lists is given in Table 13.11. Does it appear that the differences between numbers of women selected by each of the judges are due to factors other than chance?

TABLE 13.11

JUDGE	MEN	WOMEN	TOTAL
A	235	119	354
B	533	197	730
C	287	118	405
D	149	77	226
E	81	30	111
F	403	149	552
G	511	86	597

3   **Independence of Amoebas:**   An epidemic of severe intestinal disease occurred among the workers in a woodworking plant in South Bend, Indiana. Doctors attributed the illnesses to unusually large numbers of an amoeba found in the victims. (It was later concluded that the amoebas entered the plant's water supply through a faulty water main located near a leaky sewage pipe.)

In order to see how many of the men in the plant who had not fallen ill had been infected with the amoeba unawares, public health officials chose a random sample of 138 apparently well workers in the plant. The results are summarized in Table 13.12.[8] They identified two types of the amoeba—one large and one small. It is believed that the small type does not cause disease. Hence, it was wondered whether the frequency of infections is related to the presence or absence of the small and the large types of amoebas. For example, if both types are present, does the number of infections significantly decrease? Use a chi-square to test whether infection with one type of amoeba is independent of infection with the other type of amoeba.

TABLE 13.12

SMALL TYPE OF AMOEBA	LARGE TYPE OF AMOEBA		TOTALS
	PRESENT	ABSENT	
Present	12	23	35
Absent	35	68	103
Totals	47	91	138

4   The creel census based on annual samples of fishermen on the Big Piney River shows that in 1951, 92 of the 216 smallmouth bass recorded were at least 12 inches in length. In 1958, however, only 51 of the 206 recorded were at least 12 inches long. These data are summarized in Table 13.13. Is there enough sample evidence to say that the proportion of larger smallmouth caught in 1958 is different from the proportion caught in 1951?

TABLE 13.13[9]

| YEAR | LENGTH OF FISH | | TOTAL |
	AT LEAST 12 INCHES	LESS THAN 12 INCHES	
1951	92	124	216
1958	51	155	206
Totals	143	279	422

5   A remedy is proposed for the common cold and a doctor is skeptical of its value as a cure. He decides to test the new remedy by giving it to some of his patients while treating another group by the usual treatment. To get a fair comparison of patients, he tosses a coin for each cold sufferer and gives the new remedy if the coin falls heads and the standard treatment to those for whom the coin falls tails. The results are in Table 13.14. Do a chi-square test and decide what advice you would give to the doctor for treating patients with colds in the future. Suppose that the new treatment were very expensive relative to the older method. Would this affect your advice? Explain.

TABLE 13.14[10]

| TREATMENT | STATUS OF ILLNESS | | TOTAL |
	RELIEVED	NOT RELIEVED	
New	31	15	46
Old	29	25	54
Totals	60	40	100

13.3 BINOMIAL PROBABILITY

As we approach the end of this book, we take another look at perhaps the most familiar of probabilistic activities—tossing coins. It turns out that coin tossing provides an example of a very important kind of probabilistic behavior, called *binomial probability*.

Suppose we toss three coins. The *theoretical* outcomes can be shown as in Figure 13.2 (see Section 4.5).

FIGURE 13.2

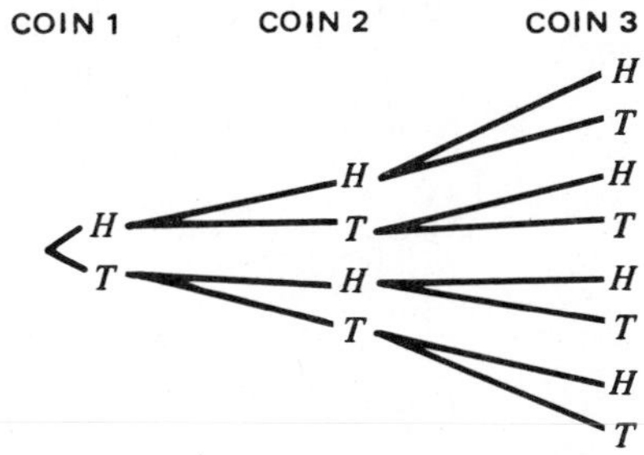

We see from this diagram that:

Coin 1 has 2 outcomes

H and *T*

Coins 1 and 2 have 4 outcomes

HH, HT, TH, and *TT*

Coins 1, 2, and 3 have 8 outcomes (count them)

HHH, HHT, HTH, HTT, THH, THT, TTH, TTT

This pattern can be generalized to tosses of N coins. That is, if N coins are tossed, then the number of possible outcomes of these N coins is 2^N

Now let's look at a way of finding how many heads may be obtained when 3 coins are tossed (Table 13.15).

TABLE 13.15

1 COIN	2 COINS	3 COINS
H (1 head)	*HH* (2 heads)	*HHH* (3 heads)
T (0 heads)	*HT* (1 head)	*HHT* (2 heads)
	TH (1 head)	*HTH* (2 heads)
	TT (0 heads)	*HTT* (1 head)
		THH (2 heads)
		THT (1 head)
		TTH (1 head)
		TTT (0 heads)

We see that the number of heads obtained follows this pattern. Do you see how this pattern is formed? In order to make it easier to see, we'll write down the outcomes as in Table 13.16.

TABLE 13.16

NUMBER OF COINS	NUMBER OF HEADS							DENOMINATOR
	0	1	2	3	4	5	6	
1	1	1						$2^1 = 2$
2	1	2	1					$2^2 = 4$
3	1	3	3	1				$2^3 = 8$
4	1	4	6	4	1			$2^4 = 16$
5	1	5	10	10	5	1		$2^5 = 32$
6	1	6	15	20	15	6	1	$2^6 = 64$

Notice that the numbers in the above table form a triangular-shaped pattern. This pattern is called *Pascal's Triangle*, named in honor of French mathematician Blaise Pascal (1623–1662). Each row of the triangle is formed by a single addition procedure from the preceding row. For example, in Row 4 we have

$$1 \quad 4 \quad 6 \quad 4 \quad 1$$

These numbers can be obtained from Row 3 as follows:

Row 3   1   +   3   +   3   +   1

Row 4   1   4   6   4   1

Check Row 5 to see how it is obtained from Row 4.

Pascal's Triangle can be used to find the theoretical probability of obtaining 0, 1, 2, or 3 heads when three coins are tossed.

We go to Row 3 and find

$$1 \quad 3 \quad 3 \quad 1$$

Now, we see from the right column (of Table 13.16) that three coins can fall in $2^3 = 8$ equally likely ways. So, 8 is the denominator for each probability and we have

$$p(0 \text{ heads}) = \frac{1}{8}$$

$$p(1 \text{ head}) = \frac{3}{8}$$

$$p(2 \text{ heads}) = \frac{3}{8}$$

$$p(3 \text{ heads}) = \frac{1}{8}$$

We also note that these fractions

$$\frac{1}{8} \quad \frac{3}{8} \quad \frac{3}{8} \quad \frac{1}{8}$$

can be obtained by multiplying the binomial

$$\frac{1}{2} + \frac{1}{2}$$

by itself three times. That is,

$$\left(\frac{1}{2} + \frac{1}{2}\right)^3 = \left(\frac{1}{2} + \frac{1}{2}\right)\left(\frac{1}{2} + \frac{1}{2}\right)\left(\frac{1}{2} + \frac{1}{2}\right)$$

$$= \left(\frac{1}{4} + \frac{2}{4} + \frac{1}{4}\right)\left(\frac{1}{2} + \frac{1}{2}\right)$$

$$= \left(\frac{1}{8} + \frac{3}{8} + \frac{3}{8} + \frac{1}{8}\right)$$

$$= 1$$

Because of the relationship between these probabilities and a binomial expression such as

$$\left(\frac{1}{2} + \frac{1}{2}\right)$$

probabilistic behavior such as tossing coins or rolling dice is often called *binomial probability*.

EXERCISES

1   Refer to the triangular (Pascal's Triangle, Table 13.16) pattern of numbers. Go through the procedure for obtaining the numbers for Row 6 from those in Row 5. Now use this same procedure to obtain the values that would be in Row 7, using the given values for Row 6.

2   Continue the method used in Exercise 1, above, to write down the numbers in Rows 8, 9, and 10.

3   Using the information from Pascal's Triangle, find these probabilities for outcomes of tossing four coins.

(a) $p(0$ heads$)$ (b) $p(1$ head$)$
(c) $p(2$ heads$)$ (d) $p(3$ heads$)$
(e) $p(4$ heads$)$

4   Toss four coins (preferably with the help of a computer) 50 times. Record the outcomes in this table and find:

(a) $P(0$ heads$)$ (b) $P(1$ head$)$
(c) $P(2$ heads$)$ (d) $P(3$ heads$)$
(e) $P(4$ heads$)$

Number of Heads	Frequency	P
0		
1		
2		
3		
4		
	50	1.00

Compare your estimated probabilities with the theoretical probabilities found in Exercise 3, above.

5   Using information from Pascal's Triangle, find these theoretical probabilities:

(a) 6 coins
    $p(0$ heads$)$; $p(1$ head$)$; $p(2\ H)$; $p(3\ H)$; $p(4\ H)$; $p(5\ H)$; $p(6\ H)$
(b) 10 coins
    $p(0\ H)$; $p(1\ H)$; $p(2\ H)$; $p(3\ H)$; $p(4\ H)$; $p(5\ H)$; $p(6\ H)$; $p(7\ H)$; $p(8\ H)$; $p(9\ H)$; $p(10\ H)$
(c) 5 coins
    $p(0\ H)$; $p(1\ H)$; $p(2\ H)$; $p(3\ H)$; $p(4\ H)$; $p(5\ H)$

6   Toss five coins, preferably with a computer, 100 times. Complete this table. Compare the estimated probabilities with the theoretical probabilities in Exercise 5 (c), above.

Number of Heads	Frequency	Probability
0		
1		
2		
3		
4		
5		
	100	1.00

SPECIAL INTEREST
FEATURE

*Statistics and
Social Security* [11]

Thomas Paine, the political thinker, wrote *The Rights of Man* in the late 1700s. In this important work, Paine presents arguments for universal education, government financial help for the aged, and other such national welfare activities. In particular, Paine requires an estimate of the number of elderly people in order that he can have a rough idea as to the cost of providing them with financial aid. Here is a quotation from his work:

> "To form some judgment of the number of those above fifty years of age, I have several times counted the persons I met in the streets of London, men, women, and children, and have generally found that the average is about one in sixteen or seventeen [who are older than fifty]. If it be said that aged persons do not come much in the streets, so neither do infants; and a great proportion of grown children are in schools and in workshops as apprentices. Taking, then, sixteen for a divisor, the whole number of persons in England of fifty years and upwards, of both sexes, rich and poor, will be four hundred and twenty thousand."

Paine has made several errors in his estimate. How many can you identify?

Postscript: There was no real census in England until 1901, but the figure of about 9,000,000 is on record as the population for England and Wales. No information about ages is available. (It is not clear from Paine's work whether Wales or Scotland are included in his estimate.) Since Paine gives 420,000 as the number of persons 50 years or older, and states that this is one-sixteenth of the total population, we have

$$\frac{\text{Total population}}{16} = 420,000$$

So, total population = 6,720,000, according to Paine.

13.4   TWO-TAILED TESTS

Let's begin by reviewing the problem about extra-sensory perception (ESP) that we discussed in Section 5.4.

■ **Example 1:**   Rob claims he has extra-sensory perception (ESP). So he doesn't have to study for tests! He just concentrates and the answers flash before him. Mary is doubtful and does an experiment. She gives Rob a 10-item true-false test on classical Greek, written in Greek. Rob takes the test, concentrating on each item, and gets 8 out of 10 of the questions correct.

What do you think of Rob's claim? Do you believe he has ESP?

Solution:

We solved this problem, following our five steps, and tossing coins.

Step 1—*Choose a model*:   We assume Rob is just guessing, that he does not have ESP. So we use a fair coin:

    Heads: Correct
    Tails: Wrong

Step 2—*Trial*:   A trial consists of tossing the coin ten times, once for each item on the test.

Step 3—*Outcome of the trial*:   We record the number of heads (number of correct answers).

Step 4—*Repeat the trial*:   We do 100 trials and get the results shown in Table 13.17.

TABLE 13.17

OUTCOME (HEADS)	TALLY	f	
0		0	
1		0	
2		2	
3		14	
4		17	
5		32	
6		14	
7		15	
8		5	} 8 or
9		0	} more
10		1	} correct
		100	

Step 5—*Probability of obtained statistic (or greater)*:   We assume Rob is guessing. So we ask how likely is it that he would get 8 *or more* questions correct by guessing. From the data of Table 13.17, we found

$$P(8 \text{ or more correct}) = .06$$

Getting 8 or more out of 10 correct by guessing is unusual. It happens less than .10 of the time. So we are doubtful that he is just guessing.

But let's notice something else about the results in Table 13.17. There is another kind of evidence that a person is not guessing. It is that

$$P(2 \text{ or less correct}) = .02$$

That is, just by chance, in the 100 trials, very low scores could happen as well. It could be that some effect (you might call it extra-sensory deception!) was at work, causing Rob to do much worse on the test than he would by chance.

Therefore, it is important to recognize that for this problem, evidence of non-chance behavior could be either a *very large* (close to 10) number of correct answers or a *very small* (close to 0) number of correct answers. Either outcome of such an experiment might be evidence that could cast doubt on the claim, or the hypotheses, which we are testing.

Figure 13.3 illustrates our discussion. It is a relative frequency graph of the data of Table 13.17.

FIGURE 13.3

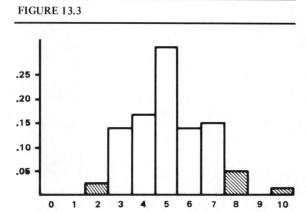

The shaded parts of the graph on the right represent the probabilities

$$P(8 \text{ or more correct}) = .06$$

The bar on the left represents

$$P(2 \text{ or less correct}) = .02$$

So we have the probability

$P(\text{number correct is greater than 8 or less than 2})$ is

$$.06 + .02 = .08 \blacksquare$$

Let's explore this idea of a two-way probability (greater than, or less than certain values) to hypothesis testing.

■ **Example 2:**   ABC Supermarkets believes that the average amount spent by shoppers on a bag of groceries in their stores is $22.00. A random sample of 25 bags of groceries is found to have a mean cost of $21.00 and a standard deviation of $2.50. Do these results cast doubt on ABC's claim?

Solution:

We follow our six steps of Chapter Twelve.

Step 1—*Choose a model:* We assume the claim is true. We are drawing samples from a population whose mean is $22.00 and whose estimated standard deviation is $2.50. Since we have a large sample (greater than 20) our estimate of the standard deviation is reliable.

Step 2—*Trial:* A trial consists of drawing a random sample of 25 "shopping bags" from the population.

Step 3—*Outcome of the trial:* We find the mean price of the sample of 25 "shopping bags."

Step 4—*Repeat the trial:* We do not actually draw samples, so we will not actually repeat the trials a large number of times. But since we have a large sample, we use the Central Limit Theorem and draw a graph of the expected shape of the distribution of the mean of hundreds of samples of size 25 drawn from a population with a mean of $22.00 and a standard deviation of $2.50 (Figure 13.4).

FIGURE 13.4

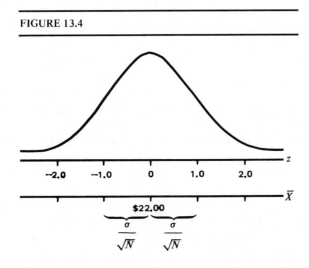

Step 5—*Find* p*(obtained statistic, or less, or greater):* Let's look back at the problem we are to solve. We are to decide if ABC's claim is acceptable. But what evidence from our sample of 25 shopping bags would cast doubt on their claim before we actually take our survey of shopping bags and find their mean price?

You see, we should decide ahead of time using the statement of the problem, what probability we are to find at this step. From a logical point of view, the mean price of the sample of shopping bags could be greater or less than the claimed (hypothesized) price of $22.00. If, from the terms of the problem, both outcomes (less than the hypothesized mean and greater than the hypothesized mean) are of interest, we should use this information to estimate the probability of the obtained sample statistic. Look at Figure 13.6.

With our sample information, we estimate the theoretical standard error as (see Figure 13.5):

$$\frac{\sigma}{\sqrt{N}} = \frac{2.50}{\sqrt{25}} = \frac{2.50}{5} = .50$$

FIGURE 13.6

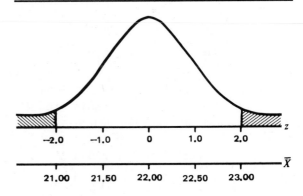

FIGURE 13.5

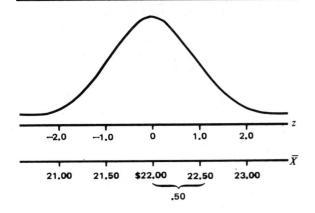

Since we are to take into account the fact that a sample mean at either end of the distribution shown in Figure 13.6 (that is, at either tail of the distribution) would cast doubt on ABC's claim, we must split up the area under the curve into two equal parts.

Now we look at our sample mean. It is, we are told, $21.00. So, we have

$$z = \frac{\overline{X} - \mu}{\sigma/\sqrt{N}} = \frac{21.00 - 22.00}{.50} = \frac{-1.00}{.50} = -2.00$$

We find from our table of normal curve areas (Appendix C) that

$$p(z < -2.0) = .02$$

But since getting a sample mean larger than the hypothesized mean would also cast doubt on ABC's claim, we must find

$$p(z > 2.0) = .02$$

Step 6—*Make a decision:* The probability of getting a sample mean which gives a z smaller than -2.0 (or larger than 2.0) is .04. This is a rare event. Its probability of occurrence is less than .10. So we have doubts that the average amount spent by shoppers is $22.00 per bag of groceries. ∎

We can use the same approach for small samples.

∎ **Example 3:**   A sample of 16 measurements of the amount of manganese in an alloy has a mean of 81.3 and a standard deviation of 2.4. Test the hypothesis that the true amount of manganese in the alloy is 80.0. In this problem, a normal distribution is reasonably assumed.

Solution:

Step 1—*Choose a model:* We assume that the amount of manganese in the deposit is 80. So we are to sample from a normal population with a mean of 80.

Step 2—*Trial:* A trial consists of drawing a sample of size 16 from a normal population with a mean of 80.

Step 3—*Outcome of the trial:* We find the mean and standard deviation of the 16 measurements and calculate the value of Student's t:

$$t = \frac{\overline{X} - \mu}{S/\sqrt{N}}$$

Step 4—*Repeat the trials:* We do not repeat the trials. But we know that if we did, we expect the distribution of Student's t, with 15 degrees of freedom (since sample size is 16).

Before we do Step 5, we look again at our problem. What sample values would cast doubt on the claim that the true amount of manganese in the alloy is 80%? Either large amounts, or small amounts. So we want to find probabilities of t's at both ends (both tails) of the t distribution.

Step 5—*Find* p*(obtained* t *in both tails of distribution):* Our sample gives us

$$t = \frac{\overline{X} - \mu}{S/\sqrt{16}} = \frac{81.3 - 80.0}{2.4/\sqrt{16}} = \frac{1.3}{.6} = 2.17$$

We find from Appendix E that for 15 degrees of freedom,

$$p(t > 2.17) \doteq .025$$

So, we have

$$p(t > 2.17 \text{ or } t < -2.17) \doteq .025 + .025 = .05$$

Step 6—*Make a decision*:   The probability of the obtained value of t (that is, .05) is less than .10. This is a rare event. We doubt that the true amount of manganese in the alloy is 80%. ∎

As a final example, consider the following.

■ **Example 4:**   A public health research group believes that a certain type of carpeting in hospital rooms may contribute to the level of airborne bacteria in the room. A sample of 25 carpeted rooms is tested by pumping air from the rooms over a substance that grows bacteria and counting the number of bacteria colonies that form. The sample results were:

Mean number of bacteria colonies = 10.3
Standard deviation =   5.4

Test the hypothesis that this bacteria count is not significantly greater than the standard level of 9.1.

Solution:

Since this is a large sample problem (size greater than 20), we use the Central Limit Theorem.

Step 1—*Model*:   We assume the claim is true. We sample from a population with a mean of 9.1 and an estimated standard deviation of 5.4.

Step 2—*Trial*:   A trial consists of taking a sample of size 25 from the population.

Step 3—*Outcome of the trial*:   We find the mean of the sample and record it.

Step 4—*Repeat the trials*:   On repeated trials, the Central Limit Theorem tells us that the distribution of sample means will look more and more like a normal distribution (Figure 13.7) with

$$\text{Mean} = 9.1$$

$$\text{Standard error} = \frac{5.4}{\sqrt{25}} = 1.08$$

FIGURE 13.7

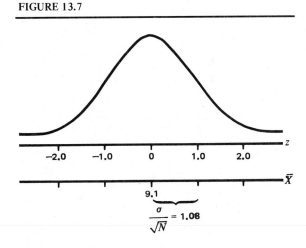

Before we do Step 5, we must determine what probability we are to find. According to the problem, the carpeted rooms are expected to yield higher bacteria counts. So, sample means larger than the claimed (population) level of 9.1 are those which could cast doubt on the claim. Those smaller, indicating lower than standard levels of bacteria in carpeted rooms would be welcome, but would not cast doubt on the stated claim of the problem. So this is a one-tailed, or one-sided, probability problem.

Step 5—*Find p(obtained statistic, or greater):*
Our obtained value of z is

$$\frac{\overline{X} - \mu}{\dfrac{\sigma}{\sqrt{N}}} = \frac{10.3 - 9.1}{\dfrac{5.4}{\sqrt{25}}} = \frac{1.2}{\dfrac{5.4}{5}} = \frac{1.2}{1.08} = 1.11$$

See Figure 13.8.

FIGURE 13.8

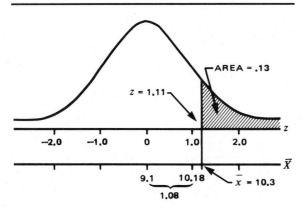

We find from the table of areas under the normal curve (Appendix C), that

$$p(z > 1.11) = .13$$

Step 6—*Make a decision:*   Getting a sample mean as large, or larger, than 10.3 from the assumed population is *not* unusual (.13 is greater than .10). So we conclude there is insufficient evidence that the amount of airborne bacteria in the carpeted rooms is greater than the standard level. ∎

EXERCISES

For each exercise, you should first decide whether it is a large sample or small sample problem. If it is a small sample problem, assume that it is known that the population from which the sample was drawn is normal. Then you must determine whether a one-tailed or two-tailed probability is to be found.

1   A sample of 9 ninth graders has a mean weight of 110 pounds with a standard deviation of 5 pounds. Do a test to help decide whether the sample differs significantly from the assumed mean weight for the total class of 114 pounds.

2   A savings and loan association claims that the average size of its savings account is $150. A random sample of 49 accounts has a mean size of $145 and a standard deviation of $25.00. Make a decision as to whether the association's claim is true.

3   A sample of 16 measurements of the percentage of manganese in a mineral deposit has a mean of 89.1 and a standard deviation of 3.2. Test the claim that the true amount of manganese in the deposit is *only* 85.

4   A school district is interested in whether student achievement in mathematics is below the nationwide mean of 85. A random sample of 49 students takes the test and gets a mean of 83 and a standard deviation of 7.3. Do a test to help decide whether student achievement is significantly below the nationwide mean.

5   A supermarket chain claims that a family of four spends an average of $95.00 per week on groceries. A consumer group surveys 25 families and finds these amounts reported spent on groceries for a week:

$ 90	$120	$114	$105	$100
$110	$102	$ 82	$130	$109
$101	$102	$100	$132	$125
$120	$ 90	$ 95	$ 87	$106
$105	$108	$110	$ 74	$132

Should the supermarket's claim be accepted?

6   A random sample of 100 fraternity men on a college campus has a scholastic average of 3.7 with a standard deviation of .25. The campus-wide average of all students is 3.6. Do the fraternity men have a significantly *higher* scholastic average than the campus in general?

7   A statistics teacher has developed a new way of teaching how to solve statistics problems by tossing coins. She wants to try it out with her mathematics students. So she uses the method on a class of 16 students. Then she gives them an exam on statistics. Her students usually get a (average) score of 58 on this examination. Her class of 16 students gets a mean score of 64 with a standard deviation of 6.6. Did the class do significantly better on the exam?

8   The length-to-width ratios of humerus bones from the same species of animal tend to be approximately the same. This ratio is used by archeologists to help identify species whose fossils of humerus bones are discovered. A certain species is known to have a mean length-to-width ratio of 8.5. Suppose that a sample of 49 fossils of humerus bones was unearthed at an archeological dig in East Africa where this species is believed to have been located. (Assume that all 49 bones are from the same species.) The statistics from this sample are:

Mean ratio = 9.2
Standard deviation = 1.12

Do a statistical test to determine whether the sample mean is significantly different from the mean ratio of 8.5 for the known species.

9   In a certain campus newspaper, the mean number of letters to the editor for the past year was 3.2 per issue. A campaign is carried out to make students more aware of the newspaper and encourage them to write letters to the editor. Following the campaign, a random sample of 9 issues of the newspaper is found to have a mean of 4.3 letters to the editor with a standard deviation of 1.2. Has the number of letters to the editor significantly increased?

13.5   RANDOMNESS

Whatever source you use for random digits—rolls of a die, numbers from a telephone book, a computer program, and so on—you may want to know the goodness of that source of random data. For example, if you produced 1,000 such digits, you would expect them to have two properties:

1.   The frequency of each of the 10 digits would be approximately equal. That is, you would expect about 100 zeroes, 100 ones, and so on.
2.   There would be no pattern in the appearance of the digits. That is, you would have no procedure for correctly predicting the next digit by using information about the preceding digits (the successive digits are independent of each other).

The chi-square test can be used to help determine whether a list of digits is random.

EQUAL FREQUENCIES

Let's say we wish to determine whether a list of 100 digits provided by a telephone directory is random. The list in Table 13.18 was produced by taking the last 2 digits of 50 successive telephone numbers taken from a page opened at random.

TABLE 13.18

DIGIT	E	O	(O − E)	(O − E)²
0	10	11	1	1
1	10	10	0	0
2	10	6	−4	16
3	10	15	5	25
4	10	8	−2	4
5	10	8	−2	4
6	10	10	0	0
7	10	5	−5	25
8	10	17	7	49
9	10	10	0	0
	100	100		124

Recall from Chapter Six that

$$\chi^2 = \text{sum of } \frac{(O-E)^2}{E}$$

where

O = obtained number of outcomes
E = expected number of outcomes

therefore

$$\chi^2 = \frac{124}{10} = 12.4$$

Since Table 13.18 has $10-1=9$ degrees of freedom, we have (from Appendix A) that

$$p(\chi^2_9 > 12.4) = .24$$

This is not an unusual event, so we have no evidence of a lack of randomness in the data. That is, the departures of digits from their expected value of 10 are those which take place fairly often by chance. (Notice from Table 13.18, for example, that seventeen 8s were produced by the telephone numbers and only six 2s were obtained.)

PRESENCE OF PATTERNS

The frequency test seems a reasonable one. But a little reflection will show that it does not tell the whole story. Consider the following sequence of digits:

0, 1, 2, 3, 4, 5, 6, 7, 8, 9, 0, 1, 2, 3, 4, ...

While it is clear that this sequence is far from random, it still will pass the frequency test with flying colors! In fact, each digit occurs exactly as often as would be predicted, and the associated chi-square would be zero.

The reason that the sequence of digits is not random, of course, is that each successive digit appears according to a pattern. A 1 always follows a 0, a 2 always follows a 1, and so on.

Although there are tests for patterns in data, we will not consider them in this book. Informally, you could try looking for such patterns. For example, you could pick any digit, such as 4. Then record the digit that immediately follows 4. Does there appear to be a digit that is more likely than any other to follow 4? Each digit following a 4 should appear about one-tenth of the time. And of course, there is nothing special about choosing 4.

PSEUDO-RANDOM DIGITS

Computers (and some calculators) can produce random digits rapidly and in very large quantities. But it would not be practical to require such devices to store great long lists of random data. It would take too much space, and leave too little space for purposes of calculating!

Therefore, formulas have been invented to compute random digits. At first, you might think this is impossible, because if you *compute* a number by a formula then you will *know* what the next digit will be and they can not be random. But to make the prediction requires knowing the formula. Knowing just the digits is not enough to make the prediction.

Digits produced by such a formula are called *pseudo-random*. They are *pseudo*-random because they are not *strictly* random. For example, they may repeat themselves after a very long string of digits has been produced. But for most purposes, such pseudo-random numbers usually work extremely well in that they appear to be random and are very convenient to use.

Some such formulas are very complicated, others are surprisingly simple. You could find out from a computer manual, or a consultant, the formula that a particular machine uses for generating its random numbers.

There is one particular easy method for producing pseudo-random numbers that you can use on a calculator.

MID-SQUARE METHOD
[suggested by the mathematician
John Von Neuman (1903⁻1957)]

Take, for example, an arbitrary four digit number (you could use the last four digits of your telephone number of social security number, for example).

Suppose we start with 63537. (This is *not* a part of your list of digits—it is only the starter.) We square this number:

$$(63537)^2 = 403(69503)69$$

We now take the middle five digits, 69503, and square:

$$(69503)^2 = 48(30667)009$$

Again, we take the middle five digits and square:

$$(30667)^2 = 940464889$$

We repeat this process as long as desired. When we write down the string of digits, for our example, we get:

69503   30667   04648

This string of digits may be regarded as a set of random digits. But the advantage is that it has been mechanically produced. In this case, we used an ordinary, inexpensive calculator to do the job.

EXERCISES

1   Table 13.19 shows 400 digits obtained from the last four digits of 100 telephone numbers from a telephone directory. Use the χ^2 test to check for equal frequencies of the digits.

TABLE 13.19

4917	6272	1009	2829
4567	6250	7753	4326
6064	0607	0098	5006
6602	6190	0899	9540
4398	4972	0361	8172
6861	0269	4317	3095
0169	4885	2750	5201
1921	2914	5642	5962
8032	2846	3846	2406
0559	6152	3465	4750
6285	2725	6872	4085
5017	0925	4596	4303
5600	5516	2918	3095
5033	4011	4745	2750
9187	3943	4685	6895
9611	8549	6329	7631
8701	6698	4504	0378
8892	5545	0674	6557
5261	5330	1938	0988
8321	2222	0615	4875
1311	2033	6106	2223
2319	4365	1439	2979
3449	5771	5614	9878
1982	3158	6693	9427
8229	6857	6300	9572

2   Explore the telephone directory data in Table 13.19 for patterns. Pick a digit, such as 1. Begin reading (for example) across the table. Record the digits that immediately follow 1. For instance, the first 5 of these digits are

$$7 \quad 0 \quad 9 \quad 8 \quad 7$$

(obtained from reading the first 5 rows of the table). Does the 1 appear to affect the frequency of occurrence of each digit?

3   Here are the results of 300 rolls of a die from a party game:

Row			
1	4 5 2 2 3	3 2 2 6 6	5 6 6 2 4
2	2 6 4 1 1	2 1 3 3 6	3 4 3 6 4
3	1 4 1 4 2	6 3 3 3 3	6 3 2 5 1
4	2 5 6 2 5	2 5 5 5 6	3 6 3 2 3
5	2 4 3 3 5	6 2 3 3 2	2 2 3 4 4
6	4 5 3 5 1	5 1 6 3 1	6 5 6 1 6
7	3 3 3 4 3	1 4 6 1 2	3 6 6 1 2
8	4 4 1 6 4	4 6 5 5 5	1 2 1 2 4
9	5 5 3 3 2	5 6 4 5 5	3 1 3 3 3
10	4 5 2 2 2	4 5 6 4 5	3 3 1 1 6
11	3 5 3 5 4	3 6 5 1 3	6 2 3 2 1
12	2 1 3 6 5	3 6 6 1 3	4 3 2 6 4
13	3 5 4 5 3	3 1 3 6 1	5 3 3 5 3
14	6 6 3 5 3	3 4 6 3 2	6 2 5 2 4
15	6 4 5 4 4	4 6 3 2 5	6 3 6 3 4
16	5 4 6 5 1	3 3 1 5 5	1 1 6 2 6
17	6 4 4 2 5	1 3 4 5 4	1 1 4 3 6
18	2 2 1 4 2	5 1 4 4 6	6 4 5 1 2
19	5 6 4 2 6	3 1 3 6 3	5 2 3 5 5
20	2 3 3 6 4	3 2 6 5 4	2 5 4 2 6

(a)   Do the χ^2 test to help you decide if the die is fair.

(b)   Explore the rolls of the die from the above table for patterns of appearance of digits.

4   Below are 400 digits generated by the mid-square method described in this section. Use the χ^2 test to help you decide how well the mid-square method works as a random number generator.

1234	5227	3215	3362	3030
1809	2724	4201	6484	4220
1780	1684	8358	8561	2907
4506	3040	2416	8370	5690
3237	4781	8579	5992	9040
5678	2396	7408	8784	1586
5153	5534	6251	7500	5625
6406	3680	1354	8333	4388
2545	4770	7529	6858	3210
1030	6090	3708	7492	1300
9876	5353	6546	8501	2670
1289	6615	7582	4867	6875
2793	8008	1280	6384	7554
6290	3956	6499	2370	6169
5650	3192	1888	5645	8660
6543	8108	7396	7008	1120
2544	4719	2689	2307	3222
3812	5313	2279	1938	7558
1233	5202	6080	3696	6604
6128	5523	5035	3512	3341

5   With the help of classmates, ask 100 persons to name a two-digit number between 00 and 99. Use the χ^2 test to help you decide if persons are naming numbers at random or tend to prefer some numbers over others. *Note:* Here, 07 is a two-digit number!

6   Generate 300 digits by using page numbers from a book. Take a book with many pages, at least 300. Turn to a page at random, perhaps by opening it without looking at where you place your finger in the book. Decide whether to read the page number for the left or right page by tossing a coin. (Why not just take the number on the right-hand page, for instance?) If the page number is less than 100, add leading zeros. For example, if you turn to page 8, the random number is 008. If you turn to page 39, the random number is 039. Use the χ^2 test to help you decide if this is a good way to produce random digits.

7   Suppose you decided to use the following method to generate random numbers. You are to use words from the daily newspaper and assign numbers to letters in some systematic way, such as: $A = 01$, $B = 02$, and so on, up to $Z = 26$. Then go to the editorial page and write down the numbers that correspond to letters appearing in an editorial. What do you think the equal-frequencies method would show as a test for this procedure? Try it as a class project.

Monte Carlo Methods

CHAPTER 14

KEY PROBLEM

How to See Through a Rock

Geologists study the materials from which rocks are made. Often, very complicated methods are needed to tell what minerals are in a rock. For example, when the astronauts returned from their moon walk, there was great interest in seeing what minerals were found in the pieces of moon rock they brought back.

Statistical methods can help scientists "look through" rocks and tell what minerals they contain. We will learn about such a method in this chapter.

14.1 WELCOME TO MONTE CARLO: COMPUTER STYLE

The Monte Carlo method is a way of solving problems by tossing coins, rolling dice, or using other ways of getting random outcomes. The name Monte Carlo was given to the method by the famous mathematician S. Ulam in the 1940s. Ulam used the method to attack problems he couldn't solve using ordinary mathematical methods. Monte Carlo methods have been used throughout this book.

In this chapter we will see how, with the help of a computer, the Monte Carlo method can be used to solve complicated problems.

When doing Monte Carlo methods, we need random outcomes. In this book, we have used coins, dice, tables of random digits, and other sources to produce these outcomes.

When we use a computer to do Monte Carlo methods, we depend upon its random number generator to give us random outcomes.

RANDOM NUMBER GENERATORS

Most computers have a function that generates random numbers. In BASIC, the function RND is designed to produce numbers randomly, with equal probability, in the open interval from zero to one. (That is, all numbers in the interval, like .334 and .5 are candidates to be chosen at random. But the endpoints 0 and 1.0 are not eligible to be picked. See Figure 14.1.)

FIGURE 14.1

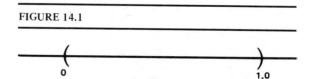

RND is usually operated by using an integer as the argument [for example, RND(1)]. Thus the statement

30 LET A=RND(1)

assigns to A a random number between 0 and 1. *Note:* There is nothing special about using a "1" in RND(1). Any digit 1 through 9 will work. On some computers, a zero or a negative value have special meaning.

PROGRAM

```
30  LET A=RND(1)
40  PRINT A;
50  GO TO 30
60  END
```

OUTPUT

```
.985444    .516529    .440934    .794276
.314788    .750572    .282484    .268628
.124840    .882872    .788047    9.06842E-3
.238821    .821371    .517585    .24723
```

Note: E−2 means multiply by 10^{-2}. So,

$$6.78473E-2 = 6.78473 \times .01 = .0678473$$

Suppose you want a random number between 0 and 2. We need to stretch the interval by two. Use

50 LET A=2*RND(1)

PROGRAM

```
50  LET A=2*RND(1)
60  PRINT A;
70  GO TO 50
80  END
```

OUTPUT

1.97089	1.03306	.881867	1.58855
.629576	1.50114	.564968	.537255
.249681	1.76574	1.57609	1.81368E-2
.477642	1.64274	1.03517	.49446

Caution: Between means just that. Instruction 50, above, will never produce the number 2, although you might get 1.999, for example. Similarly, it will never give 0. But it might give 0.00001, for example.

Then suppose you want only integers. Look at this instruction:

60 LET A=INT(2*RND(1))

The command INT cuts off the fractional part of a number. For example, INT (35.2) gives 35. Therefore, statement 60 in the program below will give the digits 0 and 1 in random order, each with equal probability.

PROGRAM

```
60  LET A=INT(2*RND(1))
70  PRINT A;
80  GO TO 60
90  END
```

OUTPUT

```
0 0 1 0 0 0 1 1 0 0 1 0 0 0 1 1 0 0 1 1
1 0 1 1 0 1 0 0 0 1 1 1 1 0 0 1 0 1 1 1
1 1 1 1 0 1 1 0 0 0 0 1 0 1 0 1 0 1 1
1 1 0 1 0 0 0 1 0 0 0 1 0 1 0 1 1 1 0
0 1 1 0 1 1 1 1 0 0 0 0 1 0 1 1 0 0 0 0
```

Now, suppose we want the random digits 2, 3, and 4. We stretch RND by 3 (to give digits 0, 1, 2) and add 2 (to give digits 2, 3, 4).

60 LET A=INT(3*RND(1)) + 2

PROGRAM

```
60  LET A=INT(3*RND(1))+2
70  PRINT A;
80  GO TO 60
90  END
```

OUTPUT

```
2 3 4 2 4 4 4 3 4 3 3 4 2 4 2 2 2 4 4 2
2 4 3 2 2 4 4 3 3 3 3 4 4 4 3 4 3 3 2 4
2 2 2 3 2 3 3 2 4 3 3 3 2 3 4 4 3 3 3 4
2 4 4 4 3 2 2 3 2 4 2 4 3 3 4
```

Let's try to get a general rule from what we have done here. We want the whole numbers in the interval from 2 to 4, inclusive. This interval contains the whole numbers: 2, 3, and 4. The number of whole numbers is 3. So we stretch the interval $(0,1)$ by 3. Since the interval we seek starts at 2, we add 2 to the result of the stretch. (We do a slide 2 units to the right.)

In general, random integers in the interval [a, b] inclusive, are produced by

70 LET X=INT((b−a+1)*RND(1)+a)

where *a*, *b* are integers.

PROGRAM

```
20    PRINT "LEFT ENDPOINT OF INTERVAL";
30    INPUT A
40    PRINT "RIGHT ENDPOINT OF INTERVAL";
50    INPUT B
70    LET X=INT((B−A+1)*RND(1)+A)
80    PRINT X;
90    GO TO 70
100   END
```

In the output, an underline shows what you reply to the computer.

OUTPUT

```
LEFT ENDPOINT OF INTERVAL ? 5
RIGHT ENDPOINT OF INTERVAL ? 9
5 6 9 5 9 9 8 7 9 7 7 8 6 8 6 6 5 9 8 5
6 9 7 6 6 9 9 7 7 7 6 9 8 9 6 9 6 6 5 8
5 5 5 7 5 6 7 5 9 7 8 8 6 7 8 8 7 7 8 9
5 9 9 9 6 6 5 7 6 9 6 9 7 8 9
```

TOSSING ONE COIN

The computer is expert at "tossing coins." If we let 1 stand for heads and 2 stand for tails, then here is a coin tosser.

PROGRAM

```
100   PRINT INT(2*RND(1))+1;
150   GOTO 10
200   END
```

This is about as simple a program as you can imagine. Here is some sample output.

OUTPUT

```
1 1 2 1 2 2 2 1 2 2 1 2 1 2 1 1 1 2 2 1
1 2 2 1 1 2 2 1 1 2 1 2 1 2 2 2 1 2 1 1 1 2
1 1 1 1 1 2 1 2 2
```

The only problem with this program (as you have already learned) is that it cannot stop. So you will need to use a stop key, or, better yet, a FOR/NEXT loop. This program will give you 25 tosses of a coin.

PROGRAM

```
90    FOR I=1 TO 25
100   PRINT INT(2*RND(1))+1;
150   NEXT I
200   END
```

OUTPUT

```
1 1 2 1 2 2 2 1 2 2 1 2 1 2 1 2 1 1 1 2 2 1
1 2 2 1 1
```

If you want a nice output, you can have the computer print H for heads and T for tails. We use what is called an array.

PROGRAM

```
5     LET A$(1)="H"
10    LET A$(2)="T"
90    FOR I=1 TO 25
100   LET J=INT(2*RND(1)+1)
120   PRINT A$(J);
150   NEXT I
200   END
```

OUTPUT

```
HHTHTTTHTTHTHTHTHHHTTHHTTHH
```

Suppose, instead, we want to count up the number of times we get heads and tails. Arrays are very useful for counting. (See line 120 below.) The resulting frequency table is printed out by lines 160–180.

PROGRAM

```
90    FOR I=1 TO 100
100   LET J=INT(2*RND(1)+1)
120   LET H(J)=H(J)+1
150   NEXT I
160   FOR K=1 TO 2
170   PRINT K,H(K)
180   NEXT K
200   END
```

(*Note:* We add one in statement 100 because zero subscripts in an array are not permitted in some forms of BASIC.)

OUTPUT

```
1              46
2              54
```

TWO COINS

If we toss two coins, what is the probability that both will be heads?

Suppose that in 40 tosses, both coins were heads a total of 12 times. Then our estimate of $P(2 \text{ heads})$ is

$$\frac{12}{40} \text{ or } .3$$

If we get tired of tossing coins, we can enlist the assistance of our infinitely patient helper, the computer.

LET A=INT(RND(1)*2) + 1

gives results of tossing coin 1 and

LET B=INT(RND(1)*2) + 1

gives results of tossing coin 2.

Then $A + B - 2$ will provide the total number of heads obtained when the two coins are tossed.

Using arrays to store the results of the tosses gives a very efficient procedure.

LET H(A+B) = H(A+B) + 1

will accumulate the number of times 0, 1, and 2 heads were obtained.

PROGRAM

```
50    FOR I=1 TO 100
100   LET A=INT(2*RND(1)+1)
110   LET B=INT(2*RND(1)+1)
120   LET H(A+B)=H(A+B)+1
130   NEXT I
140   FOR I=0 TO 2
150   PRINT I,H(I+2)
160   NEXT I
200   END
```

OUTPUT

```
0              21
1              49
2              30
```

If you want a fancier program, you can add some labels and further computation, as follows.

PROGRAM

```
10    REM PROGRAM TO FIND ESTIMATED PROBABILITY OF
15    REM GETTING NUMBERS OF HEADS
20    REM WHEN TOSSING C COINS
30    PRINT "HOW MANY COINS";
40    INPUT C
50    PRINT "HOW MANY TRIALS";
60    INPUT T
70    FOR I=1 TO T
80    LET H=0
90    FOR J=1 TO C
100   LET R=INT(2*RND(1))
110   LET H=H+R
120   NEXT J
130   LET X(H+1)=X(H+1)+1
140   NEXT I
150   REM PRINT OUT FREQUENCY TABLE
160   PRINT "N HEADS","FREQ","PROB"
200   FOR I=0 TO C
210   PRINT I,X(I+1),X(I+1)/T
220   NEXT I
300   END
```

OUTPUT

```
HOW MANY COINS? 3
HOW MANY TRIALS? 100
```

N HEADS	FREQ	PROB
0	16	.16
1	41	.41
2	31	.31
3	12	.12

EXERCISES

1   Use the random number generator in your computer to generate your own list of 500 random digits. (Suggested format: The rows have ten blocks of five digits in each row. See the programming hint for Exercise 6 of this section.)

2   Compare these BASIC instructions and explain how each works.

$$\text{LET X=INT(RND(1)+.5)}$$
$$\text{LET Y=INT(2*RND(1))}$$

3   Write BASIC instructions that will provide random digits in the (closed) intervals indicated:

(a) from 0 to 2   (b) from 1 to 2
(c) from 0 to 9   (d) from 10 to 20
(e) from −1 to +1   (f) from 0 to 100

4   Write a computer program to obtain an estimate for $P(1 \text{ head})$ on the toss of one coin. Report results for 50, 100, and 500 tosses. How do your estimates compare? Are your results identical to those obtained by classmates' programs? Why or why not?

5   Write a computer program to obtain an estimate for $P(2 \text{ heads})$ when two coins are tosses. Obtain results for 50, 100, and 500 tosses of two coins.

6   Modify your program from Exercise 4 of this section so that it gives $P(2 \text{ heads})$, $P(1 \text{ head})$, and $P(0 \text{ heads})$. Obtain estimates

based on 50, 100, and 500 tosses of two coins. From probability theory, the theoretical values are:

$$p(2 \text{ heads}) = .25$$
$$p(1 \text{ head}) = .5$$
$$p(0 \text{ heads}) = .25$$

Programming Hint: Note the effect of the following instruction:

IF I=50*INT(I/50) GOTO ____

For what values of I is this statement true? Try values of 1, 50, 75, and 100. Have you found that for multiples of 50, the equality is true? This instruction is very useful for printing out intermediate values in a FOR/NEXT loop. See what happens here:

PROGRAM

```
10   FOR I=1 TO 500
30   IF I<>50*INT(I/50) GOTO 50
40   PRINT I
50   NEXT I
60   END
```

7   Accuracy of Results: Let's look at the two coin problem again. Here are estimates for $P(2 \text{ heads})$ based on 100 tosses of two coins:

.31   .24   .21   .16   .27

The average (mean) value for $P(2 \text{ heads})$ for these five estimates is .238, or .24.

We know from our study of statistics that as we do more and more tosses, $P(2 \text{ heads})$ in tossing two fair coins will get closer and closer to the value of .25. Notice that our estimate of .238 based on 500 tosses (which is equivalent to the average of five sets of 100 tosses each) is very close to .25.

In general, the more accurate we wish to make our estimates, the more trials we will have to consider.

Modify your program of Exercise 6 in this section so that it will find $P(3 \text{ heads})$ when three coins are tossed. Obtain estimates for $P(3 \text{ heads})$ based on 100 tosses of three coins. Repeat five times and average your results. Compare your results with the theoretical value of .125 for $p(3 \text{ heads})$ as discussed in Section 13.3.

8   We learned in Chapter Nine that the graph of the proportion of heads obtained when tossing many coins can be smoothed to a normal distribution. Write a program to find the number of heads obtained when 10 coins are tossed. Do 100 trials and draw a relative frequency polygon of the outcomes. Does the graph seem to approximate a normal distribution? Probability theory tells us that the standard deviation of this distribution has the theoretical value of

$$\sqrt{10/4} = \sqrt{2.5} \doteq 1.6$$

Use a normal curve table to find the number of times we would expect to get eight or more heads when 10 coins are tossed 1000 times. Compare this with the actual number of heads obtained in the 1000 tosses. Complete this table for other outcomes as requested. Now do you believe the distribution of the outcomes of 1000 tosses of 10 coins is approximately normal?

Outcome of Toss of 10 Coins	Expected Number	Obtained Number
8 or more heads		
2 or fewer heads		
Between 3 and 7 heads, inclusive		

Note: Expected and obtained numbers are based on 1000 tosses.

9   Write and run a program to find the number of outcomes obtained in 600 rolls of a fair (not loaded) die that has six sides. Your program should print out a table like this:

Outcome	Frequency
1	
2	
3	
4	
5	
6	

Sum of Frequencies =

10   Change your program in Exercise 9, above, so that it will find the number of outcomes obtained in 400 rolls of a fair die that has eight sides, 10 sides, five sides. (*Note:* A five-sided die does not exist in the real world. But the computer does not know this—nor does it care!)

11   Write and run a program to estimate the probability of getting an ace when a card is dealt from a well-shuffled deck of regular playing cards. (A regular deck of cards consists of four suits—hearts, diamonds, clubs, spades—and 13 cards in each suit, for a total of 52 cards.) Do 100 trials.

12   Write and run a program that will deal all cards from a well-shuffled deck. (This is a challenging programming problem!)

14.2   FINDING AREAS: EASY AS PI

You can find a value for pi by throwing darts at random at a target like the one shown in Figure 14.2. We will ask the computer to throw our darts.

FIGURE 14.2

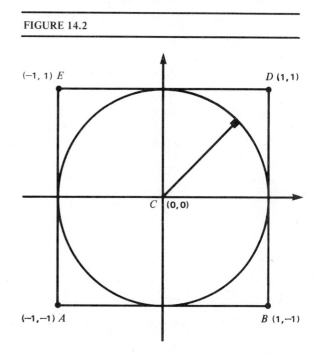

The target is the circular region in Figure 14.2. Choose a point at random in the square figure by obtaining pairs of random digits (*X, Y*) within appropriate limits. In terms of our picture, we have a circle whose center is at (0, 0) and whose radius is 1 inscribed in a square having sides of length 2. So, we need to generate values of *X* and *Y* between −1 and 1 in order to produce random points (*X, Y*). Verify that these instructions will do the job:

```
LET X=2*RND(1)−1
LET Y=2*RND(1)−1
```

How do we determine whether we have "hit the target"? You may recall that a circle is defined as the set of all points such that

$$X^2 + Y^2 = 1$$

since radius = 1.

Therefore, the expression

$$\text{If } X^2 + Y^2 \leqslant 1$$

is true for all points inside or on the circumference of our circle. We then ask the computer to keep track of how many times we hit the circular target. This is our value of $P(\text{hit})$. But

$$p(\text{hit}) = \frac{\text{area of circular target}}{\text{area of square}} = \frac{\pi r^2}{2^2} = \frac{\pi(1^2)}{2^2} = \frac{\pi}{4}$$

Since we have an estimate of $p(\text{hit})$, we solve for π and obtain

$$\pi = p(\text{hit}) \times 4 \doteq P(\text{hit}) \times 4$$

PROGRAM

```
10  FOR I=1 TO 1000
20  LET X=2*RND(1)-1
30  LET Y=2*RND(1)-1
40  IF X*X + Y*Y>1 GOTO 60
50  LET H=H+1
60  NEXT I
70  PRINT "P(HIT)=";H/1000
80  PRINT "ESTIMATED VALUE OF PI =";4*H/1000
90  END
```

OUTPUT

```
P(HIT)=.775
ESTIMATED VALUE OF PI = 3.1
```

AREA OF A CURVED REGION

The target-practice principle can be applied to finding the area of curved regions. This principle is used in the key problem for this chapter.

Consider the problem of determining the area under the graph of

$$f(x) = x^2$$

from $x = 0$ to $x = 1$.

Notice that the desired region is enclosed by the square $ABCD$ having side of length 1 unit, as in Figure 14.3.

FIGURE 14.3

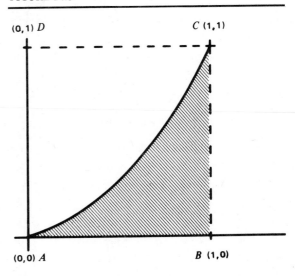

Then for a random point (x,y), we have a hit if

$$y \leqslant x^2$$

Programming Hint: This BASIC statement makes use of the DEF instruction for defining functions. Using the DEF function for $f(x)=x^2$, we might have, for example

$$10 \ \text{DEF FNA(X)=X} * \text{X}$$

When the number of hits has been determined, we have

$$p(\text{hit}) = \frac{\text{number of hits}}{\text{number of tosses}} = \frac{\text{area under } f(x)}{\text{area of } ABCD}$$

Therefore,

$$\text{area under } f(x) = p(\text{hit}) \times \text{area of } ABCD$$

The following program uses "target practice" to estimate the area under $f(x)=x^2$ as in Figure 14.3.

PROGRAM

```
10    DEF FNA(X)=X*X
20    PRINT "NUMBER OF TRIALS";
30    INPUT T
40    PRINT "LEFT END POINT FOR X";
50    INPUT E1
60    PRINT "RIGHT END POINT FOR X";
70    INPUT E2
80    PRINT "MAXIMUM VALUE OF FUNCTION IN INTERVAL";
90    INPUT M
100   FOR I=1 TO T
110   LET X1=(E2-E1)*RND(1)+E1
120   LET Y1=(E2-E1)*RND(1)+E1
130   IF Y1>FNA(X1) GOTO 150
140   LET H=H+1
150   NEXT I
160   PRINT "N HITS=";H;" P(HIT)=";H/T
170   PRINT "ESTIMATED AREA UNDER F=";M*(E2-E1)*H/T
200   END
```

OUTPUT

```
NUMBER OF TRIALS? 500
LEFT ENDPOINT FOR X? 0
RIGHT ENDPOINT FOR X? 1
MAXIMUM VALUE OF FUNCTION IN INTERVAL? 1
N HITS= 163 P(HIT)= .326
ESTIMATED AREA UNDER F= .326
```

BUFFON'S NEEDLE

One of the oldest Monte Carlo experiments was reported to the Paris Academy of Sciences in 1733 by Count Buffon. The experiment involved estimating pi by tossing sticks or needles on a surface ruled with equidistant parallel lines.

■ **Example:**   If a needle of length ℓ units is tossed randomly upon a floor of parallel planks of equal width d (greater than ℓ) units, what is the probability that the needle, once it comes to rest, will cross (or touch) a crack (assumed to be zero width) separating the planks of the floor?

Solution:

It can be shown that the probability that the needle or pin will hit a crack on the floor is

$$p(\text{hit}) = \frac{2\ell}{\pi d}$$

Note that for the special case $d = 2\ell$, when we solve for π we have

$$\pi = \frac{1}{p(\text{hit})}$$

Figure 14.4 shows a needle tossed on a ruled surface. The needle is ℓ units long and the ruled parallel marks are d units apart ($d > \ell$).

FIGURE 14.4

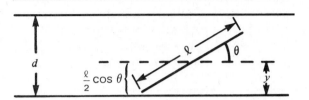

As Figure 14.4 suggests, whether a needle hits one of the parallel rulings is dependent upon (1) the distance of the center of the needle from a line and (2) the angle θ. Indeed, the condition for a hit is that this distance, which we call y, is less than or equal to

$$(\sin \theta) \cdot \frac{\ell}{2}$$

In our program below, θ is given a random value, in radians, in the interval from 0 to π (see statement 50, where X stands for θ). Y is given a random value in the interval from 0 to $d/2$ (see statement 60). ∎

PROGRAM

```
10    LET C=0
20    LET D=2
30    LET L=1
35    REM DO 500 TOSSES
40    FOR I=1 TO 500
50    LET X=RND(1)*3.14159/2
60    LET Y=RND(1)*D/2
70    IF Y>SIN(X)*L/2 GO TO 90
80    LET C=C+1
90    NEXT I
95    LET P=C/500
100   PRINT "P(HIT)=";P
110   PRINT "PI ESTIMATED AS ";(2*L)/(P*D)
120   END
```

OUTPUT

```
P(HIT)=.328
PI ESTIMATED AS 3.048
```

EXERCISES

1   Write BASIC instructions that will provide random decimal numbers in the intervals indicated.

(a)   0 to 2   (b)   −1 to 1
(c)   1 to 4   (d)   0 to 4

2   Using the target-practice method, write a program to calculate pi. Use 1 as the radius of the circle and 2 as the side of the square. Request intermediate estimates for pi every 200 tosses, up to 2000.

3   Modify your program from the preceeding problem to request values for r, the radius of the circle, and d, the side of the square containing the target. (What limitations will you want on r and d?) Experiment with relative sizes of r and d to determine if there are combinations that yield values of pi closest to the actual value of 3.14159265 . . . .

In Exercises 4 through 7, estimate the areas under the graphs for the functions given.

4    $f(x) = x$, from $x = 0$ to $x = 1$
(What should the answer be?)

5    $f(x) = x^3$, from $x = 0$ to $x = 1$

6    $f(x) = \sin(x)$, from $x = 0$ to $x = 2$ radians

7    $f(x) = (5 + 2x)^2$, from $x = 0$ to $x = 2$

In the following two exercises, the target-practice principle is extended to three dimensions. The equation

$$f(x, y) = ax + by$$

defines a surface. Given a point (x_0, y_0) in the plane, $f(x_0, y_0)$ is the distance from the plane of the surface at that point. Use the ideas of this section to write a computer program that will estimate the volume enclosed by the surfaces defined in Exercises 8 and 9. In both exercises, the given function is defined over the square with corners $(0, 0)$, $(0, 1)$, $(1, 0)$ and $(1, 1)$.

8 $f(x, y) = x^2 + y^2$

9 $f(x, y) = \dfrac{1}{1 + x + y}$

10   On Increasing Your Accuracy: Increasing the number of trials increases the accuracy of our results. Write a computer program that will help show this for estimating pi. Your program should give values for a table that shows estimates for pi using the method of this section. Print out the estimated value of pi every 100 trials. Use mean deviation (see Chapter Two) to show how sampling error is affected by increased trials.

11   Key Problem Recap—*How to See Through a Rock:*   The approach of this section can be used to estimate the mineral content of a rock sample, as suggested in the key problem. Suppose a slice of rock had the features illustrated in Figure 14.5.

FIGURE 14.5

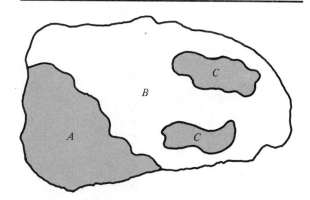

Each of the regions, *A*, *B*, and *C*, represents a different mineral. Suppose that the shape of each region can reasonably be approximated by geometric figures—such as circles or ellipses—whose equations are known. How, by throwing darts at random, could the area of each of the minerals in this slice be estimated? Then, if the thickness of the slice is measured, how is the quantity of mineral estimated?

SPECIAL INTEREST
FEATURE

Randomness

Randomness is an important idea in statistics and probability, as has been demonstrated throughout this book. Randomness is important to other fields, too, such as literature and art.[1] The novelist William Burroughs produced a book by cutting up the pages of a novel and printing them in a random order! In a tall building, random patterns can be seen at dusk as lights in the building are gradually turned on. The resulting arrays of light and dark are geometric versions of sequences of random digits.

How to produce randomness has been a subject of study for years. In 1927, L. H. C. Tippett published 41,600 random digits by taking the middle digits of the areas of parishes in England. In 1939 a table of 100,000 random digits was produced by M. G. Kendall and B. Babington Smith using a carefully fashioned roulette wheel. In 1949 the U.S. Interstate Commerce Commission took 105,000 random digits from the numbers on freight waybills.

Perhaps one of the more creative ways in which to produce random digits was devised by the statistician W. S. Gossett who worked under the pseudonymn of *Student*. In order to produce his "t" statistic (Section 12.4), Student needed a source of normally distributed random numbers. In those days, he did not have access to sources of such numbers as we do, so he used a great many measurements of lengths of criminals' left middle fingers (and also criminals' heights) as obtained from police records.

Nowadays, thanks to technology, we have ample sources of random numbers, even in some inexpensive calculators.

14.3   RADIATION

Random outcomes have been used in the study of elementary particles, such as neutrons or protons.

■ **Example:** The wall of an atomic reactor is made of lead. Neutrons enter the wall from inside the reactor and collide with atoms of lead. Determine the percentage of neutrons that penetrate the wall.

Solution:

Of those that penetrate the wall, some may also escape into the atmosphere, causing pollution and perhaps bodily injury. Figure 14.6 shows a simplified drawing of the wall of a nuclear reactor. Neutrons—atomic particles—are in continuous motion inside the reactor, bouncing against each other and colliding with the wall.

FIGURE 14.6

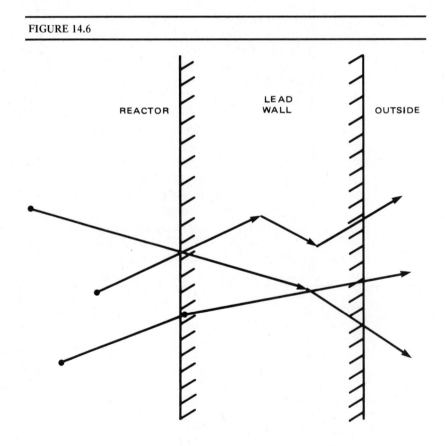

REACTOR

LEAD WALL

OUTSIDE

When neutrons penetrate the wall, they collide with lead atoms. When a neutron collides with a lead atom, there are these possibilities:

1.   The neutron can be absorbed by the atom and cease to exist (as a free neutron).

2.   The neutron can be scattered, changing direction and speed.

First, some simplifying assumptions. The wall is of constant thickness, $3d$. Neutrons enter the wall at right angles and travel a distance d before colliding with an atom of lead. After collision, the neutron rebounds in a random direction (between 0 and 360°) and travels for another distance d before the next collision, and so on. If a neutron collides 12 times with lead atoms (within the wall), it is slowed down enough to be absorbed and remains in the wall. Under these assumptions, find:

1.   Proportion of neutrons escaping from the reactor.

2.   Proportion of neutrons absorbed in the wall.

3.   Proportion of neutrons returned to inside the reactor.

We will follow one neutron as it hits the wall of the reactor. Recall that the wall has a thickness of $3d$ centimeters.

Step 1:   The neutron enters the wall at right angles at a high rate of speed. It moves a distance d before colliding with an atom of lead. After colliding, it moves away at a new direction described by the angle θ, where θ is randomly generated between 0 and 2π radians. (Think of a moving billiard ball colliding with a standing billiard ball and bouncing off in a new direction.) See Figure 14.7.

FIGURE 14.7

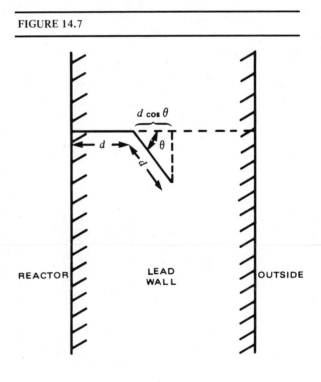

REACTOR   LEAD WALL   OUTSIDE

Step 2:   The neutron now moves in the new direction a distance d, which, in the direction perpendicular to the wall, is

$$d \cdot \text{cosine } \theta$$

The neutron then collides again, takes on a new random direction, and again moves a distance d.

Step 3:   The neutron continues until (a) it escapes the wall and goes into the atmosphere, (b) returns to the reactor, or (c) collides 12 times and is absorbed in the wall of the reactor.

Step 4:   Steps 1 through 3 are repeated for a suitably large number of neutrons, say 1000. ∎

The following program carries out the above steps.

PROGRAM

```
100   LET P=3.14159
110   REM P IS NEEDED FOR TRIG FUNCTION LINE 270
120   PRINT "HOW MANY PARTICLES ";
130   INPUT N
140   PRINT "DISTANCE TRAVELLED BY PARTICLE BEFORE COLLISION ";
150   INPUT D
160   LET I=J=0
170   REM I IS NUMBER OF PARTICLES ESCAPED
180   REM J IS NUMBER OF PARTICLES REFLECTED
190   REM N-(I+J) IS NUMBER OF PARTICLES ABSORBED
200   FOR Q=1 TO N
210   LET K=0
220   REM K COUNTS NUMBER OF COLLISIONS
230   REM D1 IS DISTANCE TRAVELLED BY PARTICLE THROUGH WALL
240   LET D1=D
250   LET K=K+1
260   IF K=12 THEN 00340
270   LET D1=D1+D*COS(P*(2*RND(1)))
280   IF D1>D*3 THEN 00310
290   IF D1<0 THEN 00330
300   GOTO 00250
310   LET I=I+1
320   GOTO 00340
330   LET J=J+1
340   NEXT Q
350   PRINT "PROPORTION OF PARTICLES"
360   PRINT "ESCAPED","REFLECTED","ABSORBED"
370   PRINT I/N,J/N,(N-I-J)/N
380   END
```

OUTPUT

```
HOW MANY PARTICLES ? 1000
DISTANCE TRAVELLED BY PARTICLE BEFORE COLLISION ? 1
       PROPORTION OF PARTICLES
   ESCAPED          REFLECTED        ABSORBED
    .284              .569             .147
```

EXERCISES

1   Write and run the program in this section. Test the program using 5000 neutrons (particles). Let the thickness of the wall be .9. (That is, $3D=.9$, where D = distance travelled by particle before collision.)

2   Now, suppose that at each collision, a neutron rebounds in a random direction restricted to lie within 90 degrees ($\pi/2$ radians) of its direction before collision. Compare your results for the motion of 5000 neutrons with this modification.

3   In the sample output, it was found that for the given thickness of the wall ($D=1.0$ cm, $3D=3.0$ cm), .28 of the neutrons escape into the atmosphere. Suppose that this level is considered unsafe and that only .10 of the neutrons should escape. Modify the program so that the thickness of the wall of the reactor can be varied independently of D, the distance travelled by a particle. Now, using the program, what is the least thickness of the wall that will provide this low a level of radiation into the atmosphere?

4   Modify the assumptions of the reactor specifications so that the thickness of the wall is five times the distance traveled by a particle before collision. Find the proportions of particles that escape and that are now absorbed.

PROJECT CORNER

The Shopping Mall Problem

Mathematical problems involving random quantities appear in an amazing variety and, at times, in surprising places. Would you believe a shopping mall? Suppose, for example, that an architect is designing a mall for some expected number of shoppers. The movement of the shoppers through the mall can be assumed to be random. What is the average (mean) distance between the shoppers?

The following problem can form the basis for solving this real-life problem whose solution requires advanced mathematics, unless the Monte Carlo method is used. Take two unit squares and place them side by side so that they form a rectangle 2 units wide and 1 unit high (Figure 14.8). Now, choose a point A at random in the first square and, again at random, a point B in the second square. Then A and B will be a certain distance apart.

Question: If we repeat the random choice of A and B many times, how far apart will they be, on the average?

From the point of view of Monte Carlo methods, the solution of this problem is simple. As an added bonus, this problem is an excellent calculator exercise.

Step 1—*Choose a model*: We need some way of generating random points A and B. To accomplish this, we could use an ordinary table of random digits or a calculator that generates random numbers.

The two unit squares are subdivided as shown in Figure 14.9.

FIGURE 14.8   FIGURE 14.9

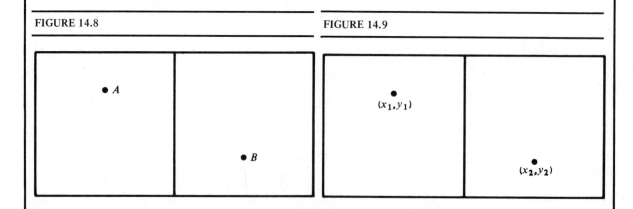

Step 2—*Trial*: To find the coordinates (x_1, y_1) for point A, generate two random digits and interpret the outcomes as decimals. For example, suppose the outcome is the pair of digits 4 and 3; then the ordered pair $(.4, .3)$ locates the random point A.

To find the coordinates (x_2, y_2) for point B, generate another pair of random digits, interpret them as decimals, and add 1 to the first to provide the value of x_2. For example, suppose the second outcome is the pair 7 and 8. Add 1 to .7 to get the coordinates $(1.7, .8)$ of the random point B. (*Note:* For calculators that generate random numbers in a single range, 0–99, a finer grid (100×100) can be used. Then generate four numbers, one for each of $x_1, y_1, x_2,$ and y_2.)

Step 3—*Outcome of the trials:* In terms of our problem, we are not interested in the points A and B, but in the distance d between them. We can use the usual distance formula here. In our example,

$$d = \sqrt{(x_1 - x_2)^2 + (y_1 - y_2)^2} = \sqrt{(0.4 - 1.7)^2 + (0.3 - 0.8)^2} = \sqrt{1.94} \doteq 1.39$$

Step 4—*Repeat trials:* Repeat Steps 2 and 3. We report the results of 10 trials. Again, as a class activity, it will not be difficult to obtain around 100 trials.

Step 5—*Calculations:* Calculate the average distance between the points A and B. Performing this experiment for only 10 trials, we found the average distance $d = 1.12$ as compared to a theoretical value of 1.0881382.[2]

It is remarkable that such a problem can be reasonably approached by a student armed with only a random number table or calculator and the ability to find squares and square roots. In working through this problem, it is worthwhile to question the limits on accuracy imposed by the grid size defined in Step 1. Even computer-generated random numbers impose the same type of constraint, except with a much finer grid. Using 500 trials, the given program yielded a very satisfactory estimate of 1.07256 for d.

PROGRAM	OUTPUT

```
10    FOR I=1 TO 500          DIST= 1.07256
30    LET X1=RND(1)
40    LET Y1=RND(1)
50    LET X2=RND(1)+1
60    LET Y2=RND(1)
70    LET C1=(X1-X2)*(X1-X2)
80    LET C2=(Y1-Y2)*(Y1-Y2)
90    LET D=C1+C2
100   LET D1=D1+SQR(D)
110   NEXT I
120   PRINT "DIST=";D1/500
130   END
```

14.4   ELECTRIC CIRCUITS

When an electric current passes through a circuit, such as in a house lighting system or in a television set, it meets resistance. The amount of current (I) is measured in amperes and the amount of resistance (R) is measured in ohms. An elementary law in physics, Ohm's Law, states that the product of the current and the resistance is the voltage (E). That is:

$$E = IR$$

Resistors are used to change electricity to heat, to change voltage, or to do many other important things in a circuit. Suppose, for example, you had a voltage supply of 12 volts and you wanted a current of .1 ampere. The resistor you would need for that circuit would be

$$R = \frac{E}{I} = \frac{12}{.1} = 120 \text{ ohms}$$

But resistors vary from their claimed value. You may buy a resistor marked 120 ohms, but its true resistance could be 100 ohms, 130 ohms, or some other value different from (but presumably close to) 120 ohms. Roughly, the more expensive the resistor, the more certain you can be that its true resistance is close to what is marked on it. In some cases, this variation may not be greatly important; but in others, the variation may be enough to produce a circuit that does not work properly. (For example, too much or too little current is allowed to pass.) The value of resistors may be thought of as random. For example, you may buy a 10% resistor which is marked 100 ohms. This means that its true resistance is likely to be between 90 ohms (10% less than 100) and 110 ohms (10% more than 100). This variation of the true resistance from the marked value on the resistor is called *tolerance.* So, for example, we can talk about resistors with 10% tolerance or 5% tolerance. Resistors are often color-coded to indicate their tolerance.

We can use the ideas of Section 14.1 to produce a random number uniformly distributed in the interval (90, 110) to be a reasonable model for pulling a resistor at random from a bin of 10% resistors. The following BASIC instruction produces such a random number.

$$50 \text{ LET } R = 90 + 20 * RND(1)$$

■ **Example:**   Suppose we have two 10% resistors, $R1$ and $R2$, each marked 100 ohms, and hook them up as shown in Figure 14.10. (The two resistors are said to be *in series.*) By doing so, their total resistance R is the sum of their individual resistances. What is the probability that $R1$ and $R2$ will be a resistance between 195 and 205 ohms?

FIGURE 14.10

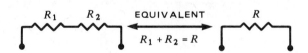

Solution:

We can use the following computer program that generates two 10% resistors in lines 50 and 60. Then, in line 100, it checks whether the sum of the resistances is between 195 and 205 ohms.

Step 1—*Model:*   We use RND, where

$$R1=90+20*RND(1)$$
$$R2=90+20*RND(1)$$

Step 2—*Trial:*   A trial consists of selecting the two values from our model in Step 1.

Step 3—*Successful trial:*   A successful trial occurs when the sum $R1 + R2$ is between 195 and 205 ohms. Successful trials are counted in lines 140 and 150.

Step 4—*Repeat trials:*   We do 500 trials.

Step 5—*P(success):*   This is the number of successes divided by N trials.

PROGRAM

```
20    PRINT "N TRIALS";
30    INPUT N
40    FOR I=1 TO N
50    LET R1=90+20*RND(1)
60    LET R2=90+20*RND(1)
70    LET R=R1+R2
100   IF R<195 OR R>205 THEN 130
110   LET P=P+1
130   NEXT I
200   PRINT "N TRIALS=";N;"PROB=";P/N
210   END
```

OUTPUT

```
N TRIALS? 500
N TRIALS= 500 PROB= .482
```

So, the probability that $R1$ and $R2$ is between 195 and 205 ohms is estimated as .48 (rounded). ∎

EXERCISES

Modify the program in this section to answer these exercises.

1   Suppose you wish to buy 5% resistors. That is, the actual resistance is likely to be within 5% of the marked value. If you buy two such resistors marked 500 ohms, what is the probability that, when hooked in series, the resulting resistance will be between 975 and 1025 ohms?

2   Suppose that in the example in Section 14.4, you use 10% resistors, one marked 100 ohms and one marked 50 ohms. You join them in series. What is the probability that the resistance of the circuit is between 145 and 155 ohms?

3   When two resistors are joined in parallel (see Figure 4.11), the resulting resistance R is given by

$$\frac{1}{R} = \frac{1}{R_1} + \frac{1}{R_2}$$

FIGURE 14.11

Find the probability that two resistors marked 100 ohms and joined in parallel will have a resistance between 45 and 55 ohms. Assume they are 10% resistors.

4   In Exercise 3, above, use 5% resistors instead of 10% resistors. Now find the required estimated probability.

5   The power P associated with a resistance R is given by

$$P = \frac{E^2}{R}$$

Suppose the voltage can be assumed to vary uniformly in a circuit. Estimate the probability that the power will be between 0 and .5, if the voltage E is a random (uniformly distributed) value in the interval $(0, 1)$ and the resistance is 1.0 ohm.

6   Refer to Appendix H to see how to generate normally distributed random values in a given interval. Use this method to solve Exercise 1 of this section when it is modified to assume that the 5% resistor has a true resistance that is normally distributed with a mean of 500 ohms and a standard deviation of 10 ohms.

7   Answer Exercise 3 of this section, assuming the resistances to be normally distributed with a mean of 100 ohms and a standard deviation of 5 ohms.

14.5   GENETICS

Genetics is the study of how characteristics of living things are passed on from one generation to another. The founder of genetics was Gregor Mendel, who published his ideas just over 100 years ago. Reference was made to his work in Chapter Thirteen. Mendel developed the idea that characteristics such as blood type or eye color are caused by factors (later called genes) in the body cells. These genes have two components, called alleles, one from each parent. If the two alleles are different, then one characteristic *dominates* over the other. If, for example, tallness dominates shortness in a certain plant, then the offspring will be tall. These two alleles are written as Tt (T is for tall, t is for short).

Mendel also assumed that the alleles were matched up in the genes in an independent way from factor to factor. So, if there are genes for eye color, blood type, and so on, the resulting match up of pairs of genes in the next generation can be demonstrated by random experiments, or Monte Carlo methods.

■ **Example:**   Persons who can roll their tongues into a U-shape have the characteristic R. Those who cannot roll their tongues into a U-shape have the characteristic r. R dominates over r.

Since traits are carried by pairs of alleles, a parent could be either one of:

RR can roll tongue
Rr can roll tongue, since R dominates over r
rR can roll tongue, since R dominates over r
rr cannot roll tongue

Suppose each parent is Rr. What proportion of their children will be tongue-rollers?

Solution:

Step 1—Model: We assume the alleles are chosen with equal likelihood. So we have

$$p(R) = .5$$
$$p(r) = .5$$

and we will use the rule for random digits:

$$0 - 4 : R$$
$$5 - 9 : r$$

Step 2—Trial: A trial consists of generating two random digits from $0 - 9$. We will use RND to do this.

PROGRAM

```
5     BASE0
10    PRINT "N TRIALS";
20    INPUT N
30    FOR I=1 TO N
35    LET X=0
40    LET Y=0
50    LET F=INT(10*RND(1))
60    IF F>4 THEN 80
70    LET X=1
80    LET M=INT(10*RND(1))
90    IF M>4 THEN 110
100   LET Y=1
110   LET H(X+Y)=H(X+Y)+1
120   NEXT I
130   PRINT "P(0)","P(1)","P(2)"
140   PRINT H(0)/N,H(1)/N,H(2)/N
150   END
```

OUTPUT

```
N TRIALS? 500
P(0)          P(1)          P(2)
.268          .498          .234
```

In line 50 of the above program, we get a random digit for the father (F). In line 80 we get a random digit for the mother (M). Suppose, for example, we get $F = 1$ and $M = 7$. According to our model, the child is Rr. Since R (tongue-rolling) is dominant over r, the child is a tongue-roller.

Step 3—Successful trial: We are to estimate the proportion of children expected to be tongue-rollers. Therefore, R determines a success.

Line 60 checks to see if the digit is greater than 4 for the father and line 70 sets $X = 1$ if we have an R. Lines 90 and 100 do the corresponding procedure for the mother.

You should verify that $X + Y$ gives the number of Rs in the child's gene.

For this characteristic, it does not matter whether the gene comes from the father or the mother. In more complicated genetic problems, the sex of the parent may be taken into account.

Step 4—Repeat trials: Using our computer, we do 500 trials.

Step 5—P(success): From the program below, the required probability is

$$P(1) + P(2)$$

since we are estimating the proportion of children who are either Rr, rR, or RR. So the estimated proportion is

$$.498 + .234 \doteq .73 \ \blacksquare$$

EXERCISES

Use a computer program and do 500 trials.

1   In the example from this section, what is the probability that, assuming mates to be chosen at random, a child will *not* have tongue-rolling capability?

2   Suppose in the example for this section, the father is *RR* and the mother is *Rr*. What proportion of their children is expected to be tongue-rollers?

3   Suppose in the example for this section, the father is *rr* and the mother is *Rr*. What proportion of their children is expected to be tongue-rollers?

4   In guinea pigs, the contrasting traits are black and white coat colors. Black (*B*) dominates over white (*b*). Suppose a *BB* male is crossed with a *bb* female. What proportion of their offsprings is expected to have black coat color?

5   The ability to taste a substance called TPC is inherited. To some people, it tastes bitter. To others, it is tasteless. Let *T* represent the ability to taste TPC. Then let *t* represent the nontasters. Suppose a woman who is *Tt* marries a man who is *Tt*. What proportion of their children is expected to be a taster (that is, *TT* or *Tt*)?

6   Suppose a blue-eyed man marries a brown-eyed woman whose father is blue-eyed. What proportion of their children would be expected to be blue-eyed? Brown is dominant over blue. (*Hint:* Since the man is blue-eyed, he is *bb*. Since the woman has a blue-eyed father, but she herself is brown-eyed, she is *Bb*.)

7   Opalescent dentine is a rare genetic disorder of the teeth that causes them to be translucent, discolored, and friable (easily broken). Geneticists believe that the condition is due to a single, rare, dominant allele, *D*. They assume the gene is *Dd*, since *DD* would require both parents having this rare disease. Thus, a child who inherits the disease has parents of whom at least one is *Dd*. The following records from the hereditary clinic at the University of Michigan are for 100 children for couples where one parent has opalescent dentine.

Affected	Normal	Total
52	60	112

Do the data support the genetic theory?

14.6 RELIABILITY OF SYSTEMS

We can think of a system as a collection of parts that work together. How well the system works depends on how well its parts work.

A television set: The tubes, transistors, loudspeaker, and so forth make up a system that produces sound and pictures.

An automobile: The engine, brakes, lights, transmission, and so on make up a system that enables the car to transport you safely.

■ **Example 1:**   Suppose the control mechanism for the photographic module of an interplanetary rocket has three components connected as shown in Figure 14.12.

FIGURE 14.12

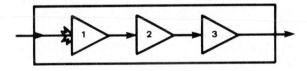

For the module to function, each component must work properly. Suppose the probability that each functions is .9 and that each functions independently of the other component. Estimate the probability that the photographic module will operate successfully.

Solution:

**Step 1—*Model:*   Use random digits:

1–9:	Works
0:	Fails to work

**Step 2—*Trial:*   Obtain three random digits, one for each component (see lines 40–60 of the following program).

PROGRAM

```
10    PRINT "N TRIALS";
20    INPUT N
30    FOR I=1 TO N
40    LET X=INT(10*RND(1))
50    LET Y=INT(10*RND(1))
60    LET X=INT(10*RND(1))
70    IF X*Y*Z=0 THEN 90
80    LET S=S+1
90    NEXT I
100   PRINT "P(SUCCESS) IN ";N;" TRIALS =";S/N
110   END
```

OUTPUT

```
N TRIALS? 500
P(SUCCESS) IN 500 TRIALS=.776
```

**Step 3—*Successful trial:*   A trial is successful if all three digits *are not zero* (see line 70 of the following program).

**Step 4—*Repeat trials:*   With the computer program below, run 500 trials.

**Step 5—*P(success):*   Number of successes is counted in line 80.

The probability that the module will operate successfully is

$$P(\text{success}) \doteq .78 \ ■$$

The reliability of a system can be increased by using multiple systems that perform the same tasks. Then, if one system fails, another takes over.

■ **Example 2:**   Suppose we designed the photographic unit so that two modules were connected, as shown in Figure 14.13.

FIGURE 14.13

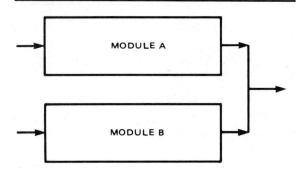

Suppose further that each module operates independently of the other and with a probability of successful operation of .75. What is the probability that the unit, consisting of two modules, will operate now?

Solution:

Step 1—*Model*:   Each module operates successfully with a probability of .75. So we use random digits:

00–74:	Module works properly
75–99:	Module does not work

Step 2—*Trial*:   Select two random numbers, one for each module (see lines 50 and 80 of the following program).

PROGRAM

```
10    PRINT "N TRIALS";
20    INPUT N
30    FOR I=1 TO N
40    LET M1=M2=0
50    LET X=INT(100*RND(1))
60    IF X>74 THEN 80
70    LET M1=1
80    LET Y=INT(100*RND(1))
90    IF Y>74 THEN 110
100   LET M2=1
110   IF M1+M2=0 THEN 130
120   LET S=S+1
130   NEXT I
140   PRINT "P(SUCCESS IN N TRIALS)=";S/N
150   END
READY.
```

OUTPUT

```
N TRIALS? 500
P(SUCCESS) IN N TRIALS=.93
```

Step 3—*Successful trial*:   A trial is a success if *one of the two* random numbers (or both) is in the interval 00–74, inclusive. (That is, at least one of the two modules operates successfully. See lines 110 and 120 of the following program.)

Step 4—*Repeat trials*:   Do 500 trials.

Step 5—*P(success)*:   As the output of our program shows,

$$P(\text{success}) = .93 \ ■$$

EXERCISES

For each exercise, use a computer program and do 500 trials.

1   Suppose that in the control mechanism in Example 1 of this section, each of the three components of the module operates with only a .8 probability of success. Now what is the probability that the module will operate? What is the probability that the unit (consisting of two modules) will operate?

2   Answer Exercise 1, above, if each of the three components of the module operates with a .95 probability of success.

3   Suppose in Example 2 of this section that the photographic unit consists of three modules, instead of only two. Suppose further that each module operates successfully with a probability of .75. What is the probability now that the unit will operate?

4   Modify the program for Example 2 so that it will estimate the probability of successful operation of a unit consisting of M modules. Each module operates successfully with probability P. The program should ask the user to specify how many modules are in the unit and with what probability the modules operate successfully.

5   Suppose that an experimental airplane is being developed with a new type of engine. The airplane can make a safe flight if at least one-half of its engines are operating. If the probability of the failure of an engine is .10, which is safer, a two-engine or four-engine plane? Assume the engines operate independently of one another.

6   A rocket has three modules, each of which must operate before lift-off can occur. Module A has a probability of failure of .1, module B a probability of failure of .15, and module C a probability of failure of .2. What is the probability of lift-off?

Appendices

A CHI-SQUARE TABLE

df*				PROBABILITY (AREA)					
	0.99	0.95	0.90	0.50	0.20	0.10	0.05	0.01	0.0001
1	0.00	0.00	0.02	0.45	1.64	2.71	3.84	6.63	10.83
2	0.02	0.10	0.21	1.39	3.22	4.61	5.99	9.21	13.82
3	0.11	0.35	0.58	2.37	4.64	6.25	7.81	11.34	16.27
4	0.30	0.71	1.06	3.36	5.99	7.78	9.49	13.28	18.47
5	0.55	1.15	1.61	4.35	7.29	9.24	11.07	15.09	20.52
6	0.87	1.64	2.20	5.35	8.56	10.64	12.59	16.81	22.46
7	1.24	2.17	2.83	6.35	9.80	12.02	14.07	18.48	24.32
8	1.65	2.73	3.49	7.34	11.03	13.36	15.51	20.09	26.12
9	2.09	3.33	4.17	8.34	12.24	14.68	16.92	21.67	27.88
10	2.56	3.94	4.87	9.34	13.44	15.99	18.31	23.21	29.59
11	3.05	4.57	5.58	10.34	14.63	17.28	19.68	24.72	31.26
12	3.57	5.23	6.30	11.34	15.81	18.55	21.03	26.22	32.91
13	4.11	5.89	7.04	12.34	16.98	19.81	22.36	27.69	34.53
14	4.66	6.57	7.79	13.34	18.15	21.06	23.68	29.14	36.12
15	5.23	7.26	8.55	14.34	19.31	22.31	25.00	30.58	37.70
16	5.81	7.96	9.31	15.34	20.47	23.54	26.30	32.00	39.25
17	6.41	8.67	10.09	16.34	21.61	24.77	27.59	33.41	40.79
18	7.01	9.39	10.86	17.34	22.76	25.99	28.87	34.81	42.31
19	7.63	10.12	11.65	18.34	23.90	27.20	30.14	36.19	43.82
20	8.26	10.85	12.44	19.34	25.04	28.41	31.41	37.57	45.31
21	8.90	11.59	13.24	20.34	26.17	29.62	32.67	38.93	46.80
22	9.54	12.34	14.04	21.34	27.30	30.81	33.92	40.29	48.27
23	10.20	13.09	14.85	22.34	28.43	32.01	35.17	41.64	49.73
24	10.86	13.85	15.66	23.34	29.55	33.20	36.42	42.98	51.18
25	11.52	14.61	16.47	24.34	30.68	34.38	37.65	44.31	52.62
26	12.20	15.38	17.29	25.34	31.79	35.56	38.89	45.64	54.05
27	12.88	16.15	18.11	26.34	32.91	36.74	40.11	46.96	55.48
28	13.56	16.93	18.94	27.34	34.03	37.92	41.34	48.28	56.89
29	14.26	17.71	19.77	28.34	35.14	39.09	42.56	49.59	58.30
30	14.95	18.49	20.60	29.34	36.25	40.26	43.77	50.89	59.70
50	29.71	34.76	37.69	49.33	58.16	63.17	67.50	76.15	86.66

(From F. Mosteller, S. Fienberg and R. Rourke: *Beginning Statistics with Data Analysis,* Menlo Park: Addison-Wesley Publishing Co., 1983, pp. 543–544.)

*df = degrees of freedom.

B SOME NOTES

Note 1–Another Formula for Finding Variance: The method used in Chapter Two for finding variance (and standard deviation) helps to show the meaning of those measures of variation in a set of data. *Variance* is an "averaged-out" squared distance of the data from their mean.

However, the method in Chapter Two has some disadvantages. For example, if you have a large set of data, it is cumbersome to first find the mean of the data, then go back and find each deviation score, square, and so on. Also, the method may introduce rounding errors first when finding the mean, then in the deviation scores.

Following is a method for finding variance that does not require you to first find the mean of the data. This method is particularly well-suited for use on a calculator or computer.

The formula is:

$$\text{Variance} = \frac{\text{sum of squares of } X - \frac{(\text{sum of } X)^2}{N}}{N}$$

Or, using symbols defined below:

$$\text{Variance} = \frac{S_2 - \frac{(S_1)^2}{N}}{N}$$

The steps to follow are:

i. Find the sum of all the numbers. Call this S_1.
ii. Square each number, then find the sum of the squares. Call this S_2.
iii. Square S_1 and divide by the number of scores, N.
iv. Subtract the result of step iii from S_2.
v. Divide the result of step 4 by N.* The result is the variance.

We will follow these five steps to find the variance of these quiz scores.

	Score	$(\text{Score})^2$
Bill	8	64
Sue	6	36
Ted	6	36
Sally	7	49
Mickey	4	16
Jose	7	49
Fran	4	16
	$S_1 = 42$	$S_2 = 266$

i. Sum of all scores. $S_1 = 42$.

ii. Square each score, then add. $S_2 = 266$.

iii. There are 7 scores. $\dfrac{42 \times 42}{7} = 252$.

iv. $266 - 252 = 14$.

v. Variance $= \dfrac{14}{7} = 2.0$.

Compare the result with the value obtained in Chapter Two, page 48.

The computer program STAT in Appendix G can be used to find the mean, variance, and standard deviation of a set of data. The number of scores, N, is entered in line 10 of the program. The data are entered beginning at line 100. It is recommended that the data be entered five to a line. This will help you keep track of the data as entered and help in checking for errors in entry.

Note 2–Continuous and Discrete Data (Refer to Chapter Nine, page 270): Discrete data have only limited, definite values that they can assume. For example, tosses of coins or rolls of dice give discrete data. You cannot roll a die, for example, and get a value of .3333 or pi.

*For an unbiased estimate of the population variance, divide by $N-1$ (see Chapter Eleven).

Continuous data, on the other hand, can take on values on a continuous scale. For example, the weight of a 17-year-old student can have any value within a certain range (such as between 50 and 250 pounds). Measurements are examples of continuous data.

Suppose C is a point on a continuous scale, as in the figures below. Then if z is a random value defined over that scale, what can be said about

$$p(z \leqslant C)?$$

In particular, we can want to know how

$$p(z \leqslant C) \text{ and } p(z < C)$$

compare.

From Figure i we want to know, for $p(z \leqslant C)$, the area under the curve **up to and including the value C.** But from Figure ii we want to know, for $p(z < C)$, the area under the curve **up to but not including the value of C.** Since C is only a point on the continuous scale, it seems reasonable that including C adds no area to the region under the curve. That is, the two areas are equal, and we have

$$p(z \leqslant C) = p(z < C)$$

Note 3—Normally Distributed Random Data: In Chapter Fourteen, it is pointed out that the BASIC statement $RND(1)$ produces random values that are uniformly distributed in the interval $(0, 1)$. Suppose, however, that you want random values that are *normally* distributed instead of uniformly distributed.

We can obtain such values using $RND(1)$ and appealing to the Central Limit Theorem (see Chapter Eleven).

$RND(1)$ defines a population of values R that are uniformly distributed in the interval $(0, 1)$. These values have a mean of .5. The variance of these values is not so obvious—it is $1/12$. While this value is not obvious, it should not surprise us that the variance of R is small. That is, on average, the squared distance of R from .5 is $1/12$.

We now use $RND(1)$ to draw random samples from $(0, 1)$. But what sample size shall we use? We choose a sample size that helps us in our search for normally distributed random values. It turns out that for a sample size of 12, the variance of the sum of these 12 values is $\sigma^2 = 1.0$ and the mean of the sum is $\mu = 6.0$.

What else can be said about the sum of these 12 values of R? By the Central Limit Theorem, we know that the sample means are approximately normally distributed about the population mean. But since the sample means are so distributed, the sum of the sample values also has the same-shaped distribution (since sum of $X = N \times \bar{X}$).

The program NORM in Appendix G does the job for us. Lines 210–220 of the program produce the sum of 12 values of $RND(-1)$. By the discussion of the preceeding paragraph, we know that the resulting value of R in line 250 is normally distributed with $\mu = 6.0$ and $\sigma = 1.0$. Subtracting 6 from R in line 150 gives a normally distributed random value with $\mu = 0$ and $\sigma = 1.0$.

Still in line 250, $R - 6$ is then stretched by the desired standard deviation S and given a slide of M units. The result is a random value N that is normally (approximately) distributed with $\mu = M$ and $\sigma = S$.

Line 260 rounds the resulting value of N to P decimal places.

FIGURE i

AREA (PROBABILITY)

C

$p(Z < C) = $ AREA

FIGURE ii

AREA (PROBABILITY)

C

$p(Z < C) = $ AREA

C NORMAL CURVE TABLE

AREAS UNDER THE
STANDARD
NORMAL CURVE
(AREAS TO THE
LEFT)

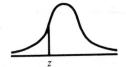

z	0	1	2	3	4	5	6	7	8	9
−3.0*	.0013	.0013	.0013	.0012	.0012	.0011	.0011	.0011	.0010	.0010
−2.9	.0019	.0018	.0017	.0017	.0016	.0016	.0015	.0015	.0014	.0014
−2.8	.0026	.0025	.0024	.0023	.0023	.0022	.0021	.0021	.0020	.0019
−2.7	.0035	.0034	.0033	.0032	.0031	.0030	.0029	.0028	.0027	.0026
−2.6	.0047	.0045	.0044	.0043	.0041	.0040	.0039	.0038	.0037	.0036
−2.5	.0062	.0060	.0059	.0057	.0055	.0054	.0052	.0051	.0049	.0048
−2.4	.0082	.0080	.0078	.0075	.0073	.0071	.0069	.0068	.0066	.0064
−2.3	.0107	.0104	.0102	.0099	.0096	.0094	.0091	.0089	.0087	.0084
−2.2	.0139	.0136	.0132	.0129	.0125	.0122	.0119	.0116	.0113	.0110
−2.1	.0179	.0174	.0170	.0166	.0162	.0158	.0154	.0150	.0146	.0143
−2.0	.0228	.0222	.0217	.0212	.0207	.0202	.0197	.0192	.0188	.0183
−1.9	.0287	.0281	.0274	.0268	.0262	.0256	.0250	.0244	.0239	.0233
−1.8	.0359	.0351	.0344	.0336	.0329	.0322	.0314	.0307	.0301	.0294
−1.7	.0446	.0436	.0427	.0418	.0409	.0401	.0392	.0384	.0375	.0367
−1.6	.0548	.0537	.0526	.0516	.0505	.0495	.0485	.0475	.0465	.0455
−1.5	.0668	.0655	.0643	.0630	.0618	.0606	.0594	.0582	.0571	.0559
−1.4	.0808	.0793	.0778	.0764	.0749	.0735	.0721	.0708	.0694	.0681
−1.3	.0968	.0951	.0934	.0918	.0901	.0885	.0869	.0853	.0838	.0823
−1.2	.1151	.1131	.1112	.1093	.1075	.1056	.1038	.1020	.1003	.0985
−1.1	.1357	.1335	.1314	.1292	.1271	.1251	.1230	.1210	.1190	.1170
−1.0	.1587	.1562	.1539	.1515	.1492	.1469	.1446	.1423	.1401	.1379
−.9	.1841	.1814	.1788	.1762	.1736	.1711	.1685	.1660	.1635	.1611
−.8	.2119	.2090	.2061	.2033	.2005	.1977	.1949	.1922	.1894	.1867
−.7	.2420	.2389	.2358	.2327	.2296	.2266	.2236	.2206	.2177	.2148
−.6	.2743	.2709	.2676	.2643	.2611	.2578	.2546	.2514	.2483	.2451
−.5	.3085	.3050	.3015	.2981	.2946	.2912	.2877	.2843	.2810	.2776
−.4	.3446	.3409	.3372	.3336	.3300	.3264	.3228	.3192	.3516	.3121
−.3	.3821	.3783	.3745	.3707	.3669	.3632	.3594	.3557	.3520	.3483
−.2	.4207	.4168	.4129	.4090	.4052	.4013	.3974	.3936	.3897	.3859
−.1	.4602	.4562	.4522	.4483	.4443	.4404	.4364	.4325	.4286	.4247
−.0	.5000	.4960	.4920	.4880	.4840	.4801	.4761	.4721	.4681	.4641

* For z ≤ −4 the areas are 0 to four decimal places.

z	0	1	2	3	4	5	6	7	8	9
.0	.5000	.5040	.5080	.5120	.5160	.5199	.5239	.5279	.5319	.5359
.1	.5398	.5438	.5478	.5517	.5557	.5596	.5636	.5675	.5714	.5753
.2	.5793	.5832	.5871	.5910	.5948	.5987	.6026	.6064	.6103	.6141
.3	.6179	.6217	.6255	.6293	.6331	.6368	.6406	.6443	.6480	.6517
.4	.6554	.6591	.6628	.6664	.6700	.6736	.6772	.6808	.6844	.6879
.5	.6915	.6950	.6985	.7019	.7054	.7088	.7123	.7157	.7190	.7224
.6	.7257	.7291	.7324	.7357	.7389	.7422	.7454	.7486	.7517	.7549
.7	.7580	.7611	.7642	.7673	.7704	.7734	.7764	.7794	.7823	.7852
.8	.7881	.7910	.7939	.7967	.7995	.8023	.8051	.8078	.8106	.8133
.9	.8159	.8186	.8212	.8238	.8264	.8289	.8315	.8340	.8365	.8389
1.0	.8413	.8438	.8461	.8485	.8508	.8531	.8554	.8577	.8599	.8621
1.1	.8643	.8665	.8686	.8708	.8729	.8749	.8770	.8790	.8810	.8830
1.2	.8849	.8869	.8888	.8907	.8925	.8944	.8962	.8980	.8997	.9015
1.3	.9032	.9049	.9066	.9082	.9099	.9115	.9131	.9147	.9162	.9177
1.4	.9192	.9207	.9222	.9236	.9251	.9265	.9279	.9292	.9306	.9319
1.5	.9332	.9345	.9357	.9370	.9382	.9394	.9406	.9418	.9429	.9441
1.6	.9452	.9463	.9474	.9484	.9495	.9505	.9515	.9525	.9535	.9545
1.7	.9554	.9564	.9573	.9582	.9591	.9599	.9608	.9616	.9625	.9633
1.8	.9641	.9649	.9656	.9664	.9671	.9678	.9686	.9693	.9699	.9706
1.9	.9713	.9719	.9726	.9732	.9738	.9744	.9750	.9756	.9761	.9767
2.0	.9772	.9778	.9783	.9788	.9793	.9798	.9803	.9808	.9812	.9817
2.1	.9821	.9826	.9830	.9834	.9838	.9842	.9846	.9850	.9854	.9857
2.2	.9861	.9864	.9868	.9871	.9875	.9878	.9881	.9884	.9887	.9890
2.3	.9893	.9896	.9898	.9901	.9904	.9906	.9909	.9911	.9913	.9916
2.4	.9918	.9920	.9922	.9925	.9927	.9929	.9931	.9932	.9934	.9936
2.5	.9938	.9940	.9941	.9943	.9945	.9946	.9948	.9949	.9951	.9952
2.6	.9953	.9955	.9956	.9957	.9959	.9960	.9961	.9962	.9963	.9964
2.7	.9965	.9966	.9967	.9968	.9969	.9970	.9971	.9972	.9973	.9974
2.8	.9974	.9975	.9976	.9977	.9977	.9978	.9979	.9979	.9980	.9981
2.9	.9981	.9982	.9982	.9983	.9984	.9984	.9985	.9985	.9986	.9986
3.0†	.9987	.9987	.9987	.9988	.9988	.9989	.9989	.9989	.9990	.9990

† For $z \geq 4$ the areas are 1 to four decimal places.
Adapted from *Probability with Statistical Applications*, second edition, by F. Mosteller, R. E. K. Rourke, and G. B. Thomas, Jr. Reading, Mass.: Addison-Wesley, 1970, p. 473.

D LINEAR INTERPOLATION

Interpolation can be used to help find a value that is not provided in a table. The procedure is called *linear* interpolation, since it assumes that the relations between the values involved are (or are nearly) *linear*, that is, are straight-line relationships. This assumption gives very good estimates of desired values for the exercises in this book.

Example 1:   Find $p(\chi^2_{10} \leqslant 20.0)$.

The chi-square table (Appendix A) does not have a value of 20.0 for a chi-square with 10 degrees of freedom. So interpolation is used.

The chi-square values of 18.31 and 23.21 are given in the table, with corresponding probabilities (areas) of .05 and .01, respectively. So, $p(\chi^2_{10} \leqslant 20.0)$ is somewhere between .05 and .01.

$$\frac{1.69}{4.90} = \frac{d}{.04}$$

So we have

$$d = \frac{1.69}{4.90}(.04) \doteq .01$$

And

$$p(\chi^2_{10} \leqslant 20.0) \doteq .05 - .01 \doteq .04$$

Example 2:   Find the value of ? in the following:

$$p(\chi^2_{12} \leqslant ?) = .03$$

The chi-square table (Appendix A) does not give probabilities of .03. However, values of .05 and .01 are given, so we use interpolation to find the required chi-square with 12 degrees of freedom.

$$\frac{d}{5.19} = \frac{.02}{.04} = \frac{1}{2}$$

So we have

$$d = \frac{1}{2}(5.19) \doteq 2.60$$

And value of chi-square is

$$21.03 + d \doteq 21.03 + 2.60 \doteq 23.62$$

so that

$$p(\chi^2_{12} \leqslant 23.62) \doteq .03$$

E *t* TABLE

STUDENT'S *t*
DISTRIBUTION

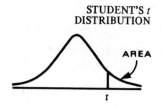

AREA

t

df	.10	.05	.025	.01	.005
			PROBABILITY (AREA)		
1	3.08	6.31	12.71	31.82	63.66
2	1.89	2.92	4.30	6.96	9.92
3	1.64	2.35	3.18	4.54	5.84
4	1.53	2.13	2.78	3.75	4.60
5	1.48	2.02	2.57	3.36	4.03
6	1.44	1.94	2.45	3.14	3.71
7	1.42	1.89	2.36	3.00	3.50
8	1.40	1.86	2.31	2.90	3.36
9	1.38	1.83	2.26	2.82	3.25
10	1.37	1.81	2.23	2.76	3.17
11	1.36	1.80	2.20	2.72	3.11
12	1.36	1.78	2.18	2.68	3.05
13	1.35	1.77	2.16	2.65	3.01
14	1.35	1.76	2.14	2.62	2.98
15	1.34	1.75	2.13	2.60	2.95
16	1.34	1.75	2.12	2.58	2.92
17	1.33	1.74	2.11	2.57	2.90
18	1.33	1.73	2.10	2.55	2.88
19	1.33	1.73	2.09	2.54	2.86
20	1.33	1.72	2.09	2.53	2.85
21	1.32	1.72	2.08	2.52	2.83
22	1.32	1.72	2.07	2.51	2.82
23	1.32	1.71	2.07	2.50	2.81
24	1.32	1.71	2.06	2.49	2.80
25	1.32	1.71	2.06	2.49	2.79
26	1.32	1.71	2.06	2.48	2.78
27	1.31	1.70	2.05	2.47	2.77
28	1.31	1.70	2.05	2.47	2.76
29	1.31	1.70	2.05	2.46	2.76
∞	1.28	1.64	1.96	2.33	2.58

Adapted from D. B. Owen, *Handbook of Statistical Tables*. Courtesy of the Atomic Energy Commission. Reading, Mass., Addison-Wesley, 1962.

Note: The last row of the table ($df = \infty$) gives values for z. For example, the table shows that $p(z > 1.28) = .10$ and $p(z > 1.64) = .05$.

F RANDOM DATA

1. COIN TOSSES
(OUTCOMES ARE H AND T. *H* STANDS FOR HEADS. *T* STANDS FOR TAILS.)

ROW										
1	HHHHT	TTTTT	THHTT	TTHHH	HTHHH	HHTTH	THHTH	TTTHH	HTTTH	HHHHH
2	HHHHT	HHTHH	HHHTT	THTHT	THTTH	TTHHT	HTHTT	TTHTH	TTHTH	HTHHT
3	HTTHT	HTHTH	HHTHT	HTTHH	HTHTH	TTHHT	TTTTH	HTHTH	HHTHT	HTTTH
4	HHHHH	THTTH	THTTT	HTTHT	HHTHT	THHTH	TTTTT	TTTTT	TTHTT	HTHTT
5	THTHT	HTHHT	THTHH	TTTHH	TTHHT	HTTHT	HHHHT	TTHTH	THTTT	HTHTT
6	TTTHT	THTTT	TTHHT	TTHHT	HHTHH	TTTHT	HTTTH	THTHH	THTHH	THHHH
7	THHHT	THTHT	THHTH	HHTTT	HHTHH	HTHTH	HHHHH	THTTT	TTHTH	HHHTH
8	HHHTT	THHTT	HHTTH	HTHTT	THTHH	THTHH	TTHHH	HTHTT	THTHT	HHTHT
9	HHHTT	HHTTH	HTHTT	THHTH	THTTT	HTTTT	HHHTT	THTTT	HTTTT	TTHTT
10	HTTHT	TTTHT	THHHH	TTHHH	HHHTT	TTTTT	HHTHH	HHTHT	THHTH	TTTTT
11	HHHTT	HTTHT	TTHTT	THHTT	THHTT	TTTTH	TTTHT	TTHTT	HHTTH	HHHTH
12	THTHT	HTTHH	THTHH	THTHT	HHTHT	HTTHT	HTHTH	TTHHT	THTTT	HHTHH
13	TTHTH	THHHH	HHHTH	HTHHT	HTTHH	TTTTT	TTHTH	HHHHH	HTHHT	TTTTT
14	THTHT	THHHH	HHTHH	TTHHH	TTTTH	HTHHT	HTHHH	HTTHH	HHTHT	HHTHT
15	HHTHH	TTTTT	THTHT	THTHH	THHHH	HHTHT	HTHHH	HTTHH	HHTHT	HHTHT
16	HTHHH	HHTHH	THHHH	HTHTT	HTTHT	HHTTH	TTTHH	TTHHT	HHTHH	TTTTT
17	HTTHT	HHTTT	TTTHT	TTHHT	HTHHH	THHTT	TTHHT	THTTH	THTHH	HTTHT
18	HHTTT	THHTH	HHHHH	HTTHH	TTHHH	THHHH	TTTTT	TTTHH	HTHTT	THHTT
19	HHHHT	TTTTH	HHTHH	TTTTT	THTTT	THHTT	HHTTH	TTHHH	HTHTT	THHTT
20	THHTH	HHHHT	THTTT	THTTT	THTTT	TTHTT	TTHTT	THHTH	HTHHT	HHTHH
21	THHTH	HHHTH	HHTTT	TTTTH	HHHTT	HHHHH	TTHTH	THHTT	HHHTT	HTHTH
22	HTTHH	THTTT	HHTHT	TTHHH	HTHHT	THTTT	TTTHH	TTTHH	HHHHT	THTTH
23	TTTHH	HTTHT	TTHHH	TTHTH	TTHHT	TTTTH	HTHTT	HTHTT	HTTTT	HHTTH
24	TTHHH	THTHT	HHTTH	HTHHT	HTHTH	HTHTH	THTHT	TTHTT	THTHT	THHTT
25	TTTHT	HTHHT	THHTT	HHTTT	HHTTH	TTTTT	HTTTT	TTHTH	THHTH	TTTHH
26	TTTTH	HHTTT	HTHHH	HHHHH	HHHHT	THTHT	TTHHH	THTHH	TTHTT	HTHHH
27	HHHHH	HHHHH	HTTHH	THTHT	TTHTT	TTHHH	TTTHT	THHTH	HTHTT	HTTHH
28	HHTTH	HHHHH	THHTT	TTHTH	TTTTH	THHTT	THTTT	TTTHH	HTTTT	TTTTH
29	HTTTT	THTHH	HTTTT	TTHTT	HHHTH	TTHHT	HTTHH	THTTH	TTHTH	HTHHT
30	THHHH	HTTHH	HHTTH	THHHT	HTTHT	HHTTT	THHHT	TTTTH	HTHTH	TTTTT
31	TTTHT	HHHHT	HTTHH	TTTHH	HHHHH	HHHTT	HHTHT	HHTHT	HHHTH	TTHHH
32	TTHHH	TTTHT	HHTHT	TTTTT	HTHTT	HTTTH	TTTTT	TTTHH	TTHHT	TTTHT
33	HHHHH	HTTTT	TTTHH	HTTHH	HTHHH	HHHTH	HHTTT	THTTH	HHHHH	THTHT
34	HHHTT	THHHT	HTHHH	HHHHT	HHTTH	TTTHH	HHHTH	HTHHT	THHTT	THHHH
35	THHTT	THTHT	THHHH	HTTTH	HHTHT	HHTTH	HTTTT	HTHHH	HTHHH	THHTT
36	TTHTH	HHTTH	THTTH	THTTH	THHHT	HTHHT	HHHTH	HTHHH	TTHTH	TTHHH
37	HTHHH	THHTH	THHTH	HHHHH	HTHTH	HTHTH	HHTTH	HHTHH	TTHTH	HHHTT
38	HTHHH	TTTTT	HTTHH	HHTTT	HHHTT	HHHHT	HHHTT	HTTHT	HTTTT	HTHTH
39	HTTTH	HTHTH	HHHTT	HTTTT	HTHHT	HHHHT	TTTHH	HHHHH	HHTHT	TTHHT
40	TTTHH	HTHHT	THTTT	HHTTT	HHHTHT	THTTH	THHHT	HTTTT	HHTHH	HHHHT
41	HTTHT	THHTH	THHTH	HHHHH	THTTT	TTHTT	HHTHT	HTTTT	HTTHT	THTTH
42	TTHHH	HTTHH	TTTTH	TTHHT	HHHTH	THTHT	TTTHH	TTHTT	HHHHT	THTTH
43	HTHHH	HTTHH	THTTH	HHHHT	THHTT	HTHTH	HTHTT	TTHHH	TTHTH	HHHTT
44	HHTHH	THHTH	THHTT	THHTH	THTTH	HTTHT	TTTTT	THTTT	HTTTH	HTTHT
45	TTTHH	HTTTH	HHHTT	TTHHH	THHHH	TTHTH	HHHHH	HHTHT	TTHTT	TTHTH

2. ROLLS OF A 4-SIDED DIE
(OUTCOMES ARE 1, 2, 3, 4.)

ROW										
1	34144	13243	23232	43444	14323	14332	21412	42141	42131	43331
2	42214	23422	11434	42423	41323	23442	43213	41234	12224	44233
3	23311	34241	41441	12134	22442	32424	21443	34413	41343	24444
4	44434	32233	22234	32234	13212	12334	13112	13241	42123	21222
5	33133	22234	33412	41312	12132	23232	13211	34321	33322	43441
6	33421	42244	43133	12124	43213	23221	11344	21222	34314	14241
7	24232	24142	34121	24142	41131	31131	21321	44442	34243	11334
8	32232	42321	12342	33424	32322	13413	12122	11212	12322	11434
9	24113	34424	13344	33423	33431	32433	21334	21333	12211	43212
10	11113	22233	43123	34132	23314	32131	24114	34111	42412	32223
11	11323	11144	21142	22232	23144	33313	12242	23323	21212	24333
12	31413	12213	32143	13413	21111	23324	11133	41132	41333	11423
13	41423	31422	43412	22144	32412	43244	43413	44234	11211	31144
14	31211	41222	42443	23244	32411	34143	33131	11114	31223	31321
15	41133	43212	43142	43213	31234	31142	42311	21443	14443	21312
16	11422	42344	34121	13222	33223	22424	44443	12122	22444	22414
17	22321	23423	33333	43321	11313	42114	41131	11334	23232	42123
18	41121	41124	23444	24331	41113	21222	11334	44322	31323	41221
19	21343	34424	44344	23443	23342	32214	22422	23114	13144	12434
20	34241	11324	11231	33232	34223	42422	42334	11312	43431	43322
21	12443	24413	11311	42121	21311	43124	31212	14122	23332	14414
22	12122	12332	22124	34431	31344	11313	33313	41111	44222	22342
23	23213	44134	33313	34342	41424	12233	43334	12322	41234	43433
24	22413	13142	23433	42111	12244	23442	21433	22323	21412	13424
25	12444	31423	41331	44212	33234	43323	23142	33232	43431	31334
26	31111	44231	32413	32143	31214	22214	21224	11313	22244	14242
27	34212	11222	11223	42431	34424	33313	22331	42341	41414	33333
28	42341	21233	11411	12421	41323	24324	43322	23241	44314	34433
29	42424	41123	44121	13112	33212	43144	14144	31142	24414	34331
30	14143	42444	31323	33411	22242	34434	43332	44413	34311	13133
31	44333	42333	32322	11412	42234	13232	21221	22343	41244	44143
32	12132	34314	42444	44343	12211	14434	13133	21241	31223	33311
33	43441	44322	22243	13144	24442	41234	44124	31321	41412	32413
34	21212	31334	23343	32132	32333	23231	44432	13321	43312	42411
35	13321	22332	21434	34413	43341	42311	43432	43444	31121	11222
36	13411	14324	33433	41443	42334	44233	43112	33131	21412	33412
37	22134	13234	21243	21431	22431	14143	14212	12134	13322	23312
38	24144	13344	24111	42413	23333	21211	34113	34343	13324	34314
39	34312	33234	42143	13233	44242	41434	14112	11212	41341	24114
40	13431	11414	42321	12232	42432	41424	43342	23341	34243	21142
41	34444	32433	34114	31332	14231	43322	23443	14241	31413	13314
42	41322	34113	11321	34221	44433	22131	31442	31132	11313	21141
43	32143	13114	23141	42412	31323	31132	22311	34134	34112	43242
44	11414	21331	24442	24413	13434	43313	14221	21242	41224	43333
45	22414	14134	42424	34312	12223	21241	11314	32432	41423	21334

3. ROLLS OF A 6-SIDED DIE
(OUTCOMES ARE 1, 2, 3, 4, 5, 6.)

ROW									
1	45652	16621	36565	26243	64223	41222	35642	16545	66563	66245
2	32221	15252	51533	61154	31413	31411	31151	66163	41566	22622
3	46143	52566	44236	24511	66641	54364	41513	15441	12414	43111
4	62313	44642	14344	55142	32256	24342	64435	52333	15236	15125
5	42241	11222	42463	26223	66523	14531	34264	32226	16262	14122
6	52643	22162	14331	24644	33613	32436	64433	25365	21264	25632
7	44115	61446	45662	51133	33656	23456	13355	61511	54566	15536
8	32356	15413	56265	33142	31551	61354	55513	14566	31544	53423
9	62335	41652	25622	54333	46562	65446	43444	53414	65614	56214
10	31225	46145	56365	52643	25543	35553	23634	33462	21166	15563
11	64263	24233	36566	42364	25643	16624	14652	33454	31531	31245
12	54666	51654	45663	66232	66213	62334	62343	55635	25111	45161
13	22531	46525	14333	26612	34365	41343	64163	66424	24425	21366
14	61563	26632	11552	24545	46221	51256	63542	54324	64642	65211
15	24322	54234	46536	65561	61113	66541	16452	43225	15456	42421
16	25511	32632	44226	65224	66631	13225	53643	25123	21113	23123
17	63545	16262	61144	53541	24642	61256	51632	56146	26654	26421
18	21446	63343	66416	16336	12326	31211	32644	16624	62631	21422
19	43211	65153	62115	51626	11226	11663	11615	31443	61623	34511
20	11362	51554	22341	51151	54415	46564	41132	25241	61422	56462
21	63564	53526	22134	51412	45512	36613	13421	54222	63351	56466
22	32345	43464	33422	36256	45666	42225	14555	32332	11233	11322
23	61243	24433	32536	55414	61225	55664	64654	55136	55632	12626
24	41415	34622	63136	51661	32324	32153	15511	53235	51461	35551
25	44555	61616	25455	63523	61526	32526	55552	35261	12625	45335
26	42344	21656	44466	35463	22252	44515	52544	21214	33436	12452
27	52553	56335	25541	66215	13211	43631	55513	13346	56561	62322
28	32645	42445	51456	21522	41552	62556	43446	41465	66565	22464
29	51632	35146	42453	62442	36353	36415	15333	46514	11445	32666
30	26126	54455	34665	23646	11213	46624	14233	53466	15216	24554
31	55542	24466	61633	65261	25512	31343	36616	66625	13411	24241
32	42421	35252	15416	13445	36213	66454	61516	31351	11261	62536
33	35134	66442	46216	56424	45416	62341	24526	16445	33532	23645
34	53156	55253	33216	66435	12313	23463	45513	54331	43552	63131
35	52655	53453	11625	11122	43144	26265	34365	41213	53253	22532
36	36333	51511	62442	66446	14655	46155	46261	12216	15411	22216
37	55336	46136	13234	55526	56445	34264	16166	34443	24223	34654
38	51533	56362	63145	31445	15444	11431	65266	62526	32355	43414
39	52313	35216	53244	62321	12645	61345	46435	15352	22331	21164
40	22642	36525	31253	64642	22422	51662	43562	63543	22431	14355
41	14615	45632	22335	36655	54614	25321	63551	42516	56142	42456
42	64213	62151	51161	22452	33415	63333	52325	36544	11662	63513
43	42412	42245	66556	12312	25254	31432	45641	16336	21234	14145
44	25662	65466	42616	32363	53133	33236	55241	26643	33125	55215
45	41464	11562	44454	36511	43615	35411	42234	51461	24331	61636

4. ROLLS OF A 10-SIDED DIE
(OUTCOMES ARE 1, 2, 3, 4, 5, 6, 7, 8, 9, 0.)

ROW										
1	58139	72299	45134	38619	57933	90550	01588	54371	75886	76924
2	89239	57597	76540	71227	99868	03424	37642	78680	84922	89389
3	84372	48677	00399	79564	72285	34756	27033	31475	56610	36043
4	97354	39847	59654	40710	73908	61113	29706	56402	51165	42020
5	90874	09571	87331	34191	23105	52514	17456	73894	59422	43192
6	88622	78438	12475	69239	22565	29443	33159	30866	20201	20026
7	25536	41685	65620	66079	99677	78399	81362	59803	98849	80672
8	71607	03440	72563	39924	63702	49261	26680	81184	70223	58010
9	36352	94005	97564	79522	07014	26087	47015	12853	12353	79106
10	56870	38586	66196	01158	59065	58937	42268	69560	68127	73022
11	34612	62755	53740	98780	85169	80311	34946	74837	42551	11083
12	23195	53667	92865	98482	33285	11042	71318	14878	26702	22314
13	81973	74878	14070	14573	85941	10097	85966	74409	84230	76907
14	32239	03851	06229	60273	19045	04673	38096	17544	16586	84405
15	70787	96762	41229	86528	16580	80861	80223	71724	40671	30428
16	40845	23593	21351	37754	74799	52534	47573	37442	51573	64000
17	55936	88983	44115	81325	72969	67040	24768	27174	89558	94219
18	35874	81739	12529	20927	93848	58960	39848	52033	69686	26906
19	70233	03194	13621	26165	12911	25322	58448	92490	16904	01562
20	21603	05760	03928	50098	21705	33835	89285	46141	98501	27972
21	78024	83509	61761	97682	73774	02331	15209	85337	83632	90720
22	63716	24699	15675	20456	31575	98560	59548	35129	53671	40061
23	80218	98435	22840	30968	16528	93342	93557	49269	79391	79223
24	43084	65484	49311	72287	89741	65295	90996	69317	28984	42819
25	79820	87404	47868	33569	83368	69647	75264	69540	73930	54260
26	70119	54953	28077	55996	25792	87169	33594	62382	15094	40131
27	80846	92783	85904	37547	63640	98319	14640	14562	19421	94666
28	12462	03837	13749	18979	04356	26617	22403	85267	21200	25648
29	19133	26729	54710	50055	27684	46961	71513	15202	93105	16161
30	38271	33657	89959	61483	82074	10304	29883	82677	00108	35764
31	15616	16844	02856	07827	99470	37692	65163	96188	50683	18319
32	01745	25767	57051	17415	10001	36224	91802	53910	60517	57683
33	62748	29668	89093	66476	09269	24763	24612	82569	79327	90904
34	99093	83381	05990	88194	83099	10513	29126	04936	64950	89433
35	67065	85216	38036	62480	61915	89617	21077	70787	21259	71007
36	51039	39172	16012	07448	76820	34958	04526	37917	97867	05129
37	46199	44042	74526	62519	03866	36468	73418	10839	77228	61770
38	12393	94813	76144	56189	44628	94361	41531	12779	14521	86059
39	92718	18696	77999	55547	28954	31996	41126	07047	89994	66437
40	86686	35542	76537	37681	29659	87981	70342	19925	38277	70873
41	63877	03511	34354	28095	33077	06105	43190	45778	95068	42281
42	97117	61271	74027	17478	01866	99892	84231	42012	23995	39131
43	34742	97707	92160	42632	14916	08779	74865	37468	27059	25676
44	48757	80066	31423	91113	28838	01783	93203	17146	30341	68400
45	38186	82830	60499	31499	37113	08398	98966	72032	90143	36614

5. ROLLS OF A 12-SIDED DIE
(OUTCOMES ARE 1, 2, 3, 4, 5, 6, 7, 8, 9, 0, E, T. *E* STANDS FOR 11. *T* STANDS FOR 12.)

ROW										
1	86135	5E370	53257	8514E	48042	342E5	13TT9	E8594	T2077	83023
2	45T24	05439	2T761	202T3	93677	E6111	84369	956T7	76E2T	TT189
3	T9256	E2922	10661	11493	52321	79731	25E26	03969	32890	17341
4	TET75	74200	11732	9TT62	8ETT6	E0020	52634	89487	08132	36470
5	17219	406T9	16E22	26E13	TE07E	E164E	15986	74141	ET44E	49E72
6	9T443	E6042	2E377	0ETE6	90T85	2E150	10ET4	455T7	860TE	472E0
7	87418	T5770	9899T	11098	85732	79997	95452	57417	3T749	72437
8	6T662	58289	E27ET	51657	65290	32352	E0948	57176	1037T	T07E8
9	5676T	99TE3	E3925	ET198	TT575	68144	890E3	94785	E2E78	35253
10	TT110	E4697	86424	T3ET1	77980	46636	57637	6TT84	T7009	2E295
11	89501	95T6T	3E199	7485E	52898	93204	53785	79523	42394	15135
12	25749	T4800	19913	6T1T8	76489	827T3	33816	77534	53030	07178
13	1950E	70T54	5757E	9ET52	69981	82359	T485E	38017	T8T2E	0E627
14	16971	8E298	5E3ET	E7832	66643	94366	21ET3	70516	E2243	54045
15	1848T	56E61	28261	4979E	02727	92625	455TT	5T727	0T0T1	5812T
16	298E7	99398	96387	0549E	80164	50053	82TT7	32662	38T0E	00056
17	T8278	71233	03893	2T1E8	21987	53778	T657T	100E3	94726	T66ET
18	732T1	74193	5T19E	T9189	12002	T5571	628E0	61481	18291	1T495
19	E2628	1440T	47T33	30E11	796T2	39422	529E7	T3173	89974	03273
20	67059	1T24E	8T063	8E1TE	T6935	6E8TT	9E5ET	2082E	731T8	1ETE3
21	02788	530E4	38325	57T77	5T920	75890	09576	05496	T46ET	3817E
22	72763	E6461	24152	9378E	E854E	03T66	643T6	84960	1T331	15E06
23	T56E9	63293	53E41	709E3	01433	T9E2E	51903	T6567	59875	T6945
24	4383E	7152T	22606	49E07	7687T	3024E	76545	591E5	06598	29E75
25	391E8	429E3	09E92	57895	758E2	E5490	40T94	5E211	08199	81E09
26	EE02E	91379	03886	68494	E711T	E2973	970T7	99760	42107	96630
27	636T4	426T2	15T3E	14702	29077	47T87	T3TT8	317E8	37886	6329E
28	53164	41422	73844	60E25	06491	76T93	72838	T4879	5E969	582TE
29	645E6	00T11	T7201	T0709	70920	00E6E	T0135	03005	36054	37678
30	ET955	65354	3E02E	E7879	T1T09	78998	64986	5T313	4E7T1	30822
31	3787E	01122	1T214	7381T	T0209	83414	92427	03330	69E12	87T57
32	5782E	T87E4	E9125	0T627	19386	9414T	68947	E146T	E9466	3345E
33	91316	16126	6T020	8E751	91E87	T1T80	T0137	93048	370TT	1E893
34	T5031	82057	48TT3	06784	69786	28758	05E6E	74471	57T2E	72467
35	62929	3E78T	46397	5E908	6773T	16713	E1076	79654	18645	E9627
36	746E0	84T16	04239	9E697	13644	EE132	4842E	55085	89E30	682E5
37	10367	76058	E6323	1T855	05E06	60126	87012	28524	T0169	61979
38	57292	25775	85689	3138T	30083	15349	49307	90525	636E7	E179E
39	ET761	3T9E5	67E49	47T45	62102	489T6	25724	9E6T7	02546	77637
40	73436	18986	56190	57007	7E001	7780E	71423	402T6	37564	94264
41	74T67	2092E	58721	T4477	T9715	7T82E	30752	52282	67551	53761
42	43T96	30683	53013	10132	E4656	T1973	E2271	65832	63488	07E91
43	33338	7E0T8	15202	E348E	68T08	8TE66	648ET	20T31	5589T	16T04
44	80717	38842	01T43	88TE3	1865E	2226T	ET9T6	0011E	94828	24E04
45	17782	05765	183ET	E2232	687T5	06907	39E12	1E600	T0E22	62471

G COMPUTER PROGRAMS

1. STAT
(FINDS THE MEAN, VARIANCE, AND STANDARD DEVIATION OF A SET OF DATA.
SEE APPENDIX B AND CHAPTER TWO.)

PROGRAM

```
10 N=7
20 FOR I=1 TO N
30 READ X
40 S1=S1+X
50 S2=S2+X*X
60 NEXT I
70 PRINT "NUMBER OF SCORES =";N
80 PRINT "  MEAN =";S1/N
85 V=(S2-S1*S1/N)/N
90 PRINT "  VARIANCE=";
95 PRINT V
96 PRINT "    STANDARD DEVIATION=";SQR(V)
100 DATA 8,6,6,7,4
110 DATA 7,4
200 END
```

OUTPUT

```
NUMBER OF SCORES = 7
  MEAN = 6
  VARIANCE= 2
    STANDARD DEVIATION= 1.41421
```

2. RANDOM
(PRODUCES RANDOM DATA AS IN APPENDIX F.)

PROGRAM

```
00120 PRINT "1. COIN TOSSES"            00330 PRINT P;". ROLLS OF A ";M;"-DIE"
00130 PRINT "ROW"                       00340 PRINT
00140 LET C$(1)="H"                     00350 PRINT "ROW"
00150 LET C$(2)="T"                     00360 LET A$(1)="1"
00160 FOR K=1 TO 10                     00370 LET A$(2)="2"
00170 PRINT K;TAB(5);                   00380 LET A$(3)="3"
00180 FOR  I=1 TO 10                    00390 LET A$(4)="4"
00190 FOR J=1 TO 5                      00400 LET A$(5)="5"
00200 PRINT C$(INT(2*RND(-1)+1));       00410 LET A$(6)="6"
00210 NEXT J                            00420 LET A$(7)="7"
00220 PRINT "  ";                       00430 LET A$(8)="8"
00230 NEXT I                            00440 LET A$(9)="9"
00240 IF K/5<>INT(K/5)GOTO 00260        00450 LET A$(10)="0"
00250 PRINT                             00460 LET A$(11)="E"
00260 NEXT K                            00470 LET A$(12)="T"
00270 FOR P=2 TO 5                      00480 FOR K=1 TO 10
00280 READ M                            00490 PRINT K;TAB(5);
00290 DATA 4,6,10,12                    00500 FOR I=1 TO 10
00300 DIM A$(12)                        00510 FOR J=1 TO 5
00310 PRINT                             00520 LET X=INT(M*RND(-1)+1)
00320 PRINT                             00530 PRINT A$(X);
                                        00540 NEXT J
                                        00550 PRINT "   ";
                                        00560 NEXT I
                                        00570 IF K/5<>INT(K/5)GOTO 00590
                                        00580 PRINT
                                        00590 NEXT K
                                        00600 PRINT
                                        00610 NEXT P
                                        00620 PRINT
                                        00630 PRINT 6;". BIRTHDAY PROBLEM DIGITS"
                                        00640 PRINT "ROW"
                                        00650 FOR K=1 TO 5
                                        00660 PRINT K;TAB(5);
                                        00670 FOR I=1 TO 10
                                        00680 PRINT INT(365*RND(-1)+1);
                                        00690 PRINT "   ";
                                        00700 NEXT I
                                        00710 PRINT
                                        00720 NEXT K
                                        00800 END
```

OUTPUT

(SEE APPENDIX F.)

3. NORM
(PRODUCES RANDOM DATA THAT ARE NORMALLY DISTRIBUTED WITH A GIVEN MEAN AND STANDARD DEVIATION. SEE APPENDIX B, NOTE 3, AND CHAPTER NINE.)

PROGRAM

```
00100 PRINT "DESIRED MEAN";
00110 INPUT M
00120 PRINT "DESIRED STANDARD DEVIATION";
00130 INPUT S
00140 PRINT "HOW MANY DECIMAL PLACES";
00150 INPUT P
00160 PRINT "ROW"
00170 FOR I=1 TO 25
00180 PRINT I;TAB(5);
00190 FOR J=1 TO 5
00200 R=0
00210 FOR K=1 TO 12
00220 R=RND(-1)+R
00230 NEXT K
00240 PRINT TAB(10*J);
00250 N=S*(R-6)+M
00260 PRINT INT(N*10**P+.5)/(10**P);
00270 NEXT J
00280 PRINT
00290 IF I/5<>INT(I/5) GOTO 00310
00300 PRINT
00310 NEXT I
00320 END
```

OUTPUT

```
DESIRED MEAN? 5
DESIRED STANDARD DEVIATION?  2
HOW MANY DECIMAL PLACES?  1

ROW
  1      2.2       5.6       7.8       4.7       5.9
  2      3.4       4.8       3.4       8.2       6.1
  3      6.2       4.0       6.5       7.8      10.2
  4      7.5       6.8       7.0       4.7       4.8
  5      3.5       3.2       4.6       5.3       6.7

  6      3.0       2.2       6.0       3.1       7.4
  7      2.2       4.3       5.7       5.0       5.6
  8      4.5       8.1       3.8       5.9       2.7
  9      4.0       3.4       2.9       3.3       1.8
 10      8.2       6.7       7.7       2.4       8.0

 11      4.7       3.3       5.3       6.9       8.1
 12      4.1       4.9       3.3       2.8       1.9
 13      5.5       5.4       5.7       4.2       5.3
```

4. CHI
(PRODUCES VALUES OF CHI-SQUARE FOR A FAIR DIE OF A GIVEN NUMBER OF SIDES. SEE CHAPTER SIX.)

PROGRAM

```
00100 BASE0
00110 A$(0)="0"
00120 A$(1)="1"
00130 A$(2)="2"
00140 A$(3)="3"
00150 A$(4)="4"
00160 A$(5)="5"
00170 A$(6)="6"
00180 A$(7)="7"
00190 A$(8)="8"
00200 A$(9)="9"
00210 DIM X(500)
00220 DIM Y1(500),Y2(500)
00230 DIM Y(500)
00240 PRINT "N SIDES OF DIE=";
00250 INPUT N2
00260 PRINT "N ROLLS OF DIE=";
00270 INPUT N1
00280 PRINT "N TRIALS =";
00290 INPUT T
00300 FOR J=1 TO T
00310 D1=D2=0
00320 FOR K=1 TO N2
00330  O(K)=0
00340 NEXT K
00350 FOR I=1 TO N1
00360  U=INT(N2*RND(-1)+1)
00370 O(U)=O(U)+1
00380 NEXT I
00390 FOR I=1 TO N2
00400  E(I)=N1/N2
00410 D2=D2+(O(I)-E(I))*(O(I)-E(I))
00420 D1=ABS(O(I)-E(I))+D1
00430 NEXT I
00440  X(D1)=X(D1)+1
00450  Y(D2)=Y(D2)+1
00460 D3=INT(10*D2/E(1)+.5)
00470 Y2(D3)=Y2(D3)+1
00480 NEXT J
00490 PRINT "D1";"    D","FREQ"
00500 FOR D1=1 TO 100
00510 IF X(D1)=0 GOTO 00530
00520 PRINT D1;D1/E(1),X(D1)
00530 NEXT D1
00540 PRINT
00550 PRINT "STEM-LEAF FOR CHI-SQUARE"
00560 FOR D2=0 TO 500
00570 IF Y(D2)=0 GOTO 00630
00580 GOTO 00600
00590 PRINT D2;D2/E(1),Y(D2)
00600  S=S+Y(D2)*D2/E(1)
00610  S1=S1+(D2/E(1))*D2/E(1)*Y(D2)
00620  F=F+Y(D2)
00630 NEXT D2
00640 PRINT "TOTAL FREQ=";F
00650 PRINT "MEAN CH SQ=";S/F
00660   PRINT"STEM","LEAF"
00670 FOR I=0 TO 500
00680 IF Y2(I)=0 GOTO 00710
00690 J=INT(I/10)
00700 Y1(J)=Y1(J)+Y2(I)
00710 NEXT I
00720 FOR J=0 TO 25
00740 PRINT J;TAB(12);
00750 FOR K=0 TO 9
00760 IF Y2(10*J+K)=0 GOTO 00800
00770 FOR L=1 TO Y2(10*J+K)
00780 PRINT A$(K);
00790 NEXT L
00800 NEXT K
00810 PRINT
00820 NEXT J
```

OUTPUT

```
N SIDES OF DIE= ? 6
N ROLLS OF DIE= ? 30
N TRIALS =? 100
D1   D          FREQ
 2   .4           2
 4   .8          10
 6  1.2           9
 8  1.6          27
10  2            21
12  2.4          20
14  2.8           4
16  3.2           3
18  3.6           3
22  4.4           1


STEM-LEAF FOR CHI-SQUARE
TOTAL FREQ= 100
MEAN CH SQ= 4.86
STEM          LEAF
 0            44888
 1            2222226666
 2            000488888888
 3            222222222666666
 4            004444444448888
 5            22222226666666
 6            00044448888888
 7            66
 8            0888
 9            2222
10            0
11            2
12
13            26
14
15
16
17
18
19
20            0
21
```

5. LINREG
(HELPS IN SEARCH FOR BEST-FITTING LINE FOR A SET OF DATA.
THIS PROGRAM WAS USED, FOR EXAMPLE, TO PRODUCE THE VALUES IN EXERCISE 10, PAGE 210.)

PROGRAM

```
00100 REM LINEAR REGRESSION PGM FOR BEST FITTING LINE
00110 REM*****    KEN TRAVERS AND PHIL HEELER   6-1-81
00120 REM    **************************************************
00130 DIM X(50),Y(50)
00140 DEF FNR(X)=INT(100*X+.05)/100
00150 S1=S2=S3=S4=S5=S6=S7=0
00160 REM READ IN NUMBER OF DATA POINTS
00170 READ N
00180 DATA 9
00190 REM
00200 REM READ IN DATA VALUES AND SUM
00210 FOR I=1TON
00220     READ X(I),Y(I)
00230     LET S1=S1+X(I)
00240     LET S2=S2+Y(I)
00250     LET S5 = S5 + X(I) * X(I)
00260     LET S6 = S6 + Y(I) * Y(I)
00270     LET S7 = S7 + X(I) * Y(I)
00280 NEXT I
00290 REM
00300 PRINT " THERE ARE ";N;" PAIRS OF DATA"
00310 PRINT "X","Y"
00320 PRINT "--------------------"
00330 FORI=1 TO N
00340     PRINT X(I),Y(I)
00350 NEXT I
00360 PRINT "MEAN OF X =";S1/N
00370 PRINT "MEAN OF Y =";S2/N
00380 LET E=S5/N - S1*S1/(N*N)
00390 PRINT "STANDARD  DEVIATION FOR X=";FNR(SQR(E))
00400 T9= SQR(S6/N-S2*S2/(N*N))
00410 PRINT "STANDARD DEVIATION FOR Y=";FNR(T9)
00420 LET C = S7/N - S1*S2/(N*N)
00430 PRINT "COVARIANCE OF X AND Y =";FNR(C)
00440 PRINT "CORRELATION BETWEEN X AND Y=";C/(SQR(E)*T9)
00450 PRINT "SLOPE OF Y ON X=";C/E
00460 PRINT
00470 PRINT "**** SEARCH FOR BEST FITTING LINE ****"
00480 PRINT
00490 PRINT "ENTER LEFT END POINT FOR SLOPE";
00500 INPUT E1
00510 PRINT "ENTER RIGHT END POINT FOR SLOPE";
00520 INPUT E2
00530 PRINT "ENTER INCREMENT";
00540 INPUT E3
```

PROGRAM (Continued)

```
00550  PRINT
00560  REM OUTPUT TABLE OF ERRORS
00570  PRINT "SLOPE","MEAN ERROR","MEAN SQUARE ERROR"
00580  FOR M=E1 TO E2 STEP E3
00590  REM COMPUTE INTERCEPT B FOR THIS SLOPE
00600     B = S2/N - M*S1/N
00610     S3 =0
00620     S4 = 0
00630     FOR I=1 TO N
00640  REM COMPUTE PREDICTED VALUE FOR Y
00650  REM AND SUM UP DIFFERENCES FROM ACTUAL VALUE
00660        Y1 = M*X(I) +B
00670        D1= Y(I) - Y1
00680        S3 = S3 + ABS(D1)
00690        S4 = S4 + D1 * D1
00700     NEXT I
00710  PRINT M,FNR(S3/N),FNR(S4/N)
00720  NEXT M
00730  PRINT
00740  PRINT "CONTINUE SEARCH 0=NO; 1=YES";
00750  INPUT P
00760  IF P = 1 THEN 00480
00770  PRINT
00780  REM OUTPUT RESULTING LINE
00790  PRINT "ENTER MINIMIZING SLOPE";
00800  INPUT M
00810  B = S2/N - M * S1/N
00820  PRINT
00830  PRINT "***EQUATION OF BEST FITTING LINE ***"
00840  PRINT " Y = ";M;"* X + ";FNR(B)
00850  PRINT
00860  PRINT "COMPARED TO THEORETICAL VALUE FOR SLOPE"
00870  PRINT "SLOPE = COVARIANCE / VARIANCE OF X"
00880  PRINT " M = ";FNR(C);"/ ";FNR(E);" = ";FNR(C/E)
00890  REM
00900  REM DATA VALUES GO HERE IN X(I),Y(I) ORDER
00910  DATA 12,19
00920  DATA 4,18
00930  DATA 8,8
00940  DATA 4,16
00950  DATA 12,26
00960  DATA 7,12
00970  DATA 5,12
00980  DATA 9,29
00990  DATA 14,31
01000  DATA 2,1
01010  DATA 46,21
01020  DATA 32,23
01030  DATA 30,18
01040  DATA 37,27
01050  DATA 35,25
01060  END
```

OUTPUT

(SEE, FOR EXAMPLE, EXERCISE 10, PAGE 210.)

```
 THERE ARE  9  PAIRS OF DATA
X                 Y
-------------------
 12               19
 4                18
 8                8
 4                16
 12               26
 7                12
 5                12
 9                29
 14               31
MEAN OF X = 8.33333
MEAN OF Y = 19
STANDARD  DEVIATION FOR X= 3.49
STANDARD DEVIATION FOR Y= 7.61
COVARIANCE OF X AND Y = 17.11
CORRELATION BETWEEN X AND Y= .642672
SLOPE OF Y ON X= 1.4

**** SEARCH FOR BEST FITTING LINE ****

ENTER LEFT END POINT FOR SLOPE? 1.0
ENTER RIGHT END POINT FOR SLOPE ? 2.0
ENTER INCREMENT ? 0.1

SLOPE          MEAN ERROR       MEAN SQUARE ERROR
1              5.25             36
1.1            5.23             35.14
1.2            5.2              34.53
1.3            5.17             34.16
1.4            5.14             34.04
1.5            5.11             34.16
1.6            5.08             34.53
1.7            5.05             35.14
1.8            5.02             36
1.9            4.99             37.1
2.             5.03             38.44

CONTINUE SEARCH 0=NO; 1=YES ? 0

ENTER MINIMIZING SLOPE? 1.4

***EQUATION OF BEST FITTING LINE ***
 Y =  1.4 * X +  7.33

COMPARED TO THEORETICAL VALUE FOR SLOPE
SLOPE = COVARIANCE / VARIANCE OF X
 M =  17.11 /  12.22  =  1.4
```

6. SAMPLE
(PRODUCES RANDOM SAMPLES OF A GIVEN SIZE FROM A NORMALLY DISTRIBUTED POPULATION HAVING A GIVEN MEAN *M* AND STANDARD DEVIATION *S*.)

PROGRAM

```
00100 LET M=50
00110 LET S=10
00120 PRINT "SAMPLING FROM A NORMALLY DISTRIBUTED POPULATION"
00130 PRINT "WITH A MEAN OF ";M;" AND STANDARD DEVIATION ";S
00140 REM SAMPLE SIZE IS N
00150 FOR I=1 TO 5
00160 READ N
00170 DATA 5,10,25,50,100
00180 PRINT
00190 PRINT "SAMPLE ";I;" HAS N=";N
00200 LET A=0
00210 LET B=0
00220 FOR K=1 TO N
00230 LET R=0
00240 FOR L=1 TO 12
00250 LET R=RND(-1)+R
00260 NEXT L
00270 LET X=INT(S*(R-6))+M
00280 PRINT X;
00290 LET A=A+X
00300 LET B=B+X*X
00310 NEXT K
00320 PRINT
00330 LET B=B-A*A/N
00340 PRINT "MEAN=";A/N,"VARIANCE(BIASED)=";B/N,
00350 PRINT "ST DEV=";SQR(B/N)
00360 NEXT I
00370 END
```

OUTPUT

SAMPLING FROM A NORMALLY DISTRIBUTED POPULATION
WITH A MEAN OF   50   AND STANDARD DEVIATION   10

SAMPLE   1   HAS N= 5
 54   54   39   46   38
MEAN= 46.2   VARIANCE(BIASED)= 48.16   ST DEV= 6.93974

SAMPLE   2   HAS N= 10
 56   61   50   63   53   59   48   36   57   57
MEAN= 54   VARIANCE(BIASED)= 55.4   ST DEV= 7.44312

SAMPLE   3   HAS N= 25
 64   44   52   36   50   52   48   51   53   44   40   51   49   43   43   45   68   34
 52   46   40   46   66   60   49
MEAN= 49.04   VARIANCE(BIASED)= 70.4384   ST DEV= 8.39276

SAMPLE   4   HAS N= 50
 50   50   26   46   36   42   40   32   37   72   41   47   53   53   74   57   42   58
 51   50   55   41   57   58   49   47   56   59   51   72   42   46   45   54   43   41
 48   51   61   64   51   49   46   62   47   59   65   66   55   41
MEAN= 50.76   VARIANCE(BIASED)= 100.102   ST DEV= 10.0051

SAMPLE   5   HAS N= 100
 55   65   42   58   60   73   52   56   55   47   38   40   55   46   57   35   33   53
 63   59   58   43   47   62   49   70   47   39   54   44   38   44   43   62   41   34
 53   54   53   43   47   33   31   53   47   45   38   48   47   51   60   54   44   51
 39   39   53   58   46   44   42   45   31   36   50   55   57   60   38   47   76   44
 45   66   47   22   55   59   52   39   38   58   48   42   33   49   51   42   35   53
 56   44   65   57   29   35   33   40   50   60
MEAN= 48.32   VARIANCE(BIASED)= 102.838   ST DEV= 10.1409

H TABLE OF SQUARES AND SQUARE ROOTS

N	N^2	$\sqrt{N}$	$\sqrt{10N}$	N	N^2	$\sqrt{N}$	$\sqrt{10N}$
1.0	1.00	1.000	3.162	5.5	30.25	2.345	7.416
1.1	1.21	1.049	3.317	5.6	31.36	2.366	7.483
1.2	1.44	1.095	3.464	5.7	32.49	2.387	7.550
1.3	1.69	1.140	3.606	5.8	33.64	2.408	7.616
1.4	1.96	1.183	3.742	5.9	34.81	2.429	7.681
1.5	2.25	1.225	3.873	6.0	36.00	2.449	7.746
1.6	2.56	1.265	4.000	6.1	37.21	2.470	7.810
1.7	2.89	1.304	4.123	6.2	38.44	2.490	7.874
1.8	3.24	1.342	4.243	6.3	39.69	2.510	7.937
1.9	3.61	1.378	4.359	6.4	40.96	2.530	8.000
2.0	4.00	1.414	4.472	6.5	42.25	2.550	8.062
2.1	4.41	1.449	4.583	6.6	43.56	2.569	8.124
2.2	4.84	1.483	4.690	6.7	44.89	2.588	8.185
2.3	5.29	1.517	4.796	6.8	46.24	2.608	8.246
2.4	5.76	1.549	4.899	6.9	47.61	2.627	8.307
2.5	6.25	1.581	5.000	7.0	49.00	2.646	8.367
2.6	6.76	1.612	5.099	7.1	50.41	2.665	8.426
2.7	7.29	1.643	5.196	7.2	51.84	2.683	8.485
2.8	7.84	1.673	5.292	7.3	53.29	2.702	8.544
2.9	8.41	1.703	5.385	7.4	54.76	2.720	8.602
3.0	9.00	1.732	5.477	7.5	56.25	2.739	8.660
3.1	9.61	1.761	5.568	7.6	57.76	2.757	8.718
3.2	10.24	1.789	5.657	7.7	59.29	2.775	8.775
3.3	10.89	1.817	5.745	7.8	60.84	2.793	8.832
3.4	11.56	1.844	5.831	7.9	62.41	2.811	8.888
3.5	12.25	1.871	5.916	8.0	64.00	2.828	8.944
3.6	12.96	1.897	6.000	8.1	65.61	2.846	9.000
3.7	13.69	1.924	6.083	8.2	67.24	2.864	9.055
3.8	14.44	1.949	6.164	8.3	68.89	2.881	9.110
3.9	15.21	1.975	6.245	8.4	70.56	2.898	9.165
4.0	16.00	2.000	6.325	8.5	72.25	2.915	9.220
4.1	16.81	2.025	6.403	8.6	73.96	2.933	9.274
4.2	17.64	2.049	6.481	8.7	75.69	2.950	9.327
4.3	18.49	2.074	6.557	8.8	77.44	2.966	9.381
4.4	19.36	2.098	6.633	8.9	79.21	2.983	9.434
4.5	20.25	2.121	6.708	9.0	81.00	3.000	9.487
4.6	21.16	2.145	6.782	9.1	82.81	3.017	9.539
4.7	22.09	2.168	6.856	9.2	84.64	3.033	9.592
4.8	23.04	2.191	6.928	9.3	86.49	3.050	9.644
4.9	24.01	2.214	7.000	9.4	88.36	3.066	9.695
5.0	25.00	2.236	7.071	9.5	90.25	3.082	9.747
5.1	26.01	2.258	7.141	9.6	92.16	3.098	9.798
5.2	27.04	2.280	7.211	9.7	94.09	3.114	9.849
5.3	28.09	2.302	7.280	9.8	96.04	3.130	9.899
5.4	29.16	2.324	7.348	9.9	98.01	3.146	9.950
5.5	30.25	2.345	7.416	10	100.00	3.162	10.000

(From Paul A. Foerster, *Algebra and Trigonometry: Functions and Applications.* Addison-Wesley, 1984.)

Glossary

A

Accuracy: The extent to which measurements are close to their true value. (p. 295)

Average: Any measure of central tendency for a set of data. Usually is the mean. (p. 26)

B

Bias: A systematic error in a measurement or set of measurements. (p. 294)

Bivariate Data: Data involving two variables, such as height and weight, or amount of smoking and measured condition of health. (p. 162)

Box-and-Whisker Plot: A way of displaying data that shows the measures of central tendency and variation. (p. 39)

C

Central Tendency (measure of): A value that tells what number is at the middle, or center, of a set of data. Different measures of central tendency include mean (p. 26), median (p. 27), and mode (p. 28).

Complementary Events: Events that are opposites of each other. If event A occurs, then the complement of A cannot occur. (p. 98)

Confidence Interval: An interval used to estimate some population value (parameter). (p. 261)

Correlation: A measure of the extent to which two variables are related. A closely related concept is covariance. (p. 221)

Covariance: A measure of how two variables vary with respect to one another. A closely related concept is correlation. (p. 217)

D

Data: Numerical information, usually about the real world. (p. 2)

Degrees of Freedom: An integer required to describe the distribution of certain statistics such as chi square or t. (p. 148)

Deviation Score: The difference between a score and the mean score for the set (p. 43)

E

Event: A result or outcome of a probabilistic experiment. (p. 88)

Expected Value (Experimental): The average value of a random quantity that has been repeatedly observed in an experiment. (p. 56)

Expected Value (Theoretical): The value that the average of a random quantity approaches when it is observed many, many times. (p. 56)

F

Frequency Polygon: A graph whose shape is a polygon. The graph shows the frequency of occurrence of data values in a set. (p. 10)

G

Grand Mean: An overall mean, such as the mean of a set of sample means. (p. 291)

H

Histogram: A bar graph representing the frequency of occurrence of data values. (p. 9)

I

Independent Events: Two events are independent if one of the events does not affect the probability of the occurrence of the other event. (p. 60)

Inverse Operations: Two arithmetic operations are inverses of each other if the effect of one undoes the effect of the other. For example, adding and subtracting are inverse operations. (p. 237)

L _____

Linear Relationships: In bivariate data (data involving two variables), if the graph of the two variables is close to that of a straight line, the two variables are said to have a linear relationship. (p. 170)

M _____

Mean: The sum of a set of numbers divided by the number of numbers in the set. What is usually meant by "average" or "arithmetic average." (p. 26)

Mean Data Point: In a graph of bivariate data (data involving two variables), the point whose coordinates are the respective means for the two variables. (p. 189)

Mean Deviation: The measure of variation that is the mean of the deviations of each score from the mean. (p. 44)

Median: The middle number in a set of numbers, when they are arranged in order. (p. 27)

Median Data Point: In a graph of bivariate data (data involving two variables), the point whose coordinates are the respective medians for the two variables. (p. 189)

Mode: The number in a set that occurs most frequently. (p. 28)

Monte Carlo Method: A method of solving problems by doing probabilistic experiments, such as tossing coins or rolling dice. (p. 58)

N _____

Negative Relationship: For a set of bivariate data (data in two variables), the variables have a negative relationship if one decreases as the other increases (that is, the slope of their regression line is negative). (p. 167)

Negative Slope: A line has a negative slope if the Y values *decrease* as the X values increase. (p. 176)

P _____

Point Estimate: An estimate of a population value using one specified value (as opposed, say, to using an interval). (p. 332)

Population: The entire collection of objects under consideration for study. (p. 102)

Positive Relationship: For a set of bivariate data (data in two variables), the variables have a positive relationship if one increases as the other increases (that is, the slope of the regression line is positive). (p. 167)

Positive Slope: A line has a positive slope if the Y values increase as the X values increase. (p. 174)

Precision: The degree of smallness in the variation of a set of measurements of some object, such as the diameter of an atom. High precision measurements have very little variation. (p. 294)

Probability (Experimental): The estimated likelihood of an event. Obtained by dividing the number of successful trials by the total number of trials. (p. 77)

Probability (Theoretical): The true probability of an event. Obtained by dividing the number of successful outcomes (outcomes of interest) by the total number of outcomes. (p. 92)

Pseudo-Random Digits: Digits produced by a formula but having the characteristics of randomness (*see* Random Digits). (p. 295)

R _____

Random: Not predictable; occurring by chance. For example, the outcome of tossing a fair coin is random. (p. 69)

Random Digits: Digits that occur in random fashion, as when produced by rolling a die or using a spinner. (p. 294)

Random Sample: A set of data chosen from a population in such a way that each member of the population has an equal probability of being selected. (p. 304)

Random Walk: The path taken by an object moving at random. (p. 69)

Range: The measure of variation that is the difference between the largest number and the smallest number (score) in a set of data. (p. 34)

Regression Equation: The equation of the regression line (*see* Regression Line). (p. 193)

Regression Line: A straight line used to estimate the relationship between two variables. (p. 193)

Relative Frequency: Same as experimental probability. (p. 76)

S _____

Sample: Same as random sample. (p. 102)

Shrink: A transformation of a set of data obtained by dividing each value in the set by a factor greater than one. This is equivalent to multiplying each value in the set by a factor less than one. (p. 236)

Slide: A transformation of a set of data obtained by adding a (positive or negative) constant to each value in the set. (p. 234)

Slope: The slope of a line is the rate of change of the Y values with respect to the rate of change of the X values. (p. 174)

Standard Deviation: A particular measure of the amount of variation in a set of data. It is the square root of the variance. (p. 47)

Standard Scores: Scores that have been transformed so that they have a specified mean and standard deviation. Examples are IQ (Intelligence Quotient) and SAT (Scholastic Aptitude Test) scores (p. 245).

Statistic: A piece of numerical information obtained from a sample. (p. 2)

Statistics: The study of the organizing, analyzing, and interpreting of data. (p. 103)

Stem-and-Leaf Plot: A display of data using certain digits (such as in the tens place) as "stems" and the remaining digit or digits (such as in the ones place) as "leaves." (p. 5)

Stretch: A transformation of a set of data by multiplying each value in the set by a factor greater than one (*see* Shrink). (p. 236)

T _____

Transformed Score: Same as standard score. (p. 248)

V _____

Variable: A quantity that varies. For example, the weights of members of a football team is a variable. Variables are usually represented by letters. (p. 162)

Variance: A particular measure of the amount of variation in a set of data. It is the mean of the squared distances of all the scores from their mean. (p. 48)

Variation (measure of): A value that tells how much a set of data varies. Different measures of variation include mean deviation (p. 44), variance (p. 47), standard deviation (p. 47), and range (p. 34).

Z _____

Z-Scores: Scores that have a mean of zero and a standard deviation of one. A particular standard score. (p. 247)

Index

Acknowledgments

CHAPTER 1

1 F. Mosteller, "Deciding Authorships," Wm. J. Tanuer, ed., *Statistics: A Guide to the Unknown*, San Francisco: Holden-Day, 1978.
2 1981/1982 *Survey Results*, International Airline Passenger Association.
3 *Historical Statistics of the United States.* Colonial Times to 1970, Bureau of the Census, Part I, p. 56.
4 Ibid, p. 55.
5 Scholastic Magazine, 1962. By permission of the National Science Teachers Association, 1742 Connecticut Avenue, NW, Washington, DC 20009.
6 *World Almanac*, 1981. p. 280.
7 *Information Please*, 1979, p. 271.
8 *World Almanac*, 1982.
9 P. Meier, "The Biggest Public Health Experiment Ever: The 1954 Field Trials of the Salk Poliomyelitis Vaccine," in J. Tanur, ed., *Statistics: A Guide to the Unknown.* San Francisco. Holden-Day, 1978.
10 D. G. Chapman, "The Plight of the Whales," in J. Tanur, ed., *Statistics: A Guide to the Unknown.* San Francisco. Holden-Day, 1978.
11 J. E. Kerrich, *An Experimental Introduction to the Theory of Probability*. Copenhagen: Jorgensen, 1946. p. 14.
12 *Time*, April 9, 1979, p. 57. © Time Inc. All rights reserved. Reprinted by permission.

CHAPTER 2

1 Adapted from A.S.C. Ehrenberg, *Data Reduction.* Chichester, England: John Wiley, 1975, p. 172.
2 National Cattlemen's Association.
3 *Consumer Reports*, 1979 Buying Guide, p. 75.
4 Union Bank of Switzerland.
5 From the Employment and Training Report to the President, 1979, p. 273.
6 Mosteller, F., "Escape-Avoidance Experiment." *Statistics by Example: Finding Models.* Menlo Park, CA: Addison-Wesley, 1973, pp. 35-39.
7 Boston: Newspaper Enterprise Association, *The World Almanac and Book of Facts*, 1969. As cited in Mosteller, F., "Collegiate Football Scores." *Statistics By Example: Exploring Data.* Menlo Park, CA: Addison-Wesley, 1972, pp. 61-74.
8 American Express Appointment Book, 1984, New York, p. 210.

9 John S. Rinehart, "Water-Generated Earth Vibrations," *Science*, 164, (27 June 1969), pp. 1515-1519. American Association for the Advancement of Science, 1515 Massachusetts Avenue, NW, Washington, DC 20005. As cited in F. Mosteller, "Transformations for Linearity." *Statistics by Example: Detecting Patterns.* Menlo Park, CA: Addison-Wesley, 1973, pp. 99-108.

CHAPTER 3

1 *The Ryatts* by Jack Elrod. © News Group Chicago, Inc. Courtesy America Syndicate.
2 M. Zelinda, "How Many Games to Complete a World Series?" *Statistics by Example: Weighing Chances.* Menlo Park, CA: Addison-Wesley, 1973, pp. 23-32.

CHAPTER 4

1 "20% Chance a Piece of Skylab Hits a City" by James Coates. Copyright, 1979, *Chicago Tribune.* Used with permission.
2 Adapted from Judith M. Tanur, ed. *Statistics: A Guide to the Unknown.* San Francisco: Holden-Day, 1978.
3 Adapted from: J. F. Kerrich, *An Experimental Introduction to the Theory of Probability.* Copenhagen. J. Jorgenson and Co., 1964, p. 14.
4 Richard G. Brown, "Predicting the Outcome of the World Series," *Statistics by Example: Detecting Patterns.* Menlo Park, CA: Addison-Wesley, 1973, pp. 1-12.
5 Champaign-Urbana *News Gazette*, February 25, 1976.
6 CBS Evening News, January 17, 1983.
7 R. Carlson, "Random Digits and Some of Their Uses," *Statistics by Example: Weighing Chances.* Menlo Park, CA: Addison-Wesley, 1973, pp. 11-22.
8 The Rand Corporation, *A Million Random Digits with 100,000 Normal Deviates.* New York: The Free Press, 1955.
9 Adapted from R. Link and M. Brown, "Classifying Pebbles: A Look at the Binomial Distribution," *Statistics by Example: Weighing Chances.* Menlo Park, CA: Addison-Wesley, 1973, pp. 33-44.

CHAPTER 5

1 *Education Week,* January 12, 1982, p. 8. Editorial Projects in Education, Inc., Washington, DC 20036.
2 *Weekend Magazine*, Toronto.
3 The Associated Press, July 23, 1982.

4 This article is reprinted with the permission of *Family Weekly*, 1515 Broadway, New York 10036, copyright 1980.

5 S. Chatterjees. "Estimating Wildlife Populations by the Capture-Recapture Method." *Statistics by Example: Finding Models*. Menlo Park, CA: Addison-Wesley, 1973, pp. 25-34.

6 C. A. Whitney, "Statistics, the Sun, and the Stars," in J. Tanur, ed. *Statistics: A Guide to the Unknown*. San Francisco: Holden-Day, 1978.

7 S. Fienberg. "The Care of the Vanishing Women Jurors," in *Statistics by Example: Detecting Patterns*. Menlo Park, CA: Addison-Wesley, 1973, pp. 87-93.

8 Downie and Starry, *Descriptive and Inferential Statistics*. New York: Harper and Row, 1977, p. 251.

CHAPTER 6

1 W. Kruskal. "Plagiarism and Probability." *Statistics by Example: Weighing Chances*. Menlo Park, CA: Addison-Wesley, 1973, pp. 1-10.

2 R. Carlson, "Introduction to the Chi-Square Procedure," *Statistics by Example: Weighing Chances*. Menlo Park, CA: Addison-Wesley, 1973, pp. 53-66.

3 R. K. Tsutakaua, "Chi-Square Distributions by Computer Simulation," *Statistics by Example: Detecting Patterns*. Menlo Park, CA: Addison-Wesley, 1973, pp. 39-62.

4 F. Mosteller, "Fractions on Closing Stock Market Prices," *Statistics by Example: Exploring Data*. Menlo Park, CA: Addison-Wesley, 1973, pp. 9-14.

5 S. Zahl. "Grocery Shopping and the Central Limit Theorem." *Statistics by Example: Detecting Patterns*. Menlo Park, CA: Addison-Wesley, 1973, pp. 33-38.

CHAPTER 7

1 Pierce, George W. *The Songs of Insects*. Cambridge, MA: Harvard University Press, 1949, pp. 12-21. Reprinted by permission.

2 *Time*, May 28, 1979, p. 49. © Time Inc. All rights reserved. Reprinted by permission from *Time*.

3 *Weekend Magazine,* Toronto.

4 Champaign-Urbana *News Gazette,* October 21, 1975.

5 FBI

6 *Education Week,* October 20, 1982. Editorial Projects in Education, Inc., Washington, DC 20036. Graph also supplied by source.

7 A.S.C. Ehrenberg, *Data Reduction*. Chichester, England: John Wiley, 1975, pp. 72-74.

CHAPTER 8

1 Robert C. Fadely. "Oregon Malignancy Pattern Physiographically Related to Hanford, Washington, Radioisotope Storage." *Journal of Environmental Health*, Volume 27, 1965, pp. 883-897.

2 Paul Baum and Ernest M. Scheuer. *Statistics Made Relevant*. New York: John Wiley, 1976, p. 147.

3 U.S. Department of Commerce, Bureau of the Census.

4 Elizabeth Whelan and Frederick Stare. *The One-Hundred-Percent Natural, Purely Organic, Cholesterol-Free, Megavitamin, Low-Carbohydrate Nutrition Hoax*. New York: Atheneum, 1983, as cited in *Reader's Digest*, October, 1983, pp. 183-141. "Sweet Truths About Sugar."

5 A. P. Shulte, "Points and Fouls in Basketball." *Statistics by Example: Exploring Data*. Menlo Park, CA: Addison-Wesley, 1973, pp. 27-32.

6 *Time*, July 26, 1982, p. 41. © Time Inc. All rights reserved. Reprinted by permission from *Time*.

7 V. Ziswiler, *Extinct and Vanishing Animals*. New York: Springer-Verlag, Heidelberg, 1967, p. vii.

CHAPTER 9

1 Ralph B. D'Agostino. "How Much Does a 40-Pound Box of Bananas Weigh?" *Statistics by Example: Detecting Patterns*. Menlo Park, CA: Addison-Wesley, 1973, pp. 29-32.

2 Frederick Mosteller. "Stock Market Fractions," *Statistics by Example: Weighing Chances*. Menlo Park, CA: Addison-Wesley, 1973, pp. 67-69.

3 Ibid.

4 Roger Carlson. "Random Digits and Some of Their Uses." *Statistics by Example: Weighing Chances*. Menlo Park, CA: Addison-Wesley, 1973, p. 16.

5 W. J. Youden, *Experimentation and Measurement*. 1962. National Science Teachers Association, 1742 Connecticut Avenue, NW, Washington, DC 20009, p. 50.

6 Roger Carlson. "Normal Probability Distributions." *Statistics by Example: Detecting Patterns*. Menlo Park, CA: Addison-Wesley, 1973, p. 17.

7 Roger Carlson. op. cit., pp. 13-28.

8 Adapted from *Probability with Statistical Applications,* sec. ed., by F. Mosteller, R.E.K. Rourke, and G. B. Thomas, Jr. Reading, MA: Addison-Wesley, 1970, p. 473.

9 W. J. Youden, op. cit., Table 18, p. 107.

10 Roger Carlson. op. cit., p. 28.

11 R. D'Agostino, "How Much Does a 40-Pound Box of Bananas Weigh?" *Statistics by Example: Detecting Patterns*. Menlo Park, CA: Addison-Wesley, 1973, pp. 27-32.

CHAPTER 10

1 John Mandel. "The Acceleration of Gravity." *Statistics by Example: Detecting Patterns*. Menlo Park, CA: Addison-Wesley, 1973, pp. 109-117.

2 W. J. Youden. *Experimentation and Measurement*. 1962. National Science Teachers Association, 1742 Connecticut Avenue, NW, Washington, DC 20009, p. 55.

3 John Mandel, op. cit., p. 110.

4 W. J. Youden, op. cit., pp. 51-52.

5 John Mandel, op. cit., p. 110.

6 W. J. Youden, op. cit., pp. 21-22.

7 Adapted from John Mandel, op. cit., pp. 109-118.

CHAPTER 11

1 S. Chatterjees, "Estimating the Size of Wildlife Populations," *Statistics by Example: Exploring Data.* Menlo Park, CA: Addison-Wesley, 1973, pp. 99-104.

2 *"Ancient Tree Rings Tell a Tale of Past and Future Droughts,"* by Walter Sullivan, September 2, 1980. Copyright © 1980 by the New York Times Company. Reprinted by permission.

3 *"The American Freshman, National Norms for Fall, 1982,"* by Alexander W. Astin. Published by Cooperative Institutional Research Program of the American Council of Education and the University of California at Los Angeles.

4 Special Interest Feature = Statistics, September 2, 1980. Copyright © 1980 by the New York Times Company. Reprinted by permission.

5 The Associated Press. May 19, 1980.

6 J. Swift, "Capture-Recapture Techniques as an Introduction to Statistical Inference." In *Applications in School Mathematics.* 1979 Yearbook, National Council of Teachers of Mathematics. Reston, Virginia, 1979, pp. 185-192.

CHAPTER 12

1 R. J. Larsen and D. F. Stroup. *Statistics in the Real World,* an excerpt from an article ("The Quest for the Perfect Figure"), pp. 44-46. Reprinted with permission of Macmillan Publishing Company. Copyright © 1976 by Macmillan Publishing Co., Inc.

2 J. Neter. *"How Accountants Save Money by Sampling."* In J. Tarver et al, eds. *Statistics: A Guide to the Unknown.* San Francisco: Holden-Day, 1972, pp. 203-212.

3 R. Larson and D. Stroup, op. cit.

4 "Tests Find No Harm to Environment from Titan Blast," October 2, 1980. Copyright © 1980 by The New York Times Company. Reprinted by permission.

5 Jon Van, "Space May Yield Nausea Cure." Copyrighted, 1980, *Chicago Tribune.* Used with permission.

6 R. Larsen and D. Stroup, op. cit. Table from p. 76.

7 Ibid, p. 51.

8 Ibid, pp. 60-63.

CHAPTER 13

1 Adapted from I. Francis, "Pie Crusts," *Statistics by Example: Finding Models.* Menlo Park, CA: Addison-Wesley, 1973, pp. 53-62.

2 R. Carlson, "Introduction to Chi Square." *Statistics by Example: Weighing Chances.* Menlo Park, CA: Addison-Wesley, 1973, pp. 53-64.

3 Adapted from R. Carlson, "The Poisson Approximations to the Binomial." *Statistics by Example: Finding Models.* Menlo Park, CA: Addison-Wesley, 1973, pp. 63-72.

4 R. Carlson. op. cit.

5 R. F. Link and M. L. Brown. "Classifying Pebbles: A Look at the Binomial Distribution." *Statistics by Example: Weighing Chances.* Menlo Park, CA: Addison-Wesley, 1973, pp. 33-44.

6 E. Jacobsen and J. Kossoff, "Opinion of Small Cars." Journal of Applied Psychology, Vol. 47 (August 1963), pp. 242-245.

7 S. Feinberg, "The Case of the Vanishing Women Jurors." *Statistics by Example: Detecting Patterns.* Menlo Park, CA: Addison-Wesley, 1973, pp. 87-89.

8 J. E. Cohen. "Independence of Amoebas." *Statistics by Example: Weighing Chances.* Menlo Park, CA: Addison-Wesley, 1973, pp. 71-74.

9 R. K. Tsutakawa. "Chi-Square Distributions by Computer Simulation." *Statistics by Example: Detecting Patterns.* Menlo Park, CA: Addison-Wesley, 1973, pp. 39-62.

10 R. Carlson, "Introduction to Chi-Square Procedure." *Statistics by Example: Weighing Chances.* Menlo Park, CA: Addison-Wesley, 1973, pp. 53-66.

11 W. H. Kruskel and R. S. Pieters. "Tom Paine and Social Security." *Statistics by Example: Exploring Data.* Menlo Park, CA: Addison-Wesley, 1973, pp. 105-112.

CHAPTER 14

1 Most of the examples of this feature are from M. Gardner, "On the Meaning of Randomness and Some Ways of Achieving It." *Scientific American*, Volume 219, July 1968, pp. 116-121. The reference to W. S. Gassett is from F. Mosteller, R. Rourke, and G. B. Thomas. *Probability with Statistical Applications.* Reading, MA: Addison-Wesley, 1970, p. 434.

Selected Answers

NOTE: Due to space limitations, graphs are not included.
See solution key in Teacher's Resource Manual.

1 CHAPTER ONE
DESCRIPTIVE STATISTICS

SECTION 1.1

2. Life Expectancy: Better health care; nutrition; physical fitness awareness

1980	73	2050	84
1990	75	2060	86
2000	76	2070	87
2010	78	2080	88
2020	79	2090	90
2030	81	2100	91
2040	83		

(Answers will vary; this is an illustration only)

7. 93 million miles; better equipment and techniques for measuring

SECTION 1.2

2. Stems for rats' weights: 18, 19, 20, 21, 22

3. Rats' weights

STEM	LEAF	FREQ
18	2,8	2
19	2,4,1,0,3,7,6,1,8,0,2,6,0,5	14
20	6,2,5,0,6,8,1,7,5,2,3,4,6,3	14
21	0,6,5,0,1,1,5,5	8
22	0,0	2

4. Stopping distances (meters):

64,68,60,61,64
75,71,73
82,80

6. Lowest and Highest Scores:
(a) 72; (b) 96; (c) 24

13. Copper Bar (Divide Stem by 10)

347	9
348	6,7
349	5,5,3,5,3,5,9,3
350	1,8,6,2,2,7,1,0,0,4
351	2,0,3,5,0,4,1
352	0,0

SECTION 1.3

3. Cases of Polio—Highest Number: 1952; 60,000

4. Cases of Polio—Lowest Number: 1938; 2,000

15. Fingerprint Ridgecounts

18	9,6,1,9,5,8	6
19	2,4,8,2	4
20	7,5,1,5,7	5
21	3,0,3,5	4
22	0	1

SECTION 1.4

2. Six-sided Die Roll: (a) $P(1) = 12/120$ or .10; (b) $P(3) = 15/120$ or .125; (c) $P(6) = 30/120$ or .25; (d) $P(\text{even}) = (21 + 18 + 30)/120$ or $69/120$ or .575; (e) $P(\text{odd}) = (12 + 15 + 24)/120$ or $51/120$ or .425

3. True Die Roll: About 50 ones; about 50 sixes; (a) $p(1) = 1/6$; (b) $p(6) = 1/6$; (c) $p(\text{even number}) = 1/2$

5. Rocket Lift-off: $P(\text{lift-off}) = 35/40$

10. Coin Toss: (a) $P(\text{heads}) = 42/100$, approximately; (b) $P(\text{tails}) = 58/100$, approximately

SECTION 1.5

1. Number of Children in Anabru Families

N	FREQ	PROPORTION	C.P.
0	3	3/50 or .06	.06
1	8	8/50 or .16	.22
2	26	26/50 or .52	.74
3	10	10/50 or .20	.94
4	2	2/50 or .04	.98
5	0	0/50 or 0	.98
6	1	1/50 or .02	1.00
	50		

P(2 or less) = .74; P(4 or less) = .98

4. Six-sided Die Roll (120): (a) P(3 or less) = .492; (b) P(5 or less) = .792; (c) P(more than 3) = .508; (d) P(even number) = .541; (e) P(1 or 2) = .3

SECTION 1.6 (none)

2 CHAPTER TWO
CENTERS AND SPREADS

SECTION 2.1

3. Defective Flash Bulbs: $\overline{D} = (1/5)(1+3+0+0+7) = (1/5)(11) = 2.2$ defectives

6. Fire Department Response Time: $\overline{T} = 1/15$ $(1+3+2+2+1+9+4+6+1+10+1+4+5+10+1) = (1/15)(60) = 4$ min. mean response time
   Median: Arrange in order and find eighth term.
   1 1 1 1 1 2 2 **3** 4 4 5 6 9 10 10;
   Median response time 3 min.

9. Ages of Grandchildren (arranged in order): 1, 3, 4, 5, 5, 5, 10; mode is 5

12. Three Coins Tossed 10 Times: Heads = 0, 1, 1, 1, 1, 2, 2, 2, 3, 3; median = 1.5
    Mean = 16/10 = 1.6

15. Test Scores: 29, 31, 31, 31, 33, 33, 40, 45; mode is 31

SECTION 2.2

4. Earnings for Mowing Lawns: Mode–$30.00; median–$28.00; mean–$23.64; Discuss "better" in terms of problem.
   Each average gives "different" picture, not "better."

7. Doctors' Charges: 20.50 X 110 = total = $2,255.00

11. Heights of Girls

	TALLY	FREQ	
50	/	1	Bimodal–54 and 56
51		0	
52	/	1	
53	//	2	
54	⊤HI	5	55 is median
55	////	4	
56	⊤HI	5	
57	///	3	Mean is 1384/25 = 55.36
58	//	2	
59	/	1	
60	/	1	Average height is about 55 in.

SECTION 2.3

1. Least calories: lowfat milk and plain lowfat yogurt, 119; most calories: strawberry whole milk yogurt, 211; range: 211–119 or 92

5. Hourly Wage Increase: Petroleum = 7.27–1.50 or $5.77 increase (therefore the greatest); textiles = 3.97–1.04 or $2.93 increase (therefore the least)

7. Temperature Ranges

CITY	H	L	RANGE
Anchorage	61	48	13
Denver	68	51	17
Flagstaff	78	39	39
Louisville	78	55	23
Edmonton	80	48	32

8. Smallest and Largest Temperature Ranges: Anchorage had smallest range; Flagstaff had largest range

SECTION 2.4

Note: Q_1 and Q_3 are given as the largest and smallest values in the middle half.

1. Home Runs (15)

STEM	LEAF
3	2,2,3,7
4	0,2,2,2,3,5,8,9
5	2,2
6	1

$Q_1 = 37$; $Q_3 = 49$; median = 42

2. Salary Data (9)

500	0,0,0,0
600	0
700	0,0
800	0
⋮	
6000	0

$Q_1 = \$5000$; $Q_3 = \$7000$; median = $6000

SECTION 2.5

1. High Temperatures

CITY	°F	DEVIATION
Atlanta	42	− 3
Bismark	12	−33
Great Falls	27	−18
Miami Beach	77	32
El Paso	67	22

$\overline{T} = (1/5)(225) = 45$; Atlanta was closest to mean

3. Collegiate Football Games

CONFERENCE	WIN	DEV.	LOSE	DEV.
Ivy	30.6	3.9	11.0	.3
Big Eight	22.2	−4.5	7.6	−3.1
Southeastern	25.7	−1.0	9.4	−1.3
Mississippi Valley	25.8	− .9	10.6	− .1
Big Ten	23.9	−2.8	12.0	1.3
Western Athletic	33.7	7.0	15.5	4.8
Pacific Eight	24.7	−2.0	8.6	−2.1

$\overline{W} = 186.6/7$ or 26.7; $\overline{L} = 74.7/7$ or 10.7

SECTION 2.6

1. Quiz Scores

	SCORE	DEVIATION
Bill	3	−3
Jane	7	1
Mary	6	0
Pat	5	−1
Phil	9	3

Mean = $(1/5)(30) = 6$; mean dev. = $8/5 = 1.6$

2. Wind Speed

	SPEED	DEVIATION
Juneau	8.4	−1.28
Chicago	10.3	.62
Boston	12.5	2.82
Nashville	8.0	−1.68
Miami	9.2	− .48

Mean = $(1/5)(8.0+8.4+9.2+10.3+12.5) = (1/5)$ (48.4) or 9.68; median = 9.2; range = 12.5−8.0 or 4.5; mean dev. = $(1/5)(6.88)$ or 1.376

4. Waterfall Vibration

WATERFALL	VIBRATIONS	DEV.
Lower Yellowstone	5	−10.7
Yosemite	3	−12.7
Canadian Niagara	6	− 9.7
American Niagara	8	− 7.7
Upper Yellowstone	9	− 6.7
Gullifoss (lower)	6	− 9.7
Firehole	19	3.3
Godofoss	21	5.3
Gullifoss (upper)	40	24.3
Fort Greeley	40	24.3

(a) Mean = 15.7; (b) median = 8½ or (8+9)/2; (c) They differ by quite a bit; mean is affected by extreme values; (d) see deviation column; mean dev. = 114.4/10 or 11.44

SECTION 2.7

1. Variance of Quiz Scores

		DEV.	(DEV.)²
Bill	3	−3	9
Jane	7	1	1
Mary	6	0	0
Pat	5	−1	1
Phil	9	3	9

Mean = 6; var. = 4; mean dev. = 1.6, smaller than variance

2. Temperature Variance

		DEV.	(DEV.)²
Juneau	8.4	−1.28	1.6384
Chicago	10.3	.62	.3844
Boston	12.5	2.82	7.9524
Nashville	8.0	−1.68	2.8224
Miami	9.2	− .48	.2304

Var. = $(1/5)(13.028) = 2.6056 \doteq 2.61$

4. Spelling Errors

ERRORS	DEV.	(DEV.)²
12	5	25
7	0	0
5	−2	4
4	−3	9

Mean = $(1/4)(28) = 7$; var. = $(1/4)(38) = 9.5$

9. Freeway Accidents

ACCIDENTS	DEV.	(DEV.)²
4	−1	1
0	−5	25
6	1	1
10	5	25
5	0	0

Mean = 5; var. = $(1/5)(52) = 10.4$; SD $\doteq 3.22$

12. Waterfall Data: Var. $= (1/10)((-10.7)^2 + \ldots$
    $24.3^2) = (1/10)(1788.10) = 178.81$; SD $\doteq 13.37$

3 CHAPTER THREE
EXPECTED VALUE

SECTION 3.1

1. Heads Per Coin Toss: 12 tosses; 100 trials;
   mean $= (12+44+80+80+150+147+144+36+11)$
   $/100 = (624)/100 = 6.24$

4. Rain Forecast: Tails meant "no rain" so $12-6.24$
   or 5.76 had no rain; *or* use heads for "had rain,"
   then 6.24 cities (on the average) had rain
   (Example—answers will vary)

SECTION 3.2

2. Boys in 4-child Family:
   Model: Coin Heads—girl, tails—boy;
   Trial: Toss coin 4 times, once for each child;
   Stat. of int.: Record number of tails (boys);
   Repeat; Mean value = _____

5. Batting Average:
   Model: Cards club—hit
       or Die 1—hit; 2,3,4—no hit; 5,6—ignore;
   Trial: Select a card 20 times (once for each time
   at bat) or roll a die 20 times;
   Stat. of int.: Record number of hits; Repeat;
   Mean value = _____

11. Model: Cards Club, Diamond, Heart—rain; Spade
    —no rain;
    Trial: Draw a card until Spade occurs;
    Stat. of int.: Record number of cards (days)
    before Spade (rain) occurs; Repeat; Mean value
    = _____

SECTION 3.3

1. Coin Toss: A trial consists of tossing 2 coins 10
   times. Possible outcomes are 0 to 10. 10 tosses of
   the 2 coins made up a trial. 50 trials were
   requested.

4. Coin Toss: M $= 1/1000 (0(26)+1(127)+2(250)$
   $+3(253)+4(184)+5(104)+6(46)+7(7)+8(2)+9(1))$
   $= 1/1000 (0+127+500+759+736+520+276+49$
   $+16+9) = 1/1000 (2992) = 2.992$

6. Traffic Lights:
   Model: Coins Heads—green; tails—not green;
   Trial: Toss 2 coins (one for each light) 10 times
   (once for each day);
   Stat. of int.: Record the number of times HH (2
   greens) occurred; Repeat; Mean value = _____

SECTION 3.4

3. Cereal Prizes:
   Model: 8-sided die
   Trial: Roll die until each side has occurred at
   least once;
   Stat. of int.: Record the number of rolls necessary
   to obtain all 8 sides; Repeat; Mean value = _____

6. World Series:
   Model: 6-sided die 1-4—win for NL; 5-6—loss
   for NL
   Trial: Toss die until 4 wins or losses occur;
   Stat. of int.: Record the number of tosses
   (games) needed; Repeat; Mean value = _____
   (Assumption of independence of games is made)

SECTION 3.5

2. Hermit Epidemic:
   Model: 8-sided die 1 side for each hermit
   Trial: Roll die to obtain 1st hermit infected, roll
   die to determine next hermit visited, continue
   until an immune hermit is visited (ignore
   successive repetitions);
   Stat. of int.: Record number of rolls before
   repeat occurs (immune is visited); Repeat;
   Mean value = _____

SECTION 3.6

4. Random Walk:
   Example of 1 trial:

COIN TOSSES	DIR.
H T	S
T T	E
H H	N
H T	S
T T	E
T H	W
H H	N
H T	S
T T	E
T T	E

   Final destination is $(3,-1)$. Distance from $(0,0)$
   $\doteq 3.2$. Repeat trials.

6. Temperature Estimation: (a) (1,3)
Example—results of 20 class trials:

3°	13
15°	4
30°	0
45°	3

so $\overline{T} = (39+60+0+135)/20 = 234/20 = 11.7$

4 CHAPTER FOUR PROBABILITY

SECTION 4.1

2. Carpet Tacks: $P(\text{up}) = 27/500$ or .054; tosses are independent, tosses are made in same manner

5. Drawing an Ace: $P(\text{ace}) = 14/100$ or .14

12. Casey's Batting Average: Over the long run, Casey Jones has had 250 hits out of 1000 times at bat.

SECTION 4.2

3. Boys in a Family:
Model: Coin Heads—girl; tails—boy
Trial: Toss coin 4 times—once for each child;
Successful trial: Occurs if 2 or more tails; Repeat;
Prob (2 or more boys) = Successes/Trials

6. Casey's Hits:
Model: Die 1-2—hit; 3-6—no hit;
Trial: Toss die 10 times, once for each time at bat;
Successful trial: Occurs if 5 or more tosses are 1 or 2; Repeat; Prob (5 or more hits) = _____

7. Strep Throat:
Model: Die 1—strep; 2-6—no strep
Trial: Toss die 4 times, once for each child;
Successful trial: Occurs if at least one toss results in a 1; Repeat; Prob (1 or more strep throats) = _____

10. Bull's Eye:
Model: Usual aim One coin: H—hit; T—no hit;
    Careful aim Two coins: HH, HT, TH—hit;
    TT—no hit
Trial: Toss coin(s) 40 times, once for each shot, using appropriate model as described in problem;
Stat. of int.: Record number of bull's eyes;
Repeat; Mean value = _____

SECTION 4.3

3. Plastic Bags:
(a) Model: 1 digit RN (RN represents Random Number) 1-2—burst; 3-9 and 0—not burst
Trial: Read six RN, one for each bag;
Successful trial: Occurs if one or more are 1 or 2 (burst); Repeat; Prob (at least one bursts) = _____
(b) Change;
Successful trial: Occurs if 2 or less are 1 or 2 (burst); Prob (two or less burst) = _____

7. Prints Developed:
Model*: 2 digit RN 01-95—developed; 96-99 and 00—not developed;
Trial: Read 12 RN, one for each picture;
Successful trial: Occurs if 2 or more are in 96-99 and 00 interval; Repeat; Prob (2 or more not developed) = _____
*Other assignments of RN are possible

9. Multiple Choice:
Model: 1 digit RN 1,2,3—correct; 4-9—wrong; 0—ignore
Trial: Read 12 RN, one for each question;
Successful trial: Occurs if 6 or more are 1 or 2 or 3; Repeat; Prob (6 or more correct) = _____

SECTION 4.4

1. Rain Experiment—example:

TRIAL	1-6 S.J. RAIN	1-4 CHI RAIN	SUCCESS
6	3 yes	6	
7	0	3 yes	
8	1 yes	5	
9	9	9	
10	9	1 yes	
11	0	6	
12	5 yes	8	
13	7	2 yes	
14	0	0	
15	4 yes	1 yes	yes

2. Rain in Two Cities: $P(\text{RR}) = 26/80$ or .325

5. Green Light:
Model: 1 digit RN first 1-5—green; 6-9 and 0—not; second 1-3—green; 4-9—not; 0—ignore;
Trial: Read 2 RN, one for each light (if digit is 0, read again);
Successful trial: Occurs if first digit is in 1-5 interval and second digit is in 1-3 interval;

Repeat; Prob (G,G) = _____ ; $p(G,G) = 1/2$
· 1/3 or 1/6

7. Safe Flight:
Model: 1 digit RN 0—failure; 1-9—success;
Trial: Read 2 1-digit RN, one for each engine;
Successful trial: Occurs if at least one number is
in 1-9 interval; Repeat; Prob (safe flight) = _____

SECTION 4.5

2. Tetrahedral Die: $p(1) = 1/4$; $p(2) = 1/4$; $p(3) =$
$1/4$; $p(4) = 1/4$; $p(e) = 2/4$ or $1/2$; p(number
less than 4) = $3/4$

5. Cards: p(black) = 26/52 or 1/2; p(heart) =
13/52 or 1/4; p(ace) = 4/52 or 1/13; p(king of
diamonds) = 1/52

9. Prime Number: p(prime) = 3/6 or $1/2$ (2, 3, and
5 are prime)

SECTION 4.6

2. Theoretical Probabilities:
$p(E) = 1/4$; p(not E) = $3/4$

6. Girl in the Family:
Model: Coin Head—girl; tail—boy;
Trial: Toss coin 4 times, once for each child;
Successful trial: Occurs if at least one head (girl)
is obtained; Repeat; Prob (at least one girl)
= _____

10. Rocket Stages: p(all) = (.9)(.8)(.7) = .504

11. No Firings: p(none) = .1 (.2) (.3) = .006

5 CHAPTER FIVE
MAKING STATISTICAL DECISIONS

SECTION 5.1

1. Sampling

	SAMPLE	POPULATION
(a)	40 high school seniors	senior class
(b)	a few pieces of rock	all rocks in the valley
(c)	a spoonful of soup	pot of soup
(d)	toe	body
(e)	5 samples of mix	truck loads
(f)	blood sample	blood of entire body

5. Art and Music: Samples consist of students
surveyed by National Assessment of Educational

Progress today and five years ago; populations
are American students today and five years ago

8. Changing Moods: Sample is 130,000 respondents
to the *Family Weekly's* survey; population
implied is all Americans—more realistically it
could be *Family Weekly's* readers

SECTION 5.2

1. 25 Coin Tosses: (a) P(16 or more heads) =
(7+5+2)/100 or .14; P(18 or more heads) =
2/100 or .02; P(20 or more heads) = 0
(b) P(12 or fewer heads) = (18+14+9+7+2+1)
/100 = .51; P(9 or fewer heads) = (7+2+1)/100
= .10; P(5 or fewer heads) = 0/100 = 0.0

5. Split Opinions: P(14 or more) = 6/100 or .06;
therefore it seems unlikely that the voters are
evenly split

6. Community Opposition: P(7 or less) = 13/100
or .13; since .13 is a fairly large probability we
could accept the conclusion that community is
evenly split

SECTION 5.3

2. Random Sample: P(13 or more girls) = (2+1)/50
= .06

5. Special Studies Poll: P(17 or more yeses) = (8+10
+5+5+2+1)/100 = .31

SECTION 5.4

1. Rob's ESP: P(8 or more) = (5+1)/100 = .06;
therefore unusual event—conclude he likely has
ESP

5. Drinking Age:
Model: Coin H—favor; T—oppose;
Trial: Toss coin 30 times, once for each call;
Outcome of trial: Record number of heads
(favors); Repeat 100 times; Prob (19 or more
favoring) = _____ ; Decision _____

6. Jury Selection:
Model: RN (1 digit) 1-5—woman; 6-9 and 0—man;
Trial: Read 12 RN, one for each juror;
Outcome of trial: Record number of digits in 6-9
and 0 interval (men); Repeat; Prob (8 or more
men) = _____ ; Decision _____

SECTION 5.5

1. Basket Shooting:
   Model: $p = .6$ (RN) 1-6—make it; 7-9 and 0—not;
   Trial: Read 12 1-digit RN for each shot
   Outcome: Record number of baskets obtained
   (1 to 6's); Repeat; Prob(8 of 12) = (10+5+3+1)/
   100; Decision: Since .19 is *not* unusual, no
   evidence his shooting has improved

3. Skin Medication:
   Model: RN 1-8—cure; 9 and 0—no cure;
   Trial: Read 20 RN, one for each patient;
   Outcome: Record the number of digits in 1-8
   interval (cures); Repeat; Prob(17 or more cures)
   = (10+5+3+1)/100 = .19; Decision: Since prob.
   of 19 is large (not unusual), conclude no evidence
   medication more effective

SECTION 5.6

1. Teaching Spelling: $P(5$ or more$) = (11+7+1)/50$
   or 29/50 or .58; not unlikely—so no evidence
   new method is different from old one

6 CHAPTER SIX
   CHI-SQUARE

SECTION 6.1

1. Sides of a Die: (a) Expected $N = 150$ (1/6) or
   25; (b) expected $N = 300$ (1/6) or 50;
   (c) expected $N = 600$ (1/6) or 100

6. Cereal Box Pens: Expected $N = (1/4)(20) = 5$;
   largest number is 20; smallest number is 0

SECTION 6.2

1. Six-sided Die Rolls:

OUTCOME	EXP.	OBT.	DIFF.	|DIFF.|
1	15	20	−5	5
2	15	16	−1	1
3	15	11	4	4
4	15	12	3	3
5	15	16	−1	1
6	15	15	0	0

Total diff. = 14

5. Four Equal Outcomes: Answers will differ—
   example:

COLOR	EXP.	OBT.	DIFF.
1	3	5	−2
2	3	3	0
3	3	2	1
4	3	2	1

6. Difference in Outcomes: Total difference = 4

9. Draft Lottery:

INT.	EXP.	OBT.	DIFF.
1-1000	5.6	5	.6
1001-2000	5.6	0	5.6
2001-3000	5.6	3	2.6
3001-4000	5.6	1	4.6
4001-5000	5.6	7	−1.4
5001-6000	5.6	8	−2.4
6001-7000	5.6	11	−5.4
7001-8000	5.6	7	−1.4
8001-9000	5.6	8	−2.4

Total diff. = 26.4

SECTION 6.3

1. Difference in Die Rolls: Total difference = 14
   (from 6.2.1); $D = 14/15$

4. Estimated Probabilities: $P(D \geqslant 1.20) = 4/30$;
   $P(D \geqslant .80) = 15/30$

9. Using Data from Table: $P(D \geqslant 1.40) = 12/50$;
   $P(D \leqslant 2.0) = 49/50$

10. Color Preferences:

COLOR	EXP.	OBT.	DIFF.
SB	15	17	− 2
PP	15	31	−16
DP	15	7	8
BO	15	10	5
BB	15	9	6
AA	15	16	− 1
		90	

$D = 38/15$ or 2.53; $P(D \geqslant 2.53) = 0$—does not
appear they are choosing at random; there is
evidence that people do prefer one color over
the other

SECTION 6.4

3. Estimating Probabilities: $P(\chi^2 \geqslant 7.8) = 4/30$;
   $P(\chi^2 \leqslant 10.0) = 38/30$

6. Pen Colors Picked: $\chi^2 \doteq 1/15$ (386) or 25.73; $P(\chi^2 \geqslant 25.73) = 0$, so random selection seems unlikely–there appears to be a preference

7. Traffic Tickets:

LOCATION	EXP.	OBT.	DIFF.	DIFF.^{2}
A	15	12	3	9
B	15	7	8	64
C	15	21	−6	36
D	15	15	0	0
E	15	11	4	16
F	15	24	−9	81

$\chi^2 = 1/15$ (206) or 13.73; $P(\chi^2 \geqslant 13.73) = 0$, so ticket distribution does not appear to be equally likely at all locations

13. Die Roll Probabilities: $P(\chi^2 \geqslant 3.0) = 21/50$; $P(\chi^2 \leqslant 6.2) = 48/50$

SECTION 6.5

1. Probabilities: $P(\chi^2 \geqslant 13.5) = 5/50$ or .10; $P(\chi^2 \geqslant 18.0) = 1/50$ or .02

2. Probabilities: $P(\chi^2 \leqslant 5.1) = 20/50$ or .40; $P(\chi^2 \leqslant 3.0) = 5/50$ or .10

4. Probabilities: $P(\chi^2 \geqslant 15.2) = 3/50$ or .06; $P(\chi^2 \leqslant 4.4) = 7/50$ or .14

6. Favorite Number:

	EXP.	OBT.	DIFF.	DIFF.^{2}
0	10	5	5	25
1	10	3	7	49
2	10	11	−1	1
3	10	10	0	0
4	10	19	−9	81
5	10	9	1	1
6	10	11	−1	1
7	10	15	−5	25
8	10	13	−3	9
9	10	4	6	36

$\chi^2 = 1/10$ (228) or 22.8; $P(\chi^2 \geqslant 22.8) = 0$, so it is unlikely that there is no preference–it seems that there are favorite numbers

SECTION 6.6

2. Probabilities (use Table 6.24): $P(\chi_5^2 \geqslant 6.0) = 12/50$; $P(\chi_5^2 \geqslant 8.0) = 6/50$

4. Probabilities (use Table 6.25): $P(\chi_9^2 \geqslant 10.0) = 17/50$; $P(\chi_9^2 \geqslant 15.0) = 4/50$

7. Telephone Numbers for Random Digits: 9 df since 10 possible digits (outcomes)

8. 20 Artists: 19 df, one less than the number of possible outcomes

SECTION 6.7

2. Probabilities: $P(\chi_5^2 \geqslant 7.4) = 8/50$ or .16; $p(\chi_5^2 \geqslant 7.4) \doteq .20$

5. Probabilities: $P(\chi_3^2 \geqslant 4.6) = 6/50$ or .12; $p(\chi_3^2 \geqslant 4.6) \doteq .20$

8. Probabilities: (a) $p(\chi_7^2 \geqslant 12.0) = .10$; (b) $p(\chi_7^2 \geqslant 14.1) \doteq .05$

11. Statistics Experiment:

	EXP.	OBT.	DIFF.	DIFF.^{2}
1	16	26	−10	100
2	16	7	9	81
3	16	17	−1	1
4	16	19	−3	9
5	16	14	2	4
6	16	13	3	9

$\chi^2 = 12.75$; $p(\chi_5^2 \geqslant 12.75) \doteq .05$, a small probability–so, conclude die does not appear to be fair

14. Sales Slips:

	EXP.	OBT.	DIFF.	DIFF.^{2}
0	17.4	11	6.4	40.96
1	17.4	10	7.4	54.76
2	17.4	10	7.4	54.76
3	17.4	34	−16.6	275.56
4	17.4	7	10.4	108.16
5	17.4	25	− 7.6	57.76
6	17.4	4	13.4	179.56
7	17.4	18	− .6	.36
8	17.4	6	11.4	129.96
9	17.4	49	−31.6	998.56
		174		

$\chi_9^2 = 1900.4/17.4$ or 109.21; $p(\chi_9^2 \geqslant 109.21) \doteq$ .001–so doubtful that final digits are equally likely

7 CHAPTER SEVEN
STATISTICS IN TWO VARIABLES

SECTION 7.1

1. Variables of Interest: (a) Time spent watching TV (in hours) and quality of school work (grades, written evaluations); (b) amount of expansion and increase in temperature; (c) vaccinated (yes/no) and get polio (yes/no); (d) amount of air (in pounds) and the mileage rate (miles per gallon)

4. Traffic Death Correlations: Time (months) following gas shortage and vehicular accident deaths, chronic lung deaths, cardiovascular disease deaths

6. Student Aid: (a) Income bracket and percentage favoring cuts in Federal aid for college students; (b) the higher the income, the higher the percentage of those favoring cuts

SECTION 7.2

1. Find the Rule: $S = 3R$; $Y = 1/2X$; $X = 2W + 2$

2. Complete the Table:

M=N+1		Y=X−3		W=2X+½	
N	M	X	Y	X	W
0	1	10	7	0	½
1	2	6	3	1	2½
3	4	5	2	1.25	3
7	8	3	0	2	4½
10	11	1	−2	3	6½

5. Medicine Dose: (a) $C = (50/150)A$ or $(1/3)A$; (b) $C = (100/150)A$ or $(2/3)A$

(a) C=(1/3)A		(b) C= (2/3)A	
A	C	A	C
1	1/3	1	2/3
2	2/3	2	4/3
3	3/3 or 1	3	6/3 or 2

8. Falling Object: $s = 16t^2$ (a) $1600 = 16t^2$, $100 = t^2$, $10 = t$, 10 seconds; (b) $s = 16(5)^2$, $s = 16(25)$, $s = 400$, 400 feet

SECTION 7.3

1. Relationship:

Y=X+2	
X	Y
−2	0
−1	1
1	3
2	4
5	7

3. Filling the Tank:

FUEL GAUGE	GALLONS NEEDED
(a) 3/4	5
(b) 1/2	10
(c) 1/4	15
(d) 1/8	17.5

5. Predicting Weight:

h	w = 3h − 90	w
43	3(43) − 90	39
50	3(50) − 90	60
55	3(55) − 90	75
60	3(60) − 90	90

SECTION 7.4

1. Line Slope: (a) 3; (b) 2; (c) 5; (d) 1/2; (e) 1/4; (f) 1; (g) 1; (h) 2; (i) −3; (j) −4; (k) −1/2

2. Y-intercept: (a) 2; (b) 3; (c) −3; (d) 6; (e) −1; (f) 1; (g) 0; (h) 0; (i) 2; (j) −3; (k) 0

4. Table of Values:

(a) Y=3X+2		(b) Y=5X−3	
X	Y	X	Y
−1	−1	−1	−8
0	2	0	−3
1	5	1	2
2	8	2	7
3	11	3	12
4	14	4	17
5	17	5	22

10. Line Equations: (a) $Y = 2X + 1$; (b) $Y = −2X + 3$; (c) $Y = 1/4X + 3$; (d) $Y = 3X + 0$; (e) $Y = 1X + 1$; (f) $Y = 1X + 0$; (g) $Y = 0X + 2$

11. Line Equations: (a) $m = 1$, $c = 2$, $Y = 1X + 2$; (b) $m = 1/3$, $c = 0$, $Y = 1/3X + 0$; (c) $m = −5/4$, $c = 5$, $Y = −5/4X + 5$

SECTION 7.5

3. Athlete's Popularity Data: $\overline{X} = 6.4$, $\overline{Y} = 6.9$, $Y' = .75X + 2$; answers will differ

SECTION 7.6

1. Slot Car Data: $\overline{X} = 3.75$ (time), $\overline{Y} = 975$ (distance); Eq. $Y' = 260X$, $Y' = 200X + 225$, $Y' = 300X − 150$; all answers will vary—examples are given

4. Slot Car Data: $Y' = 260X$

X	Y	Y'	Y−Y'
2	520	520	0
3	770	780	−10
4	1050	1040	10
6	1560	1560	0

mean error = 20/4 or 5

$Y' = 200X + 225$

X	Y	Y'	Y–Y'
2	520	625	–105
3	770	825	– 55
4	1050	1025	25
6	1560	1425	135

mean error = 1/4 (320) or 80

$Y' = 300X - 150$

X	Y	Y'	Y–Y'
2	520	450	70
3	770	750	20
4	1050	1050	0
6	1560	1650	–90

mean error = 1/4 (180) = 45

7. Birmingham Boys: $Y' = 2X - 40.2$

X	Y	Y'	Y–Y'
43	42	45.8	– 3.8
46	46	51.8	– 5.8
48	51	55.8	– 4.8
50	57	59.8	– 2.8
52	62	63.8	– 1.8
54	68	67.8	.2
55	73	69.8	3.2
58	82	75.8	6.2
59	88	77.8	10.2

mean error = 38.8/9 or 4.3; best line is
$Y' = 3.084X - 96$, since its error is 1.75

SECTION 7.7

3. Median Fit for Cylinders: Using median points
(4.5, 16) and (10.8, 32.5); $m = (32.5-16)/$
$(10.8-4.5) = 16.5/6.3$; $m = 2.6$, so 16 =
$2.6(4.5) + C$, 4.3 = C, $Y' = 2.6X + 4.3$

DIAM.	CIR.
3	10
4.5	18
5	16
6.8	21
10	32.3
10.8	32.5
13	40

8 CHAPTER EIGHT
REGRESSION AND CORRELATION

SECTION 8.1

2. Mice Data: Mean error = sum $|Y-Y'|/8 = 12/8$
or 1.5

3. Weight Loss Data: $Y' = 1.5X + C$; mean data pt.
(8.3, 19); 19 = 1.5(8.3) + C, 19 = 12.5 + C,
6.5 = C, so $Y' = 1.5X + 6.5$

7. Cricket Data: $Y' = 4X + 12$

X	Y	Y'	Y–Y'
20	89	92	–3
16	72	76	–4
20	93	92	1
18	84	84	0
17	81	80	1
16	75	76	–1
15	70	72	–2
17	82	80	2
15	69	72	–3
16	83	76	7
15	80	72	8
17	83	80	3
16	81	72	9
17	84	80	4
14	76	68	8

Mean error = 56/15 ≐ 3.73

SECTION 8.2

1. Mice Data: $Y' = X + 17.13$; mean error =
25/8 or 3.125

2. Mice Data: $Y' = 2X + .5$; mean error = 1.5, so
this equation gives better fit—smaller error

7. Mice Data: Best slope $m = 2$; mean data point is
(16.625, 33.75); 33.75 = 2(16.625) + C, .5 = C
Eq. is $Y' = 2X + .5$

9. Cricket Data: Best slope $m = 2.8$; mean data
point is (16.6, 80.1); 80.1 = 2.8(16.6) + C,
80.1 = 46.48 + C, 33.6 = C
Eq. is $Y' = 2.8X + 33.6$

SECTION 8.3

1. Mice Data: $Y' = 1.5X + 8.8$; MSE = 55.9688/8
or 7.00

2. Mice Data: $Y' = 1.9X + 2.16$; MSE = 37.9988/8
or 4.75

3. Mice Data: (b) is better—MSE is smaller

12. Mice Data: m is 2.1, mean data point (16.625,
33.75 = 34.91 + C; –1.16 = C
Eq. is $Y' = 2.1X - 1.16$

SECTION 8.4

7. Push-ups: $Y' = -X + 40$

X	Y	Y'	$Y-Y'$	$(Y-Y')^2$
21	10	19	-9	81
25	8	15	-7	48
22	11	18	-7	49
28	6	12	-6	36
30	7	10	-3	9
38	15	2	13	169
22	9	18	-9	81
27	6	13	-7	49
44	4	-4	8	64
48	3	-8	11	121
35	8	5	3	9
48	5	-8	13	169

ME = 96/12 or 8, MSE = 886/12 or 73.8

8. Push-ups Data: Best slope is $-.2$, mean data point (32.3, 7.6); $7.6 = -.2(32.3) + C$, $7.6 = -6.46 + C$, $14.06 = C$; $Y' = -.2X + 14.1$

SECTION 8.5

3. Mice Data: $\overline{X} = 16.625$, $\overline{Y} = 33.75$, covar. = 13.406, var.$(X) = 6.23$

4. Mice Data: $m = 13.406/6.23 = 2.15$, very close!

10. Radiation Data: $\overline{X} = 4.99$, $\overline{Y} = 159.8$, covar. = 833/8 = 104.2, var.$(X) = 87.4/8$ or 10.9

11. Radiation Data: $m = 104.2/10.9 = 9.6$; $Y' = mX + C$; $159.8 = 9.6(5) + C$, $159.8 - 48 = C$, $111.8 = C$; $Y' = 9.6X + 111.8$

SECTION 8.6

1. Correlation: (a) Strong neg.; (b) weak pos.; (c) strong pos.; (d) weak neg.

   These are possible answers—good justification can be given for other answers—encourage discussion

5. Mice Data: Covar. = 13.406, $S_X = 2.50$, $S_Y = 5.76$, $r = 13.406/((2.50)(5.76)) \doteq .93$

7. Radiation Data: Covar. = 104.2, $S_X = 3.30$, $S_Y = 33.99$, $r = 104.2/((3.30)(33.99)) = .929$

10. Mice Data: $r = m (S_X)/S_Y$, using $m = 2.15$; $r = 2.15 (2.50)/5.76$; $r \doteq .93$

SECTION 8.7

2. Dietary Fat: Weight and calorie intake; less calories cause weight loss; in this case, causal relationship seems reasonable

9 CHAPTER NINE
WORKING WITH DATA

SECTION 9.1

2. Quiz Scores: Old mean 42/6 or 7, new mean 12; old range 9, new range 9

4. Calculator Prices:

OLD PRICE	SLIDE FACTOR	NEW PRICE
15.50	-1.65	13.85
16.25	-1.65	14.60
14.95	-1.65	13.30
15.80	-1.65	14.15
16.50	-1.65	14.85
15.80	-1.65	14.15

Old price range = $16.50 - 14.95 = 1.55$; new price range = $14.85 - 13.30 = 1.55$

8. Shrinking Data:

OLD	SHRINK FACTOR	NEW
24	$\times 1/3$ or $\div 3$	8
18	$\times 1/3$ or $\div 3$	6
15	$\times 1/3$ or $\div 3$	5

11. Record Prices:

OLD PRICE	SHRINK	SALE PRICE
$6.50	$\times 1/2$	$3.25
$6.25	$\times 1/2$	$3.13
$7.00	$\times 1/2$	$3.50
$5.95	$\times 1/2$	$2.98
$5.50	$\times 1/2$	$2.75

SECTION 9.2

1. Quiz Scores:

   (a) old scores

OLD SCORE	DEV.	(DEV.)²
8	1	1
5	-2	4
7	0	0
5	-2	4
8	1	1
9	2	4
42		14

$\overline{X} = (1/6)(42) = 7$; var. = $(14)/6 = 2.3$; range = $9 - 5 = 4$; SD = $\sqrt{2.3}$ or 1.53

(b) new scores

NEW SCORE (OLD + 10)

$$
\begin{array}{r}
18 \\
15 \\
17 \\
15 \\
18 \\
\underline{19} \\
102
\end{array}
$$

Mean = 17; range = 4; SD $\doteq$ 1.53

3. Quiz Scores:

NEW SCORE (5 × OLD + 10)

$$
\begin{array}{r}
50 \\
35 \\
45 \\
35 \\
50 \\
55
\end{array}
$$

Mean = 5(7) + 10 = 45; range = 5(4) = 20;
SD $\doteq$ 5(1.53) = 7.65

4. Quiz Score Data: If transformation, (old + 10) × 5 is used, mean is (7 + 10) × 5 or 85; transformation in problem 3 gives mean of 45, therefore order of transformation affects results

6. Quiz Scores:

(a) old ÷ 2: 4, 2.5, 3.5, 2.5, 4, 4.5; (b) mean = 1/2(7) = 3.5, range = 1/2(4) = 2, SD = 1/2(1.53) $\doteq$ .77

9. Quiz Scores: $\overline{X}$ = 50, SD = 10; new $\overline{X}$ = 2(50) + 10 = 110, new SD = 2(10) = 20

13. Quiz Scores: Stretch × 3

SECTION 9.3

2. Changing Data: Change each value, X, to corresponding z-score; $Z = (X-18.1)/4.2$

4. Changing Data: Find transformed score, T; $T = 20Z + 50$ or $T = 20(X-18.1)/4.2 + 50$

7. Barb's Test Scores: $Z_1 = (40-34)/5.3 \doteq 1.132$; $Z_2 = (44-48.3)/6.8 \doteq -.632$; She did better on Test I

8. Barb's Test Scores: $T_1 = 1.132 \times 100 + 200 = 113.2 + 200 = 313.2$; $T_2 = -.632(100) + 200 = -63.2 + 200 = 136.8$; Barb did better on Test I

9. Quiz Scores:

OLD	DEV.	(DEV.)^2	Z	T*
5	$-$ .3	.09	$-$ .21	48
7	1.7	2.89	1.20	62
3	$-$2.3	5.29	$-$1.62	34
4	$-$1.3	1.69	$-$ .92	41
5	$-$ .3	.09	$-$ .21	48
5	$-$ .3	.09	$-$ .21	48
6	.7	.49	.49	55
4	$-$1.3	1.69	$-$ .92	41
8	2.7	7.29	1.90	69
6	.7	.49	.49	55

*Rounded.

$\overline{X}$ = 5.3; var. = 20.1/10 = 2.01; SD = 1.42; $Z = (X-5.3)/1.42$; $T = 10Z + 50$

SECTION 9.4

3. Phone Book Numbers:

DIGIT	FREQ.	REL. FREQ.
1	77	.077
2	93	.093
3	106	.106
4	123	.123
5	94	.094
6	88	.088
7	117	.117
8	104	.104
9	92	.092
0	106	.106

$P(1) = .077$, $P(5) = .094$, $P(\text{even}) = .514$; $p(1) = .10$, $p(6) = .10$, $p(\text{even}) = .50$, $p(0) = .10$

9. Cork Toss:

OUTCOME	FREQ.	REL. FREQ.
top	109	.109
bottom	653	.653
side	$\underline{238}$	.238
	1000	

$P(\text{top}) = .109$, $P(\text{bottom}) = .653$, $P(\text{side}) = .238$ —horizontal line seems unlikely!

13. Carlson's Assertion:

DIGIT	EXP.	f	$E-O$	$(E-O)^2$
0	1000	1026	26	676
1	1000	1107	107	11449
2	1000	997	$-$ 3	9
3	1000	966	$-$ 34	1156
4	1000	1095	75	5625
5	1000	933	$-$ 67	4489
6	1000	1107	107	11449
7	1000	972	$-$ 28	784
8	1000	964	$-$ 36	1296
9	1000	853	$-$147	21609

Exp. $= 1000$; $\chi^2 = 58.542$; $p(\chi^2_9 > 58.542)$ is very small! so we doubt that each number is equally likely—accept Carlson's assertion

SECTION 9.5

2. Stem and Leaf Table:

STEM	LEAF
4	8,5,5,3,5,7,0,0,4,6,4,9,9,0
5	3,2,0,1,3,1,5,0,3,0,3,0,0,8,1
6	2

3. Heights: Mean $= (1/30)(1467) = 48.9$; var. $= 1/30$ $(766.7) = 25.56$; SD $\doteq 5.06$

5. Pennies:

STEM ($\times 100$)	LEAF	FREQ.	REL. FREQ.
300	1	1	.025
301		0	0
302	4	1	.025
303	7,4,9,3	4	.100
304	1,2,2	3	.075
305	9,5	2	.05
306	6,6,3,1	4	.100
307	3,0,3	3	.075
308	1,7	2	.05
309	2,9	2	.05
310	2	1	.025
311	2,2,1	3	.075
312	8,0,5,5,7,5	6	.15
313		0	0
314		0	0
315	7,4	2	.05
316	4,4,0	3	.075
317	1,5,4	3	.075
		$\overline{40}$	

7. Pennies: $P(w < 3.10) = 18/40 = .45$

SECTION 9.6

1. Normal Curves: (a) $p(z < 1.96) = .9750$; (b) $p(z < -1.96) = .0250$; (c) $p(z < 1.0) = .8413$; (d) $p(z < -1.0) = .1587$; (e) $p(z < 1.5) = .9332$; (f) $p(z < -.5) = .3085$; (g) $p(z < 0) = .5000$

7. Missing Value: (a) $p(z < .47) = .68$, pos.; (b) $p(z < -.99) = .16$, neg; (c) $p(z < .84) = .80$, pos; (d) $p(z < -.84) = .20$, neg; (e) $p(z > .84) = .20$, pos; (f) $p(z > 1.04) = .15$, pos; (g) $p(z > -.47) = .68$, neg; (h) $p(z > -2.05) = .98$, neg; (i) $p(z > 0) = .5$, zero

9. Finding Value: (a) $p(z < 2.38) = .9913$ (have z); (b) $p(z > -1.65) = 1 - .0495 = .9505$ (have z); (c) $p(z < -2.57) = .005$ (have probability); (d) $p(z > -1.96) = .975$ (have probability)

SECTION 9.7

1. Teachers' Ages: $\overline{X} = 40$, SD $= 6$, $X = 45$; $z = (45-40)/6 = .83$; $p(z > .83) = 1 - .7967 = 2033$; approximately 20%

SECTION 9.8

2. Deaths from Horse Kicks:

K	f	REL. f
0	109	.545
1	65	.325
2	22	.110
3	3	.015
4	1	.005
	$\overline{200}$	

$P(K = 0) = .545$; $P(K > 1) = .11 + .015 + .005 = .13$

5. Probabilities: (a) $P(R = 0) = 50/200$ or $.25$; (b) $P(R < 2) = (50+24)/200 = .37$; (c) $P(R = 12) = 55/200 = .275$

10 CHAPTER TEN MEASUREMENT

SECTION 10.1

2. Measurement of Human Hair: Mean $= 1/30$ $(1487) = 49.6$; var. $= 3.512$; SD $\doteq 1.874$

4. Cement:

STEM	LEAF
35	89
36	76,26
37	71,77,75
38	10,42,25,89,88,71,48
39	48,22,14,07
40	81,84,30,72,18,05,46,17,84,56
41	54,23,35,72,80,26,34,35,20,41,28,81,30
42	28,41,51
43	00,34,39
44	66,09,47
45	24

$\overline{X} = 4057.68$; SD $= 210.01$; $P(\text{strength} > 4358) = 4/50$ or $.08$

SECTION 10.2

4. Scales: Scale A, $\overline{X}$ = 76.4, SD $\doteq$ 1.74;
   Scale B, $\overline{X}$ = 75.0, SD $\doteq$ 3.71; Scale A appears
   to have less error (there is less variation in the
   measurements)

SECTION 10.3

1. New Pennies—Weights:

WEIGHT	f
28.56	1
29.25	1
29.37	1
29.43	5
29.62	1
29.81	1

   $\overline{W}$ = 29.276 and is good estimate of true weight;
   SE = .416

2. New Pennies—Weights: $\overline{W}$ = 29.57; SE = .610

SECTION 10.4

1. Thermometer Error + .5:

STEM	LEAF
78	7,9
79	7,5,0,2
80	9,6,4
81	1

   $\overline{X}$ = 79.8; SD = .84; Correct, $\overline{X}$ = 79.8−.5 = 79.3

2. Another Thermometer:

STEM	LEAF
79	7
80	7,2,7
81	7,9,3,0,2,1

   $\overline{X}$ = 80.95; SD = .63; Second is more precise (less
   variation); if it is the same green house as prob. 1,
   this thermometer seems biased (too high)

SECTION 10.5

2. Sewing Thread:

STEM	LEAF
48	84,84,52
49	48,80,52,16,28,68,48,72,68,80,84,48,96,08
50	28,12,48

   $\overline{X}$ = 49.552 and is good estimate of true length;
   SE = .484

SECTION 10.6 (none)

11 CHAPTER ELEVEN
ESTIMATION

SECTION 11.1

1. Sample, Statistic, Population, and Parameter:
   (a) Temp. (parameter) in Los Angeles (population)
   is estimated as 73 (statistic) at the airport
   (sample); (b) percent opposed (parameter) from
   a large college (population) is estimated to be
   63% (statistic) based on 64 freshmen (sample);
   (c) pollen count (parameter) in Champaign-
   Urbana (population) is estimated to be 228
   (statistic) based on sample of one cubic yard of
   air; (d) amount of moisture (parameter) in a
   truck load of corn (population) is estimated to
   be 35% (statistic) found in sample of 250 grams

2. Cola Taste: $\overline{X}$ = (1/20)(66) = 3.3; range is 5−2 =
   3; $\overline{X}$ could be used to estimate the mean
   preference of the entire population—all those
   at the shopping center, or in the community

5. College Freshmen: Sample 188,000 students who
   entered college in fall 1982; population could
   be *all* students who entered college in fall 1982
   Example—in sample 43.8% preferred living in a
   dorm, statistic could estimate that 43.8% of *all*
   college students prefer living in a dorm

7. Telephone Bill: .15 (total) = 112.82, so total is
   estimated as 112.82/.15 or $752.13

SECTION 11.2

1. Population Sample: μ = 50, σ = 10
   (a) N = 4, $P(48 \leqslant \overline{X} \leqslant 52.9)$ = (1/100)(9+9+5+3+10)
   = .36; (b) N = 16, $P(48 \leqslant \overline{X} \leqslant 52.9)$ = (1/100)
   (14+14+16+13+11) = .68; (c) N = 36, $P(48 \leqslant \overline{X}$
   $\leqslant 52.9)$ = (1/100)(20+25+29+10+9) = .83

   Note: Interval is from 48.0 to 52.9

4. Batteries: μ = 45; σ = 12; N = 100;
   $P(43 \leqslant \overline{X} < 47.9)$ = (1/100)(92) = .92

   Note: Interval is from 43.0 to 47.9

SECTION 11.3

1. Army Recruits: N = 10

(a) $\bar{X} = 128/10$ or 12.8

	$X - \bar{X}$	$(X - \bar{X})^2$
15	2.2	4.84
10	−2.8	7.84
12	− .8	.64
14	1.2	1.44
14	1.2	1.44
16	3.2	10.24
12	− .8	.64
11	−1.8	3.24
13	.2	.04
		33.90

(b) Var. = 33.9/10; var. = 3.39, biased;
var. = 33.9/9; var. = 3.73, unbiased
(c) S(sample standard deviation) $\doteq 1.93$

SECTION 11.4

1. Random Samples: Means of samples are close to mean of population.

2. Theoretical Standard Error: (a) $N = 4$—theoretical standard error = 10/2 or 5; (b) $N = 16$—theoretical standard error = 10/4 or 2.5; (c) $N = 36$—theoretical standard error = 10/6 or 1.7

3. Estimated Standard Error: (a) $N = 4$—ESE = 5.3, close; (b) $N = 16$—ESE = 2.5, exact!; (c) $N = 36$—ESE = 1.7, exact (within rounding)

4. Random Samples: 2/3 of population is within one standard dev., so interval is 63.5 ± 12/6 or 61.5 to 66.5

5. Random Samples: 63.5 ± 12/5; 63.5 ± 2.4; Interval is 61.1 to 65.9

8. Random Samples: For 95%, $z = \pm 1.96$, score = z(st. error)+mean
(a) $N = 36$, interval is 59.58 to 67.42
(b) $N = 25$, interval is 58.80 to 68.20
(c) $N = 4$, interval is 51.74 to 75.26
(d) $N = 100$, interval is 61.15 to 65.85

SECTION 11.5

1. Bong Show: $\bar{X} = 31$, SD = 4.3, $N = 100$, CI (confidence interval) = 99%
Find z for 99% CI: $z = \pm 2.54$; $T = z$(SD)+mean; $T = \pm 2.54(.43)+31$; $T = \pm 1.19+31$; so 99% CI is 29.81 to 32.19

4. Length of Micro-organism: (a) $\bar{X} = 27.5$, SD = 3.2, $N = 64$, CI = 90%
$z = \pm 1.64$; $T = \pm 1.64(3.2/8)+27.5$; $T = \pm .656 + 27.5$; so CI is 26.88 to 28.16
(b) CI = 95%; $z = \pm 1.96$; $T = \pm 1.96(3.2/8)+27.5$; $T = \pm .784+27.5$; so interval is 26.72 to 28.28
(c) CI = 99%; $z = \pm 2.58$; $T = \pm 2.58(3.2/8)+27.5$; $T = \pm 1.032+27.5$; so interval is 26.47 to 28.53

12 CHAPTER TWELVE HYPOTHESIS TESTING

SECTION 12.1

1. Batteries: $P(\text{mean} \leqslant 14.5) = (15+20)/100 = .35$; A large probability—do not reject claim

3. Iron Ore: $P(\text{mean} \geqslant 14) = 3/100$; It is reasonable to doubt the claim

5. Gasoline Pumps: $P(\text{mean} \leqslant 230.3) = 18/100$; No evidence to doubt accuracy of pumps

SECTION 12.2

1. Population Data: Since $p(\bar{X} \leqslant 13.5) = .04$, do not reject claim if .01 is defined as unusual

3. Iron Ore from Section 12.1, Exercise 3: $z = (14-12)/(6.1/\sqrt{40}) = 2.07$; $p(z \geqslant 2.07) = .0192$ or about .02 which is close to .03—same decision made

6. Groceries: $z = (86-80)/(9/\sqrt{64}) = 6/1.125$ or 5.33; $p(z \geqslant 5.33)$ is very small, so claim is doubtful

9. Nite-Nite Bulbs: $\bar{X} = 48.3$, SD = 4.19, $N = 36$ so $z = (48.3-50)/(4.19/6) = -1.7/.6983 = -2.43$; $p(z \leqslant -2.43) = .0075$; so doubtful that bulb life is 50 hours

SECTION 12.3

3. Groceries: Type I error would occur if claim that average expenditure for groceries is $80.00 is rejected when it really is true; Type II error would occur if claim of $80 average expenditure was accepted when, in fact, it was not true.

SECTION 12.4

1. Estimated Probabilities: (a) $P(t > 1.8) = .04$;
   (b) $P(t > 2.3) = .02$; (c) $P(t > 1.4) = .13$;
   (d) $P(t < -1.8) = .03$; (e) $P(t < -2.1) = 0$;
   (f) $P(t < -1.4) = .05$

3. Theoretical Probabilities: (a) $df = 11$, $p(t > 1.80) = .05$; (b) $df = 6$, $p(t < -1.44) = .10$;
   (c) $df = 21$, $p(t > 1.32) = .10$; (d) $df = 24$,
   $p(t < -2.06) = .025$

SECTION 12.5

2. Student IQ's: $\overline{X} = 115$, SD $= 11.2$, $N = 9$,
   $df = 8$, $t = 2.31$ for 95% CI
   CI $= \pm2.31\,(11.2/3) + 115 = \pm8.62 + 115$; CI is
   from 106.38 to 123.62

4. Gold in Vein: $\overline{X} = 12.3$, SD $= 2.5$, $N = 10$,
   $df = 9$, $t = 1.83$, 90% CI
   CI $= \pm1.83\,(2.5/\sqrt{10}) + 12.3$; CI $= \pm1.447 + 12.3$,
   CI is from 10.85% to 13.75%

SECTION 12.6

1. Ninth Graders Weight: $\overline{X} = 118$, $S = 5$, $N = 12$,
   $df = 11$, $t = (118 - 114)/(5/\sqrt{12})$, $t = 2.77$
   $p(t \geq 2.77) \doteq .01$; Doubtful that total class has
   mean weight of 114 pounds—probably weighs
   more

4. Transylvania Effect: $\overline{X} = 13.3$, $S = 5.50$, $N = 12$,
   $\mu = 11.2$, $t = (13.3 - 11.2)/(5.50/\sqrt{12})$, $t = 1.32$
   $p(t \geq 1.32) \doteq .10$; Depends on definition of
   "rare" or "unusual," but since probability is
   slightly more than .10, claim that admissions
   averaged 11.2 would not be rejected—do not
   have sufficient evidence of more admissions
   during full moon

13 CHAPTER THIRTEEN
ADVANCED TOPICS

SECTION 13.1

2. Mendel's Experiment:

		EX.	OB.	$E - O$
Round & yellow	9/16	313	315	-2
Wrinkled & yellow	3/16	104	101	3
Round & green	3/16	104	108	-4
Wrinkled & green	1/16	35	32	3

$\chi^2 = 4/313 + 9/104 + 16/104 + 9/35 = .012 + .087 + .154 + .257$; $\chi^2 \doteq .51$; $p(\chi_3^2 \geqslant .51) \approx .90$;
Since prob. is large, no evidence model is
*in*appropriate—doubt is *not* cast on his theory

SECTION 13.2

2. Judges' Selections:

JUDGE	MEN	WOMEN	TOTAL
A	$O = 234$; $E = 261.7$	$O = 119$; $E = 92.3$	354
B	$O = 553$; $E = 539.6$	$O = 197$; $E = 190.4$	730
C	$O = 287$; $E = 299.3$	$O = 118$; $E = 105.6$	405
D	$O = 149$; $E = 167.1$	$O = 77$; $E = 58.9$	226
E	$O = 81$; $E = 82.0$	$O = 30$; $E = 29.0$	111
F	$O = 403$; $E = 408.0$	$O = 149$; $E = 144$	552
G	$O = 511$; $E = 441.3$	$O = 86$; $E = 155.7$	597
	2199	776	2975

Note: $261.7 \doteq (2199)(354)/2975$
$\chi^2 = (-26.7)^2 / 261.7 + (26.7)^2/92.3 + .081 + .229 + .505 + 1.456 + 1.961 + 5.562 + .012 + .012 + .061 + .061 + 11.001 + 31.20$; $\chi^2 \doteq 65.526$
$p(\chi_6^2 > 65.526)$ is very small, so doubtful that
number of women selected and the judges'
selections are independent

5. Cold Remedy:

	RELIEVED	NOT RELIEVED	TOTALS
New	$O = 31$; $E = 27.6$	$O = 15$; $E = 18.40$	46
Old	$O = 29$; $E = 32.4$	$O = 25$; $E = 21.6$	54
	60	40	100

$\chi^2 = .419 + .628 + .357 + .535$; $\chi^2 = 1.939$
$p(\chi_1^2 > 1.939) \doteq .17$ so conclude better results
due to chance—no evidence to suggest changing
to new treatment

SECTION 13.3

2. Pascal's Triangle:

   Row 8: 1 8 28 56 70 56 28 8 1
   9: 1 9 36 84 126 126 84 36 9 1
   10: 1 10 45 120 210 252 210 120 45 10 1

5. Theoretical Probabilities: (a) 6 coins, $2^6 = 64$;
   $p(0\,H) = 1/64$; $p(1\,H) = 6/64$; $p(2\,H) = 15/64$;
   $p(3\,H) = 20/64$; $p(4\,H) = 15/64$; $p(5\,H) = 6/64$;
   $p(6\,H) = 1/64$
   (b) 10 coins, $2^{10} = 1024$; $p(0\,H) = 1/1024$;
   $p(1\,H) = 10/1024$; $p(2\,H) = 45/1024$; $p(3\,H) = 120/1024$; $p(4\,H) = 210/1024$; $p(5\,H) = 252/1024$; $p(6\,H) = 210/1024$; $p(7\,H) = 120/1024$;

$p(8\,H) = 45/1024$; $p(9\,H) = 10/1024$; $p(10\,H) = 1/1024$

(c) 5 coins, $2^5 = 32$; $p(0\,H) = 1/32$; $p(1\,H) = 5/32$; $p(2\,H) = 10/32$; $p(3\,H) = 10/32$; $p(4\,H) = 5/32$; $p(5\,H) = 1/32$

SECTION 13.4

2. Large Sample—Two-tailed: $\overline{X} = \$145$, $S = \$25$, $N = 49$, $\mu = \$150$, $z = (145-150)/(25/\sqrt{49}) = -1.4$
$p(z < -1.4 \text{ or } z > 1.4) = .0808 + .0808 = .1616$; Do not have evidence to doubt association's claim

9. Small Sample—One-tailed: $\overline{X} = 4.3$, $S = 1.2$, $N = 9$, $df = 8$, $\mu = 3.2$, $t = (4.3-3.2)/(1.2/\sqrt{9}) = 2.75$

$p(t > 2.75) \doteq .01$; Doubtful that number of letters has remained same, probably has increased

SECTION 13.5

1. Equal Frequency Test—Phone Numbers:

DIGIT	E	O	$E-O$	$(E-O)^2$
0	40	46	-6	36
1	40	37	3	9
2	40	44	-4	16
3	40	35	5	25
4	40	36	4	16
5	40	48	-8	64
6	40	47	-7	49
7	40	30	10	100
8	40	35	5	25
9	40	42	-2	4

$\chi^2 = 8.6$; $p(\chi_9^2 > 8.6) \doteq .50$; Hypothesis that digits are equally distributed seems reasonable —phone numbers appear to be a good source of random digits

3. Die Rolls:

(a)

DIGIT	E	O	$E-O$
1	50	36	14
2	50	46	4
3	50	68	-18
4	50	47	3
5	50	50	0
6	50	53	-3

$\chi^2 = 11.08$; $p(\chi_5^2 > 11.08) \doteq .05$; Die likely not fair

(b) Answer will depend on the pattern checked—following 1 or 2 or . . . .

14 CHAPTER FOURTEEN MONTE CARLO METHODS

None given for this chapter.